New Directions of Modern Cryptography

New Directions of Modern Cryptography

Zhenfu Cao

CRC Press
Taylor & Francis Group
Boca Raton London New York

CRC Press is an imprint of the
Taylor & Francis Group, an **informa** business

CRC Press
Taylor & Francis Group
6000 Broken Sound Parkway NW, Suite 300
Boca Raton, FL 33487-2742

First issued in hardback 2019

© 2013 by Taylor & Francis Group, LLC
CRC Press is an imprint of Taylor & Francis Group, an Informa business

No claim to original U.S. Government works

ISBN-13: 978-1-4665-0138-6 (hbk)

Visit the Taylor & Francis Web site at
http://www.taylorandfrancis.com

and the CRC Press Web site at
http://www.crcpress.com

Contents

Preface

The study of cryptography is motivated by security requirements in the real world and also driven forward by security requirements. Regardless of its real applications, cryptography can be viewed as a branch of mathematics. All the new directions of modern cryptography introduced in this book, including proxy recryptography, attribute-based cryptography, batch cryptography, and noncommutative cryptography, have arisen from the requirements. In this book, we focus on the fundamental definitions, precise assumptions, and rigorous security proofs of cryptographic primitives and related protocols, as well as how they developed from the security requirements and how they are applied.

As we know, modern cryptography has evolved dramatically since the 1970s. Nowadays, the field of cryptography encompasses much more than secure communication. It covers, for example, authentication, digital signature, key establishment and exchange, zero-knowledge, secure multiparty computation, electronic auction and election, digital cash, access control, etc. Modern cryptography is concerned with security problems that arise in a variety of distributed environments where attacks may come from either internal or external forces. Instead of giving a rigid and perfect definition of modern cryptography, we here say that, from the viewpoint of applications, modern cryptography is the science and technology that focuses on defending digital information storage, transportation, and distributed computation via modern communication networks.

These networks consist of but are not limited to wired or wireless telecommunication networks, satellite communication networks, broadcast and TV networks, computer networks (including organization-wide intranet and the Internet), and all newly emerging networks, such as the internet of things, cloud computing, social networks, and named data networks. Another important difference between classic cryptography and modern cryptography is related to who is using it. Historically, the major users of cryptography have been military and intelligence organizations. Today, however, cryptography is required everywhere in our lives. Security mechanisms that rely on cryptography become an essential ingredient of information systems. For example, cryptographic methods are used to enforce access control to all web sites and to prevent adversaries from extracting business secrets from stolen laptops. In view of increasing demands on the network security, this book presents some application paradigms and general principles regarding new directions of modern cryptography.

In short, modern cryptography has gone from an art that deals with secret communication for the military and governments to the science and technology that help ordinary people to set up secure systems. This also means that cryptography becomes a more and more central topic within computer science.

In fact, with the rise of new network architectures and services, the security requirements have changed significantly, that is, from single-user communication (each side is of a single user) to multiuser communication (at least one side is of multiple users). The public-key cryptosystem proposed by Diffie and Hellman in 1976 is not sufficient to satisfy the security requirements in multiuser settings. Under the new environments of "one sender vs. multiple receivers" (one-to-many), "multiple senders vs. one receiver" (many-to-one), "multiple senders vs. multiple receivers" (many-to-many), it is a tendency to design and analyze the multiuser-oriented cryptographic algorithms. They aim to solve ciphertext access control problems, trust problems, efficiency problems in the multimessage cryptology, and challenging problems of quantum and biological computing etc.

In the last 10 years, I, as a founder and a director of Trusted Digital Technology Laboratory (TDT Lab) of Shanghai Jiao Tong University, have witnessed the progressively increased demands on cryptographic techniques. I am proud of having been engaged in cryptography research for over 30 years. The TDT Lab is one of the earliest groups focusing publicly on "Trusted-X Technology" in the world. The TDT Lab focuses on the research of cryptology and trusted digital technology. In cryptology, our research interests mainly include authorized cryptography (proxy cryptography, proxy re-cryptography), attribute-based cryptography (identity-based cryptography, spatial cryptography, functional cryptography), post-quantum cryptography (noncommutative cryptography, lattice-based cryptography), collaboration cryptography (aggregated cryptography, batch cryptography), biologic cryptography (DNA cryptography, biometric feature based cryptography), commitment and zero-knowledge proof, as well as cryptanalysis. In trusted digital technology, we mainly study trusted computing, trusted networks, secure storage/access, secure e-commerce/e-government, key management, and so on. The TDT Lab tries to pursue original innovation in fundamental research, acquire intellectual properties in key technology, and promote industrial development inspired by our academic results. During these years, I have instructed 2 post-doctorals, supervised 19 Ph.D.s and 50 masters, and obtained a number of interesting results (along with my students and colleagues). This book can be viewed as a part of the collection of these results.

Audience

The goal of this book is to be of interest to cryptographers and practitioners of network security. In particular, it is aimed at the following readers: For students who complete first degree courses in computer/information science or applied mathematics and plan to pursue a degree or career in network security, this book may serve as an advanced course in applied cryptography. Fresh Ph.D. candidates beginning their research in cryptography or information/network security would appreciate the new directions introduced in this book. For security researchers and engineers who are interested in the cloud computing security, e-health security, vehicular ad-hoc network security, RFID security, delay tolerant network security, network coding security, and other wired/wireless network security, and who are responsible for designing and developing secure network systems, this book may help them have a solid understanding of the security principles and correctly deploy the applications. This book is also designed to serve as a reference for graduate courses in cryptography, computer sciences, and mathematics, or as a general introduction suitable for people who want to learn cryptography and security themselves.

Acknowledgments

I would like to acknowledge all those who helped in the development of this book.

First, I am deeply grateful to Professor Xiaolei Dong for devoting her time to improving this book with invaluable comments, criticisms, and suggestions.

Second, I would like to thank my students, especially Licheng Wang, Jun Shao, Lifei Wei, Zhen Liu, Rong Ma, and Le Chen et al. for the preparation and copyediting of the earlier versions of this material.

Third, I would like to thank the members of TDT lab, especially Zongyang Zhang, Qingshui Xue, Jun Zhou, and Jiachen Shen for their comments and literature services.

Finally, I am grateful to Professor Hao Shen and Professor Zhengjun Cao for their invaluable suggestions.

The project would never have happened without the strong support and encouragement from CRC Press, including Ruijun He and his colleagues. We also thank them for their help in getting the book into its final form.

Comments and Errata

It seems that errors or typos in this book are inevitable. Any errors that remain are solely the author's responsibility. I am always more than happy to receive feedback on this book, especially constructive comments on how to improve this book. You can email your comments and suggestions to zfcao@cs.sjtu.edu.cn.

Shanghai, May, 2012.   *Zhenfu Cao*

Chapter 1

Introduction

In this chapter, we briefly review some security problems arisen in the network environments and present the main idea about how to use modern cryptographic techniques to solve these problems. We want to emphasize that cryptography is the building block of most solutions to security problems.

1.1 Trust Problem

1.1.1 Trusted Domains Transfer Problem

We first discuss what the trusted domains transfer problem is. Suppose Alice and Bob belong to two trusted domains CA_1 and CA_2, respectively, and they want to build trust relationship. In the public key infrastructure (PKI), CA_1 and CA_2 are two certificate authorities. Every user in a trusted domain has a public key certificate, which is a signature signed by the certificate authority to bind a public key with an identity. A certificate can be used to verify that a public key belongs to an individual.

However, Alice can only verify and trust the certificates from the certificate authority CA_1, and so can Bob from the certificate authority CA_2. In the scenario as shown in

1

Figure 1.1, how to build trust relationship between Alice in the trusted domain CA_1 and

Bob in CA_2 is a practical problem.

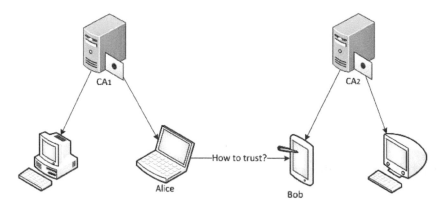

Figure 1.1: Trusted Domains Transfer Problem

Solution: The main idea to solve this problem is to set a transfer server, called proxy

that is allowed to transform certificates from CA_1 to CA_2, as shown in Figure 1.2. How-

ever, the proxy cannot generate new certificates in CA_1 or CA_2 by itself. We may require

extra abilities of the proxy in some concrete applications.

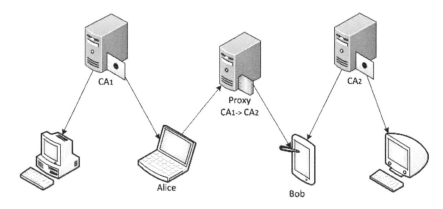

Figure 1.2: A Solution to Trusted Domains Transfer Problem

Sometimes it is desired that certificates in CA_1 can be transformed into ones in

CA_2, while certificates in CA_2 cannot be transformed into ones in CA_1, as shown in the

Figure 1.3. This requires that the proxy can only do the *unidirectional* transformation. This new ability of proxy required in this case is authorized only by CA_2. If the ability of proxy is authorized by both CA_1 and CA_2, then the proxy could do the *bidirectional* transformation.

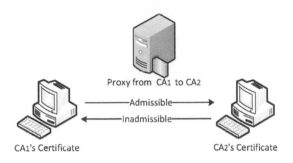

Figure 1.3: Unidirectional Transfer

On the other hand, the requirements in the trusted domains transfer problem could be further extended. As shown in Figure 1.4, certificates in CA_1 can be transformed into the ones in CA_2 via $Proxy_1$, and further transformed into the ones in CA_3 via $Proxy_2$. If the process can continue multiple times, the method of trusted domains' transfer is *multiuse*. Otherwise, it is *single-use*.

Fortunately, we have a new cryptographic primitive called proxy re-signature [29] to solve the above problem. We will give a comprehensive introduction of proxy re-signature in Chapter 2 and discuss more applications of it in Chapter 6.

1.1.2   Trusted Server Problem

Cloud computing is drawing more and more attention from the information and communication technology community, since it can significantly reduce the costs of hardware and software resources in computing infrastructure.

In cloud computing, the cloud storage server is responsible for storing users' data

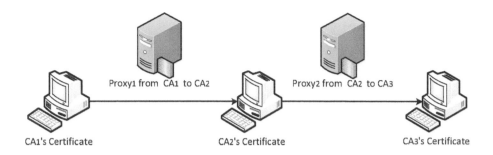

Figure 1.4: Multiuse Transfer.

including the sensitive data, and the cloud access control server is responsible for exert-
ing control over who can access the data stored in the cloud storage server. It is usually
required that the cloud access control server is fully trusted.

However, this requirement cannot be met in practice for two reasons. One is that the
provider(s) of cloud access control service cannot be assumed to be fully trusted, because that he/she could become corrupted in some situations. The other is that intruders
could break the cloud access control server even if the provider(s) is absolutely trusted.
Hence, trusted server problem in cloud computing should be solved to put forward the
development of cloud computing.

A possible solution is to store the encrypted plaintexts (i.e., ciphertexts) in the cloud
storage server. If the ciphertexts are merely used by the encryptor himself/herself, the
trusted server problem is easily solved. When the ciphertexts need to be shared by others and the access control server has no right to perform decryption, it is a challenging
problem. Under this situation, we can conceive a following solution: let the encryptor
authorize the access control server the right to transform the ciphertexts so that the delegated users can decrypt the resulting ciphertexts, but the access control server cannot
decrypt the ciphertexts. This paradigm is shown in Figure 1.5.

The above conceived solution could be implemented if the access control server un-

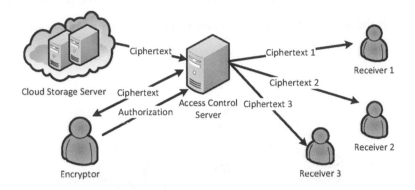

Figure 1.5: A Solution to Trusted Server Problem

der the authorization of the encryptor can transform the ciphertexts stored in the cloud storage server into a new form with the same plaintexts that can only be decrypted by the designated receivers. If we regard the access control server, encryptor, designated receivers, and authorization messages as the proxy, delegator, delegatees, and re-encryption keys, respectively, the above solution becomes a particular case of proxy re-encryption.

Proxy re-encryption was proposed by Blaze et al. [29] and has many applications. According to the concrete applications, proxy re-encryption should satisfy other properties. We will discuss them in Chapters 2 and 6.

1.2 Ciphertext Access Control Problem

Assume that data owner intends to store a private message that is accessed by a specific set of users in a storage server. The current solution is that the data owner stores the data in plaintext form in the storage server, and the user's access rights are specified by access control lists that are created by the data owner and performed by the access control server as shown in Figure 1.6. The users specified by the access control lists and verified by the access control server, can access the message. However, as mentioned in the above section, in practice, trust and security issues of the servers are always serious.

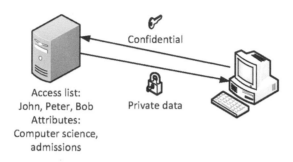

Figure 1.6: Access Control Model

A trivial method would be to store the data in ciphertext form in the servers. However current encryption systems cannot allow the ciphertext to be efficiently shared among a group of users. It becomes urgent to develop an efficient and flexible method to share data directly based on ciphertexts, which includes the access control policy. Fortunately, Bethencourt et al. [27] proposed such a cryptographic primitive called ciphertext-policy attribute-based encryption (CP-ABE), which initiates a new direction in solving the ciphertext access control problem.

The concept of attribute-based encryption (ABE) was first proposed by Goyal et al. [128] in 2005. They also constructed a concrete scheme called key-policy attribute-based encryption (KP-ABE). In KP-ABE as shown in Figure 1.7, each ciphertext is labeled by the encryptor with a set of descriptive attributes, and each private key is associated with an access structure that specifies which type of ciphertexts the key can decrypt. While in CP-ABE as shown in Figure 1.8, each private key is associated with a set of attributes and each ciphertext is associated with an access structure.

For example in Figure 1.8, the access structure of ciphertext File 1 is "{(Computer

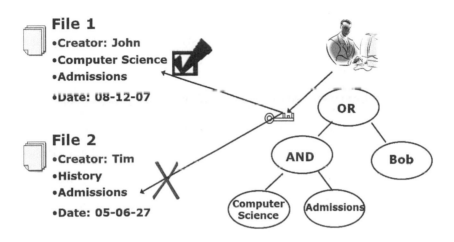

Figure 1.7: Key-Policy ABE

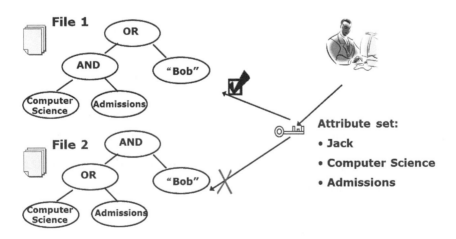

Figure 1.8: Ciphertext-Policy ABE

science) AND (Admissions)} OR (Bob)" representing that the qualified user should be an enrolled student in department of computer science or Bob. As Jack's attribute set {Jack, Computer science, Admissions} satisfies the ciphertext access structure, he can decrypt File 1 successfully by using his private key. Moreover, he can not decrypt ciphertext File 2 since his attribute set does not satisfy the access structure "{(Computer science) OR (Admissions)} AND (Bob)" of File 2.

Recall the scenario discussed in Section 1.1.2, where the proxy can merely transform a ciphertext into another one that can only be decrypted by *a designated user*. But in practice, it is more usually required that the proxy is able to transform a ciphertext into another one that can be decrypted by *a group of designated users*. We propose a new concept of attribute-based proxy re-encryption [191], called ABPRE, which is a more powerful cryptographic primitive that can settle this problem efficiently.

To better understand the concept of ABPRE, we demonstrate an application scenario of personal information system in a university. In this system, there are some confidential records of grades of every student. These records are encrypted into a ciphertext under the access structure "((AGE > 40) AND (Tenure))." Professors who are older than 40 and have a tenure position are qualified to retrieve the confidential records by using their own *different private keys*. Nevertheless, when these professors are on vacation, it is necessary to find some trustworthy delegatees who are able to decrypt the ciphertext in time. Therefore, ABPRE allows a qualified professor to authorize a proxy (administrator) who can transform a ciphertext into another ciphertext encrypted with a different access structure so that the corresponding delegatees can retrieve the records. For example, the delegated access structure can be defined by "(Secretary) AND (EXP $\geq$ 2)" that represents secretaries with at least 2 years working experience. Therefore, even if no qualified professor is available, some experienced secretaries can open the confidential records with the help of the authorized administrator.

A more general relationship between users and ciphertexts is shown in Figure 1.9.

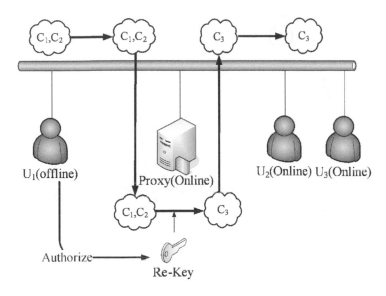

Figure 1.9: Attribute Based Proxy Re-encryption with Delegating Capabilities

Suppose there are three users and three ciphertext sets in this system, where user U_1 is able to decrypt any ciphertext in sets C_1 and C_2 encrypted under access structures AS_1 and AS_2, while users U_2 and U_3 are able to decrypt ciphertexts in C_3 corresponding with AS_3. Then, U_1 authorizes a proxy with a re-key that can be used to transform the ciphertext of C_1 and C_2 into that of C_3. In this way, even if U_1 is offline, U_2 and U_3 could still retrieve the information encrypted in C_1 and C_2 with the help of U_1's proxy.

1.3 Efficiency Problems in Multi-Message Cryptology

With the development of networks, there are many scenarios related to multiple requests processing as shown in Figure 1.10. The server receives multiple encrypted requests from different clients using the server's public key. In this scenario, is it necessary for the server to decrypt all the requests one by one? Generally, it is. But it is very

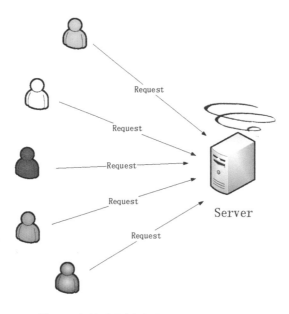

Figure 1.10: Multiple Requests Processing

inefficient when the number of requests becomes large. A similar scenario appears in the key exchange among the many clients and the central server.

Another scenario is that a famous star signs his books or discs for selling as shown in Figure 1.11. He has to meet a lot of his fans and sign his name many times, which leaves him exhausted. Digital signature continuously faces similar scenarios in the e-government or e-business. In these scenarios, is it necessary for them to sign one by one? Generally, they can sign only one by one, which results in low efficiency when the number of files to be signed is large.

A similar scenario exists in the current broadcast/multicast authentication, where the signatures need to be verified one by one. In this case, is it necessary to verify in such an inefficient way? Generally, it is. However, verifiers become exhausted due to excessive signature verification. The situation goes even worse in the case of the energy-constrained networks such as wireless sensor networks and mobile ad hoc networks.

Taking into account the above scenarios, it has become critical to introduce a new

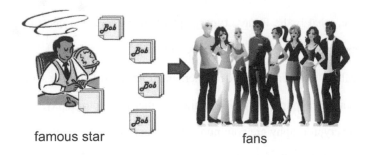

famous star   fans

Figure 1.11: Signing Multiple Files

direction of modern cryptography—batch cryptography, which includes batch decryption, batch key agreement, and batch verification to solve the multi-message processing and their security and efficiency problems in real world. Readers can refer to Chapters 4 and 6 for more details.

1.4 The Challenges from Quantum and Biological Computing

In 1994, Shor [257] proposed a polynomial time algorithm for solving integer factorization problem [1] and the discrete logarithm problem using quantum computers. Meanwhile, most of modern cryptographic schemes are based on the hardness of integer factorization problem and discrete logarithm problem. If quantum computers become available, the modern cryptography based on the above two problems should be collapsed.

Biological computing, which refers to DNA computing [2] at present, is considered with super computing capability possibly more than quantum computing. If a biological computer appears, we believe that it will bring the same challenges to modern cryptog-

---

[1]For the detail remarks of Shor's algorithm, please refer the footnote on page 234 of this book.

raphy.

Therefore, what is the future cryptography? To answer this question, we suggest that one gets an insight into the history of cryptography.

Historically, stream cipher, block cipher, and related work use the same key for both encryption and decryption, which form the symmetric cryptography. The milestones in symmetric cryptography are communication theory of secrecy systems proposed by Shannon [243, 244] in 1949 and data encryption standard (DES) [222] in 1975.

In 1976, Diffie and Hellman [99] proposed the new direction of cryptography — asymmetric encryption in the network environment. Two different keys are required in their work: one is public for encryption and the other is private for decryption.

Although they did not propose any concrete encryption schemes, it paved the way for asymmetric encryption. After that, many researchers have proposed some concrete schemes such as RSA [233] based on integer factorization problem and ElGamal [107] based on discrete logarithm problem. Following this method, researchers have obtained many fruitful results, which form the asymmetric cryptography.

Both the symmetric cryptography and asymmetric cryptography are based on commutative algebraic structures. We call these kinds of cryptography commutative cryptography. As we have seen the transition of cryptography from "symmetric" to "asymmetric," the next transition of cryptography might be from "commutative" to "noncommutative." We refer to the cryptography based on the noncommutative algebraic structures as noncommutative cryptography.

Some computing hard problems based on noncommutative algebraic structures might be essential to resist quantum computing or biological computing. Thus, one possible direction of modern cryptography is to securely bridge commutative cryptography to noncommutative cryptography. Another possible direction is to explore noncommutative algebraic structures from the quantum theory and biological technology. We will discuss this detailedly in Chapters 5 and 6.

1.5 Organization

The theme of this book is to introduce some new directions of modern cryptography. The main idea we want to express is: cryptography is brought into birth and driven by application requirements. We will cover four kinds of modern cryptography: proxy re-cryptography, attribute-based cryptography, batch cryptography, and noncommutative cryptography. This book consists of the following Chapters. Chapter 1 introduces the background and motivation to write this book. Chapters 2, 3, 4, and 5, respectively, introduce proxy re-cryptography, attribute-based cryptography, batch cryptography, and noncommutative cryptography, including the fundamental definitions, security models, concrete schemes and security proof. In Chapter 6, we present some applications and challenging problems related to the above cryptography.

Chapter 2

Proxy Re-Cryptography

2.1 Introduction

Proxy re-cryptography can be used to solve the trust problem in the scenarios mentioned in Chapter 1. Besides, there are many promising applications.

Many companies have developed digital rights management (DRM) technologies, which can prevent illegal redistribution of digital content. With DRM systems, the digital content can only be played in a specified device (regime). For example, a song playable in device (regime) A cannot be played in device (regime) B. However, it is reported that 86% of the consumers prefer to pay twice the price for a song that can run on any device than that with one single device [155]. Most of current interoperability architectures require to change the existing DRM systems significantly [176]; this modification cannot be adopted due to business reasons. At ACM DRM 2006, Taban *et al.* [271] proposed a new interoperability architecture, which does not change the existing DRM systems too much but maintains the DRM systems' security. In their architecture, only a new module called domain interoperability manager (DIM) is introduced. DIM applies a single signature scheme and a single public key encryption scheme that can transform licences and content in regime A into ones in regime B, but it cannot generate valid licenses or content either in regime A or in regime B. It is easy to see that the traditional signature and public key encryption cannot support transformation, which, however, can be easily implemented by proxy re-cryptography.

In this chapter, we will give a comprehensive introduction of proxy re-cryptography, especially the security models of proxy re-cryptography. Furthermore, we introduce one proxy re-signature scheme and one proxy re-encryption scheme for illustration.

2.2   Proxy Re-Signature

Proxy re-signature (PRS), introduced by Blaze et al. [29] at Eurocrypt 1998, and formalized by Ateniese and Hohenberger [11] at ACM CCS 2005, allows a semi-trusted proxy to transform a delegatee's (Alice) signature into a delegator's (Bob) signature on the same message by using some additional information (a.k.a. re-signature key). The proxy, however, cannot generate arbitrary signatures on behalf of either the delegatee or the delegator.

2.2.1   Properties and Definition

Before giving the formal definition of PRS, we would like to give the desired properties of PRS [11].

- *Unidirectional:* In this scheme, a re-signature key allows the proxy to transform Alice's signature to Bob's but not vice versa. On the other hand, in a bidirectional scheme, the re-signature key allows the proxy to transform Alice's signature to Bob's as well as Bob's signature to Alice's.

- *Multiuse:* In a multiuse scheme, a transformed signature can be re-transformed again by a proxy. While in a single-use scheme, a proxy can transform only the signatures not yet transformed.

- *Private proxy:* A proxy can keep the re-signature key as a secret in a private proxy scheme, but *anyone* can recompute the re-signature key by observing the re-signature process passively in a public proxy scheme.

- *Transparent:* In a transparent scheme, users may not even know the existence of a proxy.

- *Key-optimal:* In a key-optimal scheme, a user is required to protect and store only a small constant amount of secrets no matter how many signature delegations the user gives or accepts.

- *Noninteractive:* The delegatee is not required to participate in the delegation process. Bidirectional PRS cannot be noninteractive, since a delegator is also a delegatee.

- *Non-transitive:* A re-signature key cannot be generated from two other re-signature keys. For example, the re-signature key from Alice to Bob cannot be generated from the re-signature keys from Alice to Tina and Tina to Bob.

- *Temporary:* A re-signing right is temporary. This can be done by either revoking the right [11] or expiring the right.

- *Collusion-resistant:* The delegator can delegate the signing rights to the delegatee via the proxy, while keeping decryption rights for the same public key.

Now we give the formal definition of PRS.

Definition 2.2.1 (Proxy Re-Signature). A PRS scheme consists of the following five probabilistic polynomial time (p.p.t.) algorithms:

- KeyGen: It takes as input the security parameter λ, and returns a verification key pk and a signing key sk.

- ReKeyGen: It takes as input delegatee Alice's key pair (pk_A, sk_A) and delegator Bob's key pair (pk_B, sk_B), and returns a re-signature key $rk_{A \to B}$ for the proxy. If the PRS scheme is unidirectional, the delegatee's signing key is not included in the input. If the PRS is bidirectional, the proxy can easily obtain $rk_{B \to A}$ from $rk_{A \to B}$. In many bidirectional schemes [11, 29, 250], $rk_{A \to B} = 1/rk_{B \to A}$.

- Sign: It takes as input a signing key sk, a positive integer ℓ, and a message m from the message space, and returns a signature σ at level ℓ. If the PRS scheme is single-use, then $\ell \in \{1, 2\}$.

- ReSign: It takes as input a re-signature key $rk_{A \to B}$ and a signature σ_A on a message m under pk_A at level ℓ, and returns the signature σ_B on the same message m under pk_B at level $\ell + 1$ if $\mathtt{Verify}(pk_A, m, \sigma_A, \ell) = 1$, or $\mathtt{reject}$ otherwise. If the PRS scheme is single-use, then $\ell = 1$.

- Verify: It takes as input a verification key pk, a message m from the message space, a signature σ, and a positive integer ℓ, and returns 1 if σ is a valid signature under pk at level ℓ or 0 otherwise.

Correctness. The following property must be satisfied for the correctness of a PRS scheme: For any message m in the message space and any two key pairs (pk_A, sk_A) and (pk_B, sk_B), let $rk_{A \to B} \leftarrow \mathtt{ReKeyGen}(pk_A, sk_A, pk_B, sk_B)$, the following two equalities must hold:

$$\mathtt{Verify}(pk_A, m, \sigma_A, \ell) = 1,$$

where σ_A is a signature on message m under pk_A at level ℓ from $\mathtt{Sign}$. If the PRS scheme is single-use, then $\ell \in \{1, 2\}$, or $\ell \geq 1$ otherwise.

$$\mathtt{Verify}(pk_B, m, \mathtt{ReSign}(rk_{A \to B}, pk_A, m, \sigma'_A, \ell - 1), \ell) = 1.$$

If the PRS scheme is single-use, σ'_A is a signature on message m under pk_A from Sign with $\ell = 2$; if it is multiuse, σ'_A could be a signature on message m under pk_A from Sign with $\ell = 2$ or from ReSign with $\ell > 2$.

Remark 2.2.2 (Two Types of Signatures). In all existing unidirectional PRS schemes, a signature manifests in two types: the *owner-type* (i.e., the first-level defined in [11], and $\ell = 1$ in this chapter) and the *nonowner-type* (i.e., the second-level signatures [11], and $\ell > 1$ in this chapter). An *owner-type* signature can be computed only by the owner of the signing key via Sign, while a *nonowner-type* signature can be computed not only by the owner of the signing key via Sign, but also by collaboration between his/her proxy and delegatee via ReSign.

If there is only one signature type in a PRS scheme, the parameter ℓ in all algorithms can be omitted.

2.2.2 Related Work

Though PRS has many applications as we mentioned in Chapter 1, it has a rather simple history. The first PRS scheme, which is multiuse, public proxy and bidirectional, was proposed by Blaze et al. [29] at Eurocrypt 1998. However, there was no follow-up until the work published by Ateniese and Hohenberger [11] at ACM CCS 2005. One of the reasons is that the definition of proxy re-signatures [29] was informal and could be easily confused with other signature variations. Ateniese and Hohenberger [11] first formalized the definition of security for PRS, referred as the AH model in this chapter, and then proposed three PRS schemes with security proofs. The first one is multiuse, private proxy and bidirectional; the second one is single-use, public proxy and unidirectional; and the third one is single-use, private proxy and unidirectional. Later, by using Waters' identity-based signature [289], we [250] proposed a multiuse, private proxy and bidirectional PRS scheme, and Chow and Phan [85] proposed a single-use, private proxy and unidirectional PRS scheme. Libert and Vergnaud [193] proposed a multiuse, private proxy and unidirectional scheme. All the PRS schemes [11, 85, 193, 250] are proven secure in the AH model.

Recently, we [251] found that the AH model is suitable for almost all proxy re-signatures except the private proxy and unidirectional PRS. To deal with this problem, we [251] proposed an improvement of the AH model (AH$^+$ model for short). Fortunately, the previous schemes proven secure in the AH model can still be proven secure in the AH$^+$ model. Following this, we [252] further extended the AH$^+$ model to the ID-based setting, and proposed a unidirectional, single-use, private proxy, and ID-based PRS.

Some PRS schemes in terms of the satisfied properties are summarized in Table 2.1.

Table 2.1: Properties of some PRS schemes

Property[a]	P1	P2	P3	P4	P5	P6	P7	P8	P9
BBS [29]	✗	✓	✗	✓	✓	✗	✗	✓	✗
AH05a [11]	✗	✓	✓	✓	✓	✗	✗	✓	✗
AH05b [11]	✓	✗	✗	✓	✓	✓	✓	✓	✓
AH05c [11]	✓	✗	✓	✓	✓	✓	✓	✓	✓
SCWL07a [250]	✗	✓	✓	✓	✓	✗	✗	✓	✗
SCWL07b [250]	✗	✓	✓	✓	✓	✗	✗	✓	✗
CP08 [85]	✓	✗	✓	✓	✓	✓	✓	✓	✓
LV08 [193]	✓	✓	✓	✓	✓	✓	✓	✓	✓
SWLX11 [252]	✓	✓	✓	✓	✓	✓	✓	✓	✓

[a] P1,···,P9 denote unidirectional, multiuse, private proxy, transparent, key-optimal, non-transitive, temporary, and collusion-resistant, respectively.

In the rest of this section, we first introduce the security model of PRS—the AH model. Then we introduce the multiuse, private proxy and bidirectional PRS scheme [250] proposed at Indocrypt 2007. In what follows, we show the incompleteness of the AH model, and give an improvement, AH$^+$ model [251].

2.2.3   Security Model: The AH Model

The AH model mainly deals with the unforgeability of signatures; it contains external security and internal security. These two parts contain four security games between an adversary $\mathcal{A}$ and a challenger $\mathcal{C}$. Before the games start, the adversary should decide which users (delegatee/delegator) are to be corrupted. Furthermore, all the verification keys in the games are generated by the challenger. If the adversary obtains the signing key of a user via $\mathcal{O}_{sk}$, then the user is declared corrupted. We assume that the adversary never invokes a query twice. If so, the challenger simply returns the previous value.

The following oracles could be queried in the four games:

- *Verification key query* $\mathcal{O}_{pk}$: On receiving an index i from the adversary $\mathcal{A}$, the challenger $\mathcal{C}$ responds by running $\texttt{KeyGen}(1^\lambda)$ to get a key pair (pk_i, sk_i), and forwards the verification key pk_i to the adversary. Finally, the challenger records (pk_i, sk_i) in Table T_k, which is initialized as empty and used to record the key pairs of users. In the following oracles, sk_i denotes as the signing key corresponding to the verification key pk_i.

- *Signing key query* $\mathcal{O}_{sk}$: On receiving a verification key pk_i from the adversary $\mathcal{A}$, the challenger $\mathcal{C}$ responds with sk_i which is the associated value with pk_i in Table T_k, if pk_i is corrupted; otherwise the challenger $\mathcal{C}$ responds with $\texttt{reject}$.

- *Re-signature key query* $\mathcal{O}_{rk}$: On receiving two verification keys pk_i, pk_j ($pk_i \neq pk_j$) from the adversary $\mathcal{A}$, the challenger $\mathcal{C}$ responds with $\texttt{ReKeyGen}(pk_i, sk_i, pk_j, sk_j)$. Note that for bidirectional PRS, we consider queries with (pk_i, pk_j) and (pk_j, pk_i) are identical.

- *Signature query* $\mathcal{O}_s$: On receiving a verification key pk_i and a message m_i from the adversary $\mathcal{A}$, the challenger $\mathcal{C}$ responds with $\texttt{Sign}(sk_i, m, 1)$.

- *Re-signature query* $\mathcal{O}_{rs}$: On receiving two verification keys pk_i, pk_j ($pk_i \neq pk_j$), a message m_i, and a signature σ_i at level ℓ from the adversary $\mathcal{A}$, the challenger $\mathcal{C}$ responds with $\texttt{ReSign}(\texttt{ReKeyGen}(pk_i, sk_i, pk_j, sk_j), pk_i, m_i, \sigma_i, \ell)$.

Now, we introduce the four security games as follows:

External Security: This security protects a user from outside adversaries other than the proxy and any delegation parties. The game goes as follows:

- *Queries*: The adversary $\mathcal{A}$ can make queries to oracles $\mathcal{O}_{pk}$, $\mathcal{O}_{rs}$, and $\mathcal{O}_s$ adaptively.

- *Forgery*: The adversary $\mathcal{A}$ outputs a signature σ^* on message m^* on behalf of pk^* at level ℓ^*. The adversary is declared the winner of the game, if all of the following requirements are satisfied:

  - $\text{Verify}(pk^*, m^*, \sigma^*, \ell^*) = 1$;

  - The adversary $\mathcal{A}$ never obtains a signature on a message m^* under the verification key pk^* by querying $\mathcal{O}_s$ with (pk^*, m^*), or by querying $\mathcal{O}_{rs}$ with $(\cdot, pk^*, m^*, \cdot, \cdot)$.

A PRS scheme has external security if and only if for security parameter λ and all p.p.t. algorithms $\mathcal{A}$, $\Pr[\mathcal{A} \text{ wins}]$ is *negligible*.

Internal Security: This security protects a user from inside adversaries who can be any parties, i.e., the proxy, the delegatee, or the delegator, in a PRS scheme. It can be classified into the following three types:

Limited Proxy: In this case, only the proxy is a potential adversary $\mathcal{A}$. We must be sure that the proxy cannot produce signatures on behalf of either the delegator or the delegatee except the signatures produced by the delegatee and delegated to the proxy to re-sign. The game goes as follows:

- *Queries:* The adversary $\mathcal{A}$ can make queries to oracles $\mathcal{O}_{pk}$, $\mathcal{O}_{rk}$, $\mathcal{O}_{rs}$, and $\mathcal{O}_s$ adaptively.

- *Forgery:* The adversary $\mathcal{A}$ outputs a signature σ^* on message m^* on behalf of pk^* at level ℓ^*. The adversary is declared the winner of the game, if all of the following requirements are satisfied:

- Verify$(pk^*, m^*, \sigma^*, \ell^*) = 1$;

- The adversary $\mathcal{A}$ never obtains a signature on message m^* under the verification key pk^* by querying $\mathcal{O}_s$ with (pk^*, m^*), or by querying $\mathcal{O}_{rs}$ with $(\cdot, pk^*, m^*, \cdot, \cdot)$;

- If the PRS scheme is single-use, and the adversary $\mathcal{A}$ obtains a re-signature key from pk_i to pk^* directly from $\mathcal{O}_{rk}$, then the adversary $\mathcal{A}$ never obtains a signature on message m^* under the verification key pk_i by querying $\mathcal{O}_s$ with (pk_i, m^*);

- If the PRS scheme is multiuse, and the adversary $\mathcal{A}$ obtains re-signature keys corresponding to (pk_{i_1}, pk_{i_2}), (pk_{i_2}, pk_{i_3}), ..., (pk_{i_t}, pk^*), then the adversary $\mathcal{A}$ never obtains a signature on message m^* under the verification key pk_i ($i \in \{i_1, \ldots, i_t\}$) by querying $\mathcal{O}_s$ with (pk_i, m^*), or by querying $\mathcal{O}_{rs}$ with $(\cdot, pk_i, m^*, \cdot, \cdot)$.

A PRS scheme has limited proxy security if and only if for security parameter λ and all p.p.t. algorithms $\mathcal{A}$, $\Pr[\mathcal{A} \text{ wins}]$ is *negligible*.

Delegatee Security: In this case, the proxy and delegator may collude with each other. This security guarantees that their collusion cannot produce any signatures on behalf of the delegatee.

The game goes as follows:

- *Queries*: The adversary $\mathcal{A}$ can make queries to oracles $\mathcal{O}_{pk}$, $\mathcal{O}_{sk}$, $\mathcal{O}_{rk}$, $\mathcal{O}_{rs}$, and $\mathcal{O}_s$ adaptively.

- *Forgery*: The adversary $\mathcal{A}$ outputs a signature σ^* on message m^* on behalf of pk^* at level ℓ^*. The adversary is declared the winner of the game, if all of the following requirements are satisfied:

  – pk^* is uncorrupted;

  – pk^* has not been a delegator in this game;

  – $\texttt{Verify}(pk^*, m^*, \sigma^*, \ell^*) = 1$;

  – The adversary $\mathcal{A}$ never obtains a signature on message m^* under the verification key pk^* by querying $\mathcal{O}_s$ with (pk^*, m^*).

A PRS scheme has delegatee security if and only if for security parameter λ and all p.p.t. algorithms $\mathcal{A}$, $\Pr[\mathcal{A} \text{ wins}]$ is *negligible*.

Delegator Security: In this case, the proxy and delegatee may collude with each other. This security guarantees that their collusion cannot produce any owner-type signatures on behalf of the delegator.

The game goes as follows:

- *Queries*: The adversary $\mathcal{A}$ can make queries to oracles $\mathcal{O}_{pk}$, $\mathcal{O}_{sk}$, $\mathcal{O}_{rk}$, $\mathcal{O}_{rs}$, and $\mathcal{O}_s$ adaptively.

- *Forgery*: The adversary $\mathcal{A}$ outputs a signature σ^* on message m^* on behalf of pk^* at the first level. The adversary is declared the winner of the game, if all of the following requirements are satisfied:

  – pk^* is uncorrupted;

  – pk^* has not been a delegatee in this game;

  – $\texttt{Verify}(pk^*, m^*, \sigma^*, 1) = 1$.

A PRS scheme has delegator security if and only if for security parameter λ and all p.p.t. algorithms $\mathcal{A}$, $\Pr[\mathcal{A} \text{ wins}]$ is *negligible*.

Remark 2.2.3. Since a delegator is also a delegatee in bidirectional PRS, we do not consider the delegatee security and delegator security for bidirectional PRS with only one signature type. While we have a weaker security for protecting the delegatee (delegator) in this kind of PRS, which is collusion resistance. It guarantees that the signing key of a delegatee (delegator) cannot be revealed by the collusion of the delegator (delegatee) and the proxy. On the other hand, it is easy to show that the delegatee security and delegator security imply the collusion resistance. In particular, if the adversary cannot generate a valid forgery, it definitely cannot obtain the corresponding signing key.

2.2.4 Multiuse, Private Proxy and Bidirectional Scheme

In this section, we [250] introduce the multiuse, private proxy and bidirectional PRS scheme (denoted as S_{mb}) proposed at Indocrypt 2007. The security is proved in the standard model based on the computational Diffie–Hellman (CDH) assumption, that is, it is hard to compute g^{xy} given (g, g^x, g^y) where g is a random element in a cryptographic group $\mathbb{G}$ with a prime order q, and x, y are random numbers in $\mathbb{Z}_q^*$.

The system parameter of scheme S_{mb} is $(\mathbb{G}, \mathbb{G}_T, q, g, g_2, e, H_w)$, where $\mathbb{G}$ and $\mathbb{G}_T$ are bilinear groups with prime order q; g and g_2 are two random elements of $\mathbb{G}$; e is an admissible pairing, $e : \mathbb{G} \times \mathbb{G} \to \mathbb{G}_T$; and H_w is a Waters' function [289], $H_w(m) = u' \cdot \prod_{i \in \mathcal{U}} u_i$, where m is an n_m-bit message, $u', u_1, u_2, \ldots, u_{n_m}$ are random elements in $\mathbb{G}$, $\mathcal{U} \subset \{1, \ldots, n_m\}$ is the set of indices i such that $m[i] = 1$, and $m[i]$ is the i-th bit of m.

- KeyGen: It outputs the key pair $(pk, sk) = (g^a, a)$, where a is a random number in $\mathbb{Z}_q^*$.

- ReKeyGen: On input two signing keys $sk_A = a$ and $sk_B = b$, output the re-signature key $rk_{A \to B} = b/a \bmod q$. It is easy to see that $rk_{B \to A} = 1/rk_{A \to B} = a/b \bmod q$.

  (Note that the re-signature key can be obtained by the following method [11]: (1) the proxy sends a random number $r \in \mathbb{Z}_q^*$ to the delegatee Alice, (2) then Alice sends r/a to the delegator Bob, (3) Bob sends rb/a to the proxy, (4) finally, the proxy obtains b/a. The communications between any two entities are via private and authenticated channels.)

- Sign: On input a signing key $sk = a$ and an n_m-bit message m, output the signature $\sigma = (\sigma_1, \sigma_2) = (g_2^a \cdot H_{\mathtt{w}}(m)^r, g^r)$, where r is chosen randomly from $\mathbb{Z}_q^*$.

- ReSign: On input a re-signature key $rk_{A \to B}$, a verification key pk_A, a signature $\sigma_A = (\sigma_{A,1}, \sigma_{A,2})$, and an n_m-bit message m, check that Verify(pk_A, m, σ_A) = 1. If σ_A is invalid, output reject; otherwise, choose a random number $v \in \mathbb{Z}_q^*$ and output $\sigma_B = (\sigma_{A,1}^{rk_{A \to B}} \cdot H_{\mathtt{w}}(m)^v, \sigma_{A,2}^{rk_{A \to B}} \cdot g^v)$.

Note that we have:

$$
\begin{aligned}
(\sigma_{A,1}^{rk_{A \to B}} \cdot H_{\mathtt{w}}(m)^v, \sigma_{A,2}^{rk_{A \to B}} \cdot g^v) &= (g_2^b \cdot H_{\mathtt{w}}(m)^{rb/a} H_{\mathtt{w}}(m)^v, g^{rb/a} g^v) \\
&= (g_2^b H_w(m)^{r'}, g^{r'})
\end{aligned}
$$

where $r' = rb/a + v \bmod q$.

- Verify: On input a verification key pk, an n_m-bit message m, and a signature $\sigma = (\sigma_1, \sigma_2)$, output 1, if $e(pk, g_2)e(\sigma_2, H_{\mathtt{w}}(m)) = e(\sigma_1, g)$, or 0 otherwise.

Theorem 2.2.4. *In the standard model, the bidirectional PRS scheme S_{mb} is correct and existentially unforgeable under the CDH assumption in $\mathbb{G}$.*

Proof. The correctness property is easily observable.

To prepare the simulation, simulator $\mathcal{B}$ first sets $\ell_m = 2(q_s + q_{rs})$, and randomly chooses a number k_m, such that $0 \le k_m \le n_m$, and $\ell_m(n_m + 1) < q$. $\mathcal{B}$ then chooses $n_m + 1$ random numbers x', $x_i(i = 1, \ldots, n_m)$ from $\mathbb{Z}_{\ell_m}$. Lastly, $\mathcal{B}$ chooses $n_m + 1$ random numbers y', $y_i(i = 1, \ldots, n_m)$ from $\mathbb{Z}_q^*$. q_s and q_{rs} are the number of queries to $\mathcal{O}_s$ and $\mathcal{O}_{rs}$, respectively.

To make expression simpler, we use the following notations:

$$F(m) = x' + \sum_{i \in \mathcal{U}} x_i - \ell_m k_m \text{ and } J(m) = y' + \sum_{i \in \mathcal{U}} y_i.$$

Now, $\mathcal{B}$ sets the public parameters:

$$g_2 = g^b, \ u' = g_2^{r' - \ell_m k_m} g^{y'}, \ u_i = g_2^{x_i} g^{y_i} \quad (1 \le i \le n_m).$$

Note that for any message m, there exists the following equality:

$$H_{\mathbf{w}}(m) = u' \prod_{i \in \mathcal{U}} u_i = g_2^{F(m)} g^{J(m)}.$$

In the following, we focus on external security and internal security (limited proxy) of scheme S_{mb}.

External Security:

- *Queries:* $\mathcal{B}$ builds the following oracles:

  - $\mathcal{O}_{pk}$: On input index i, $\mathcal{B}$ chooses a random $x_i \in \mathbb{Z}_q^*$, and guess whether $pk_i = pk^*$. If it is, it outputs $pk_i = (g^a)^{x_i}$; otherwise, it outputs $pk_i = g^{x_i}$. Finally, $\mathcal{B}$ records (pk_i, x_i) in Table T_k.

  - $\mathcal{O}_s$: On input (pk_i, m), $\mathcal{B}$ obtains (pk_i, x_i) from Table T_k. If $pk_i \ne pk^*$, it performs Sign with x_i; otherwise, it performs as follows.

    * If $F(m) \ne 0 \bmod q$, $\mathcal{B}$ picks a random $r \in \mathbb{Z}_q^*$ and computes the signature as,

    $$\begin{aligned} \sigma &= (((g^a)^{-J(m)/F(m)}(g^{J(m)} g_2^{F(m)})^r)^{x_i}, ((g^a)^{-1/F(m)} g^r)^{x_i}) \\ &= (\sigma_1, \sigma_2) \end{aligned}$$

For $\tilde{r} = (r - a/F(m)) \cdot x_i$, we have that

$$((g^a)^{-J(m)/F(m)}(g^{J(m)}g_2^{F(m)})^r)^{x_i}$$
$$= (g_2^a(g_2^{F(m)}g^{J(m)})^{-a/F(m)}(g^{J(m)}g_2^{F(m)})^r)^{x_i}$$
$$= (g_2^a(g_2^{F(m)}g^{J(m)})^{r-a/F(m)})^{x_i}$$
$$= g_2^{ax_i}H_w(m)^{\tilde{r}},$$

and

$$((g^a)^{-1/F(m)}g^r)^{x_i} = (g^{r-a/F(m)})^{x_i}$$
$$= g^{\tilde{r}},$$

which shows that σ has the correct signature as in the actual scheme.

* If $F(m) = 0 \bmod q$, $\mathcal{B}$ is unable to compute the signature σ and outputs failure.

- $\mathcal{O}_{rs}$: On input (pk_i, pk_j, m, σ). If $\text{Verify}(pk_i, m, \sigma) \neq 1$, $\mathcal{B}$ outputs reject. Otherwise, $\mathcal{B}$ queries $\mathcal{O}_s$ with (pk_j, m) and returns the resulting value.

• *Forgery:* If $\mathcal{B}$ does not output failure in any query above, $\mathcal{A}$ will, with probability at least ε, return a message m^* and a valid forgery $\sigma^* = (\sigma_1^*, \sigma_2^*)$ on behalf of pk^*. It is easy to see that $\mathcal{B}$ guesses the right pk^* with $1/q_{pk}$ at least, where q_{pk} is the number of queries to $\mathcal{O}_{pk}$. If $F(m^*) \neq 0 \bmod q$, $\mathcal{B}$ outputs failure. Otherwise, the forgery must be of the form, for some $r^* \in \mathbb{Z}_q^*$,

$$\sigma^* = (g^{abx^*}(g_2^{F(m^*)}g^{J(m^*)})^{r^*}, g^{r^*})$$
$$= (g^{abx^*+J(m^*)r^*}, g^{r^*})$$
$$= (\sigma_1^*, \sigma_2^*).$$

To solve the CDH instance, $\mathcal{B}$ outputs $((\sigma_1^*) \cdot (\sigma_2^*)^{-J(m^*)})^{1/x^*} = g^{ab}$, where x^* is the corresponding value in Table T_k with pk^*.

At last, we need to bound the probability that $\mathcal{B}$ completes the simulation without outputting `failure`. We require that all signature and re-signature queries on a message m along with pk^* have $F(m) \neq 0 \bmod q$, and that $F(m^*) = 0 \bmod q$. Let $m_1, \ldots, m_{q_Q}$ be the messages appearing in signature or re-signature queries not involving the message m^*. Clearly, $q_Q \leq q_s + q_{rs}$. We define the events E_i, E_i', and E^* as:

$$E_i : F(m_i) \neq 0 \bmod q, \quad E_i' : F(m_i) \neq 0 \bmod \ell_m, \quad E^* : F(m^*) = 0 \bmod q.$$

From $\ell_m(n_m + 1) < q$ and $x', x_i (i = 1, \ldots, n_m) \in \mathbb{Z}_{\ell_m}$, we have $0 \leq l_m k_m < q$ and $0 \leq x' + \sum_{i \in \mathcal{U}} x_i < q$. Then $F(m) = 0 \bmod q$ implies $F(m) = 0 \bmod \ell_m$. Hence, $F(m) \neq 0 \bmod \ell_m$ implies $F(m) \neq 0 \bmod q$. Since $k_m, x', x_i (i = 1, \ldots, n_m)$ are chosen randomly, we have

$$\begin{aligned}
\Pr[E^*] &= \Pr[F(m^*) = 0 \bmod q \wedge F(m^*) = 0 \bmod \ell_m] \\
&= \Pr[F(m^*) = 0 \bmod \ell_m]\Pr[F(m^*) = 0 \bmod q | F(m^*) = 0 \bmod \ell_m] \\
&= \frac{1}{\ell_m} \cdot \frac{1}{n_m + 1}
\end{aligned}$$

and

$$\begin{aligned}
\Pr[\textstyle\bigwedge_{i=1}^{q_Q} E_i' | E^*] &= 1 - \Pr[\textstyle\bigvee_{i=1}^{q_Q} \neg E_i' | E^*] \\
&\geq 1 - \textstyle\sum_{i=1}^{q_Q} \Pr[E_i' | E^*] \\
&= 1 - \frac{q_s + q_{rs}}{\ell_m} \\
&= 1/2
\end{aligned}$$

The probability of $\mathcal{B}$ not outputting `failure` is

$$\Pr[\neg \texttt{failure}] \geq \Pr[\bigwedge_{i=1}^{q_Q} E_i \wedge E^*]$$

$$\geq \Pr[\bigwedge_{i=1}^{q_Q} E_i' \wedge E^*]$$

$$= \Pr[E^*]\Pr[\bigvee_{i=1}^{q_Q} \neg E_i'|E^*]$$

$$\geq \frac{1}{2\ell_m(n_m+1)}$$

Internal Security: Since a delegator in scheme S_{mb} is also a delegatee and only one signature type exists in scheme S_{mb}, we only consider limited proxy security here.

- *Queries:* $\mathcal{B}$ builds the following oracles:

  - $\mathcal{O}_{pk}$: On input index i, $\mathcal{B}$ chooses a random $x_i \in \mathbb{Z}_q^*$, and guesses whether $\mathcal{A}$ will issue the re-signature key queries with $(pk_i, pk_{i_1}), (pk_{i_1}, pk_{i_2}), \cdots,$ (pk_{i_j}, pk^*) or $pk_i = pk^*$. If it is, $\mathcal{B}$ sets $\theta_i = 0$ and outputs $pk_i = (g^a)^{x_i}$; otherwise, it sets $\theta_i = 1$ and outputs $pk_i = g^{x_i}$. Finally, $\mathcal{B}$ records (pk_i, x_i, θ_i) in Table T_k.

  - $\mathcal{O}_s$: On input (pk_i, m), $\mathcal{B}$ obtains (pk_i, x_i, θ_i) from Table T_k. If $\theta_i = 1$, $\mathcal{B}$ returns $\texttt{Sign}(x_i, m)$; otherwise, $\mathcal{B}$ proceeds as follows:

    * If $F(m) \neq 0 \bmod q$, $\mathcal{B}$ picks a random $r \in \mathbb{Z}_q^*$ and computes the signature as,

    $$\sigma = (((g^a)^{-J(m)/F(m)}(g^{J(m)}g_2^{F(m)})^r)^{x_i}, ((g^a)^{-1/F(m)}g^r)^{x_i})$$

    $$= (\sigma_1, \sigma_2).$$

    * If $F(m) = 0 \bmod q$, $\mathcal{B}$ is unable to compute the signature σ and outputs `failure`.

- $\mathcal{O}_{rk}$: On input (pk_i, pk_j), if $\theta_i = \theta_j$, $\mathcal{B}$ returns $rk_{i \to j} = (x_j / x_i) \mod q$; else, $\mathcal{B}$ outputs `failure`.

- $\mathcal{O}_{rs}$: On input (pk_i, pk_j, m, σ). If $\mathtt{Verify}(pk_i, m, \sigma) \neq 1$, $\mathcal{B}$ outputs `reject`; otherwise, $\mathcal{B}$ queries $\mathcal{O}_s$ with (pk_j, m) and returns the resulting value.

- *Forgery:* If $\mathcal{B}$ does not output `failure` in any query above, $\mathcal{A}$ will, with probability at least ε, return a message m^* and a valid forgery $\sigma^* = (\sigma_1^*, \sigma_2^*)$ on behalf of pk^*. If $F(m^*) \neq 0 \mod q$ or $\theta^* = 1$, $\mathcal{B}$ outputs `failure`. Otherwise, the forgery must be of the form, for some $r^* \in \mathbb{Z}_q^*$,

$$
\begin{aligned}
\sigma^* &= (g^{abx^*}(g_2^{F(m^*)} g^{J(m^*)})^{r^*}, g^{r^*}) \\
&= (g^{abx^* + J(m^*)r^*}, g^{r^*}) \\
&= (\sigma_1^*, \sigma_2^*).
\end{aligned}
$$

To solve the CDH instance, $\mathcal{B}$ outputs $((\sigma_1^*) \cdot (\sigma_2^*)^{-J(m^*)})^{1/x^*} = g^{ab}$, where x^* is the corresponding value in Table T_k with pk^*.

At last, we need to bound the probability that $\mathcal{B}$ completes the simulation without outputting `failure`. We require the followings:

R1 $\theta_i = \theta_j$ in $\mathcal{O}_{rk}$.

R2 $\theta^* = 0$.

R3 $F(m) \neq 0 \mod q$ with the corresponding $\theta = 0$ in $\mathcal{O}_s$ and $\mathcal{O}_{rs}$.

R4 $F(m^*) = 0 \mod q$.

It is easy to see that requirements R1 and R2 will be satisfied with the probability $1/2^{q_{rk}}$ and $1/q_{pk}$ at least, respectively, where q_{rk} is the number of queries to $\mathcal{O}_{rk}$.

Furthermore, we can use the similar analysis for external security to conclude that re-

quirements R3 and R4 will be satisfied simultaneously with the probability $\frac{1}{2\ell_m(n_m+1)}$

at least.

This completes the proof.   □

2.2.5   Incompleteness of the AH Model

The AH model is designed for all kinds of PRS, and almost all existing PRS schemes

are proven secure in the AH model. However, as we will see later, the AH model is

not complete. In particular, the AH model is not suitable for the unidirectional and pri-

vate proxy PRS. In this section, we propose a private proxy and unidirectional scheme,

named S_{ins}, which is proven secure in the AH model, but it cannot provide all the re-

quired security properties. This fact shows that the AH model is not complete.

The public parameter of scheme S_{ins} is $(q, g, \mathbb{G}, \mathbb{G}_T, e, H)$, where $\mathbb{G}$ and $\mathbb{G}_T$ are

bilinear groups with prime order q, g is a random element of $\mathbb{G}$, e is an admissible

pairing, $e : \mathbb{G} \times \mathbb{G} \to \mathbb{G}_T$, and $H\colon \{0,1\}^* \to \mathbb{G}$ is a cryptographic hash function.

- KeyGen: It selects a random number $a \in \mathbb{Z}_q^*$, and outputs the key pair $(pk, sk) = (g^a, a)$.

- ReKeyGen: On input the delegatee's verification key $pk_A = g^a$ and the delegator's signing key $sk_B = b$, it outputs the re-signature key

$$rk_{A \to B} = (rk_{A \to B}^{(1)}, rk_{A \to B}^{(2)}, rk_{A \to B}^{(3)}) = (r', (pk_A)^{r'}, H(g^{a \cdot r'} || 2)^{1/b}),$$

  where r' is a random number in $\mathbb{Z}_q^*$ determined by the delegator.

- Sign: On input a signing key $sk = a$, a message m in message space and an
  integer $\ell \in \{1, 2\}$,

– if $\ell = 1$, it outputs an *owner-type* signature

$$\sigma = (A, B, C) = (H(m\|0)^r, g^r, H(g^r\|1)^a),$$

– if $\ell = 2$, it outputs a *nonowner-type* signature

$$\sigma = (A, B, C, D, E) = (H(m\|0)^{r_1}, g^{r_1}, H(g^{r_1}\|1)^{r_2}, g^{r_2}, H(g^{r_2}\|2)^{1/a}).$$

- ReSign: Given an *owner-type* signature σ at level 1, a re-signature key $rk_{A \to B} = (rk_{A \to B}^{(1)}, rk_{A \to B}^{(2)}, rk_{A \to B}^{(3)})$, a verification key pk_A, and a message m, this algorithm first checks $\mathtt{Verify}(pk_A, m, \sigma, 1) = 1$. If it does not hold, outputs $\mathtt{reject}$; otherwise, outputs

$$
\begin{aligned}
\sigma' \ &= \ (A', &B', \quad &C', &D', &E') \\
&= \ (A, &B, \quad &C^{rk_{A \to B}^{(1)}}, &rk_{A \to B}^{(2)}, &rk_{A \to B}^{(3)}) \\
&= \ (H(m\|0)^r, &g^r, \quad &H(g^r\|1)^{ar'}, &(pk_A)^{r'}, &H((pk_A)^{r'}\|2)^{1/b}) \\
&= \ (H(m\|0)^{r_1}, &g^{r_1}, \quad &H(g^{r_1}\|1)^{r_2}, &g^{r_2}, &H(g^{r_2}\|2)^{1/b})
\end{aligned}
$$

Note that we set $r_1 = r \bmod q$ and $r_2 = ar' \bmod q$.

- Verify: On input a verification key pk, a message m at level $\ell \in \{1, 2\}$, and a signature σ,

– if σ is an *owner-type* signature $\sigma = (A, B, C)$ (i.e., $\ell = 1$), it checks

$$e(pk, H(B\|1)) \stackrel{?}{=} e(g, C) \ \text{ and } \ e(B, H(m\|0)) \stackrel{?}{=} e(g, A).$$

If the two equalities both hold, it outputs 1; otherwise, outputs 0.

– if σ is a *nonowner-type* signature $\sigma = (A, B, C, D, E)$ (i.e., $\ell = 2$), it checks

$$e(g, H(D||2)) \stackrel{?}{=} e(pk, E),$$

$$e(D, H(B||1)) \stackrel{?}{=} e(g, C),$$

$$e(B, H(m||0)) \stackrel{?}{=} e(g, A).$$

If all the equalities hold, it outputs 1; otherwise, outputs 0.

Correctness   The correctness is due to the following equalities:

- Owner-type signature:

$$e(pk, H(B||1)) = e(g^a, H(B||1)) = e(g, H(B||1)^a) = e(g, C),$$

$$e(B, H(m||0)) = e(g^r, H(m||0)) = e(g, H(m||0)^r) = e(g, A).$$

- Nonowner-type signature:

$$e(pk, E) = e(g^b, H(D||2)^{1/b}) = e(g, H(D||2)),$$

$$e(D, H(B||1)) = e(g^{r_2}, H(B||1)) = e(g, H(B||1)^{r_2}) = e(g, C),$$

$$e(B, H(m||0)) = e(g^{r_1}, H(m||0)) = e(g, H(m||0)^{r_1}) = e(g, A).$$

Security Analysis of Scheme S_{ins}   The security proof of scheme S_{ins} is based on extended computational Diffie–Hellman (eCDH) assumption. That is, it is hard to compute g^{uv} or $g^{u/v}$ given $(q, g, \mathbb{G}, \mathbb{G}_T, e, g^u, g^v, g^{1/v})$, where g is a random element in $\mathbb{G}$, $\mathbb{G}, \mathbb{G}_T$ are bilinear groups with a prime order q, e is a bilinear map, $\mathbb{G} \times \mathbb{G} \to \mathbb{G}_T$, and u, v are random numbers in $\mathbb{Z}_q^*$.

Theorem 2.2.5. *Scheme S_{ins} is secure in the* AH *model if the eCDH problem is hard, and hash function H is treated as a random oracle.*

Proof. We show that if adversary $\mathcal{A}$ can break scheme S_{ins} in the AH model, we can build another algorithm $\mathcal{B}$ that can solve the eCDH problem. Given

$$(q, g, \mathbb{G}, \mathbb{G}_T, e, g^u, g^v, g^{1/v}),$$

$\mathcal{B}$ aims to output g^{uv} or $g^{u/v}$. The PRS security game goes as follows:

External Security:

- *Random oracle $\mathcal{O}_h$:* On input string R, $\mathcal{B}$ first checks whether $(R, R_h, r_h, *)$ is in Table T_h. If yes, $\mathcal{B}$ returns R_h and terminates; otherwise, $\mathcal{B}$ chooses a random number $r_h \in \mathbb{Z}_q^*$, and proceeds as follows:

  - The input string R satisfies the format $m\|0$, where $m \in \mathcal{M}$. $\mathcal{B}$ guesses whether m is the target message m^*. If yes, $\mathcal{B}$ outputs $R_h = (g^u)^{r_h}$; otherwise, $\mathcal{B}$ outputs $R_h = g^{r_h}$.

  - The input string R satisfies the format $m\|1$ or $m\|2$, where $m \in \mathbb{G}$. $\mathcal{B}$ outputs $R_h = (g^u)^{r_h}$.

  - The input string R does not satisfy any of the above formats. $\mathcal{B}$ outputs $R_h = g^{r_h}$.

  Finally, $\mathcal{B}$ records $(R, R_h, r_h, \perp)$ in Table T_h.

- $\mathcal{O}_{pk}$: On input index i, $\mathcal{B}$ first chooses a random number $x_i \in \mathbb{Z}_q^*$, and guesses whether it is pk^*. For $pk_i \neq pk^*$, it sets $pk_i = g^{x_i}$; for $pk_i = pk^*$, it sets $pk_i = (g^v)^{x_i}$. Finally, $\mathcal{B}$ records (pk_i, x_i) in Table T_k.

- $\mathcal{O}_s$: On input (pk_i, m_i), $\mathcal{B}$ proceeds as follows:

  - If $m_i = m^*$, then $pk_i \neq pk^*$, $\mathcal{B}$ chooses a random number $r \in \mathbb{Z}_q^*$, and outputs
    $$\sigma = (A, B, C) = (H(m^*\|0)^r, g^r, H(g^r\|1)^{x_i}).$$

  - If $m_i \neq m^*$, then $\mathcal{B}$ chooses a random number $r \in \mathbb{Z}_q^*$, and checks whether $((g^v)^r\|1, *, *, *)$ is in Table T_h. If it exists, $\mathcal{B}$ outputs `failure` and aborts; otherwise $\mathcal{B}$ chooses a random number $r_1 \in \mathbb{Z}_q^*$, and records $((g^v)^r\|1, g^{r_1}, r_1, r)$ in Table T_h. Then $\mathcal{B}$ checks whether $(m_i\|0, \star_1, \star_2, \perp)$

is in Table T_h. If it exists, then $\mathcal{B}$ sets $r_2 = \star_2$; otherwise, $\mathcal{B}$ chooses a random number $r_2 \in \mathbb{Z}_q^*$ and records $(m_i||0, g^{r_2}, r_2, \perp)$ in Table T_h. Finally, $\mathcal{B}$ outputs

$$\sigma = (A, B, C) = ((g^v)^{rr_2}, (g^v)^r, pk_i^{r_1}) = (H(m_i||0)^{vr}, g^{vr}, H(g^{vr}||1)^{vx_i}).$$

- $\mathcal{O}_{rs}$: On input $(pk_i, pk_j, m, \sigma, 1)$, where $\sigma = (A, B, C)$. If $\mathtt{Verify}(pk_i, m, \sigma, 1) = 1$, then $\mathcal{B}$ proceeds as follows; otherwise, outputs $\mathtt{reject}$.

  - If $pk_j \neq pk^*$, $\mathcal{B}$ uses x_j, associated to pk_j in Table T_k, to run $\mathtt{ReKeyGen}$ and $\mathtt{ReSign}$, and gets the required re-signature.

  - If $pk_j = pk^*$, then $m \neq m^*$, and $\mathcal{B}$ chooses a random number $r \in \mathbb{Z}_q^*$, and checks whether $((g^v)^{x_i r}||2, *, *, *)$ is in Table T_h. If it exists, $\mathcal{B}$ outputs $\mathtt{failure}$ and aborts; $\mathcal{B}$ chooses a random number $r_1 \in \mathbb{Z}_q^*$, and records $((g^v)^{x_i r}||2, g^{r_1}, r_1, x_i r)$ in Table T_h. $\mathcal{B}$ searches $(B||1, \star_1, \star_2, \star_3)$ in Table T_h, and outputs

$$
\begin{aligned}
\sigma' \;\; = (\;\;\; & A', & B', & \;\;C', & D', & E') \\
= (\;\;\; & A, & B, & \;\;(g^v)^{x_i \star_2 r}, & (g^v)^{x_i r}, & (g^{1/v})^{r_1/x_t}) \\
= (\;\;\; & A, & B, & \;\;H(B||1)^{x_i vr}, & g^{x_i vr}, & H(g^{x_i vr}||2)^{1/(vx_t)}),
\end{aligned}
$$

- *Forgery:* At some point, the adversary must output a forgery (pk^*, m^*, σ^*).

Now, we show how $\mathcal{B}$ gets the eCDH solution from the forgery as follows.

- If σ^* is an owner-type signature, such that $\sigma^* = (A^*, B^*, C^*)$, then we have the following analysis:

  - If any owner-type signature corresponding to pk^* from $\mathcal{O}_s$ is not of the form $(*, B^*, C^*)$, then $\mathcal{B}$ finds $(B^*||1, \star_1', \star_2', \star_3')$ in Table T_h, and gets the solution of the eCDH problem:

$$(C^*)^{1/(x^* \star_2')} = (H(B^*||1)^{vx_t})^{1/(x^* \star_2')} = (g^{u \star_2' vx_t})^{1/(x^* \star_2')} = g^{uv}.$$

Note that $pk^* = (g^v)^{x^*}$.

- Once an owner-type signature corresponding to pk^* from $\mathcal{O}_s$ is of the form $(*, B^*, C^*)$, then $\mathcal{B}$ finds out $(m^*||0, \star_1, \star_2, \star_3)$ and $(B^*||1, \star_1', \star_2', \star_3')$ in Table T_h, and gets the solution of the eCDH problem:

$$(A^*)^{1/(\star_2 \star_3')} = (H(m^*||0)^{v\star_3'})^{1/(\star_2 \star_3')} = (g^{uv\star_2 \star_3'})^{1/(\star_2 \star_3')} = g^{uv}.$$

Note that $B^* = (g^v)^{\star_3'}$.

- If σ^* is a nonowner-type signature, such that $\sigma^* = (A^*, B^*, C^*, D^*, E^*)$, then we have the following analysis:

  - If any signature corresponding to pk^* from $\mathcal{O}_{rs}$ is not of the special form $(*, *, *, D^*, E^*)$, $\mathcal{B}$ finds $(D^*||2, \star_1'', \star_2'', \star_3'')$ in Table T_h and gets the solution of the eCDH problem:

$$(E^*)^{x_t/\star_2''} = (H(D^*||2)^{1/(vx_t)})^{x_t/\star_2''} = (g^{u\star_2''/(vx_t)})^{x_t/\star_2''} = g^{u/v}.$$

Note that $pk^* = (g^v)^{x_t}$.

  - If at least one signature corresponding to pk^* from the output of oracles is of the form $(*, *, *, D^*, E^*)$, but any signature corresponding to pk^* from $\mathcal{O}_{rs}$ is not of the form $(*, B^*, C^*, D^*, E^*)$, $\mathcal{B}$ finds out $(D^*||2, \star_1'', \star_2'', \star_3'')$ and $(B^*||1, \star_1', \star_2', \star_3')$ in Table T_h and gets the solution of the eCDH problem:

$$(C^*)^{1/(\star_3'' \star_2')} = (H(B^*||1)^{v\star_3''})^{1/(\star_3'' \star_2')} = (g^{u\star_2' v\star_3''})^{1/(\star_3'' \star_2')} = g^{uv}.$$

Note that $D^* = (g^v)^{\star''_3}$.

– Once an owner-type signature corresponding to pk^* from $\mathcal{O}_{rs}$ is of the form $(\ast, B^*, C^*, D^*, E^*)$, then $\mathcal{B}$ finds out $(m^*||0, \star_1, \star_2, \star_3)$ and $(B||1, \star'_1, \star'_2, \star'_3)$ in Table T_h, and gets the solution of the eCDH problem:

$$(A^*)^{1/(\star_2 \star'_3)} = (H(m^*||0)^{v\star'_3})^{1/(\star_2 \star'_3)} = (g^{uv\star_2 \star'_3})^{1/(\star_2 \star'_3)} = g^{uv}.$$

Note that $B^* = (g^v)^{\star'_3}$.

Notice that $\mathcal{B}$ guesses the right target verification key with the probability $1/q_{pk}$ at least, and $\mathcal{B}$ outputs failure in $\mathcal{O}_s$ and $\mathcal{O}_{rs}$ with the probabilities $(q_h + q_s)/q$ and $(q_h + q_{rs})/q$ at most, respectively. Here, q_{pk}, q_h, q_s, and q_{rs} are the maximum numbers that $\mathcal{A}$ can query to $\mathcal{O}_{pk}$, $\mathcal{O}_h$, $\mathcal{O}_s$, $\mathcal{O}_{rs}$, respectively. As a result, $\mathcal{B}$ solves the eCDH problem with a non-negligible probability.

Internal Security: Internal security includes limited proxy, delegatee security, and delegator security.

Limited Proxy:

- $\mathcal{O}_h$: Identical to that in external security.

- $\mathcal{O}_{pk}$: On input index i, $\mathcal{B}$ first chooses a random number $x_i \in \mathbb{Z}_q^*$, and then outputs $pk_i = (g^v)^{x_i}$. Finally, $\mathcal{B}$ records (pk_i, x_i) in Table T_k.

- $\mathcal{O}_s$: On input (pk_i, m_i), $\mathcal{B}$ proceeds as follows:

  – If $m = m^*$ and $pk_i \neq pk^*$ $\mathcal{B}$ chooses a random number r, and checks whether $(g^r||1, \ast, \ast, \ast)$ is in Table T_h. If it exists, $\mathcal{B}$ outputs failure and aborts; otherwise, $\mathcal{B}$ chooses a random number r_1 and records

$$(g^r||1, g^{r_1}, r_1, \perp)$$

in Table T_h. Finally, $\mathcal{B}$ outputs $(H(m^*||0)^r, g^r, pk_i^{r_1})$.

- If $m \neq m^*$, then $\mathcal{B}$ performs the same as that in external security.

- $\mathcal{O}_{rk}$: On input (pk_i, pk_j), $\mathcal{B}$ chooses a random number $r \in \mathbb{Z}_q^*$, and checks whether $((g^v)^{x_i r}||2, *, *, *)$ is in Table T_h. If it exists, $\mathcal{B}$ outputs failure and aborts; otherwise, $\mathcal{B}$ chooses a random number $r_1 \in \mathbb{Z}_q^*$, records

$$((g^v)^{x_i r}||2, g^{r_1}, r_1, x_i r)$$

in Table T_h and outputs $(r, (g^v)^{x_i r}, (g^{1/v})^{r_1/x_j})$.

Note that $pk_i = (g^v)^{x_i}$ and $pk_j = (g^v)^{x_j}$.

With the similar analysis in the external security, $\mathcal{B}$ solves the eCDH problem with a non-negligible probability.

Delegatee Security: Compared to the limited proxy, $\mathcal{B}$ needs to change $\mathcal{O}_{pk}$, $\mathcal{O}_{sk}$, $\mathcal{O}_s$, and $\mathcal{O}_{rk}$ as follows.

- $\mathcal{O}_{pk}$: For the target delegatee, set the verification key as $(g^v)^{x_0}$, and for all other users, g^{x_i}, where $x_i(i = 0, \ldots, n) \in \mathbb{Z}_q^*$.

- $\mathcal{O}_{sk}$: On input pk_i, $\mathcal{B}$ returns the corresponding x_i.

- $\mathcal{O}_s$: On input (pk_0, m_i), $\mathcal{B}$ performs the same as in $\mathcal{O}_s$ with input (pk^*, m_i) in external security, where pk_0 is treated as pk^*.

- $\mathcal{O}_{rk}$: On input (pk_i, pk_j), where $pk_j \neq pk_0$, $\mathcal{B}$ performs the same as in the real execution since it knows x_j corresponding to pk_j.

With the similar analysis in the external security, $\mathcal{B}$ solves the eCDH problem with a non-negligible probability.

Delegator Security: Compared to the limited proxy, $\mathcal{B}$ needs to change $\mathcal{O}_{pk}$, $\mathcal{O}_s$, and $\mathcal{O}_{rk}$ as follows:

- $\mathcal{O}_{pk}$: For the target delegator, set the verification key as $(g^v)^{x_0}$, and for all other users, g^{x_i}, where $x_i(i = 0, \ldots, n) \in \mathbb{Z}_q^*$.

- $\mathcal{O}_{sk}$: On input pk_i, $\mathcal{B}$ returns the corresponding x_i.

- $\mathcal{O}_s$: On input (pk_0, m_i), $\mathcal{B}$ performs the same as in $\mathcal{O}_s$ with input (pk^*, m_i) in external security, where pk_0 is treated as pk^*.

- $\mathcal{O}_{rk}$: On input (pk_i, pk_j), $\mathcal{B}$ proceeds as follows:

  - If $pk_j \neq pk_0$, $\mathcal{B}$ gets x_j from Table T_k, and uses x_j to run ReKeyGen(pk_i, pk_j). In the end, $\mathcal{B}$ outputs the result of ReKeyGen.

  - If $pk_j = pk_0$, $\mathcal{B}$ chooses a random number $r \in \mathbb{Z}_q^*$ and checks whether $(pk_i^r||2, *, *, *)$ is in Table T_h. If it exists, $\mathcal{B}$ outputs failure and aborts; otherwise, $\mathcal{B}$ chooses a random number $r_1 \in \mathbb{Z}_q^*$, records $(pk_i^r||2, g^{r_1}, r_1, \bot)$ in Table T_h, and outputs $(r, pk_i^r, (g^{1/v})^{r_1/x_0})$.

With the similar analysis in the external security, $\mathcal{B}$ solves the eCDH problem with a non-negligible probability. Note that in this case, the forgery σ^* is an owner-type signature.

This completes the proof.   □

An Attack on Scheme S_{ins}**:**   Although the security proof of scheme S_{ins} in the AH model is given before, there exists the following attack. Let us now consider the case: Alice $\rightarrow$ Proxy $\rightarrow$ Bob, i.e., Bob delegates his signing rights to Alice via Proxy. First, Alice can produce an owner-type signature on m: $\sigma_a = (H(m||0)^r, g^r, H(g^r||1)^a)$, where r is known to Alice. Then Proxy can transform σ_a into Bob's signature $\sigma_b = (H(m||0)^r, g^r, (H(g^r||1)^a)^{rk_{a \rightarrow b}^{(1)}}, rk_{a \rightarrow b}^{(2)}, rk_{a \rightarrow b}^{(3)})$. In this case, Alice can generate signatures on any message simply by changing m to m', since she knows r. This shows that scheme S_{ins} is insecure. Hence, the AH model is not suitable for private proxy and unidirectional scheme. Since most of the existing unidirectional PRS schemes are private proxy and unidirectional schemes, it is desired to propose a new security model to solve this problem.

Remark 2.2.6. In fact, the AH model does not consider the attack that the delegatee tries to forge a valid nonowner-type signature of the delegator without colluding with the proxy.

The main reason why scheme S_{ins} is insecure is that algorithm ReSign does not affect the value A containing the message m in the owner-type signature. However, the schemes in [11, 193] do not have this flaw.

2.2.6   AH$^+$ **Model**

In this section, we propose a new security model for private proxy and unidirectional PRS. Due to its simplicity, it is easy to verify its completeness.

Before giving our security model, we will first define several terms.

1. If user A delegates his signing rights to user B via proxy P, then both users A and B are said to be in a *delegation chain*, denoted as (B,A). User B is called user A's *delegation predecessor*. The combination of the proxy and a user, either the delegatee B or the delegator A, is called a *delegation pair*. Therefore, user A and proxy P is a *delegation pair*, so is user B and proxy P.

2. If one of the parties in a delegation pair is corrupted, the delegation pair is *corrupted*; otherwise, it is *uncorrupted*.

3. A user can be treated as the smallest *delegation chain*.

4. If two users A and B are in a delegation chain and B is A's delegation predecessor, B's signature can be transformed by a proxy or proxies into A's signature.

5. A delegation chain is its own *subchain*.

6. *(Only for multiuse.)* If user A delegates his signing rights to user B via a proxy P, and user B delegates his signing rights to user C via a proxy P', users A and C are said to be in a *delegation chain* too. User C is also called user A's *delegation predecessor*. In this case, users A, B, and C are in a delegation chain (C, B, A). The delegation chains (B, A) and (C, B) are *delegation subchains* of the delegation chain (C, B, A). The delegation chain (C, B, A) can be extended if C delegates his signing rights to another user via another proxy.

Existential Unforgeability of Private Proxy and Unidirectional Proxy Re-Signature: The existential unforgeability of private proxy and unidirectional PRS is defined by the following adaptively chosen-message attack game played between a challenger C and an adversary A. We consider the security in a static situation, that is, before the game starts, the adversary should decide which users and proxies are to be corrupted. Furthermore, the challenger generates all the verification keys in the following game.

Queries: The adversary adaptively makes a number of different queries to the challenger. Each query can be one of the following:

- $\mathcal{O}_{pk}$: On input an index i, the challenger first runs KeyGen(1^λ) to get a key pair (pk_i, sk_i), and then forwards the verification key pk_i to the adversary. Finally, the challenger records (pk_i, sk_i) in Table T_k. In the following oracles, sk_i is the signing key corresponding to pk_i.

- $\mathcal{O}_{sk}$: On input a verification key pk_i, the challenger responds with sk_i which is the associated value with pk_i in Table T_k, if pk_i is corrupted; otherwise, the challenger responds with reject.

- $\mathcal{O}_{rk}$: On input two verification keys pk_i, pk_j ($pk_i \neq pk_j$), the challenger responds with ReKeyGen(pk_i, pk_j, sk_j).

- $\mathcal{O}_s$: On input a verification key pk_i and a message m_i, the challenger responds with Sign($sk_i, m, 1$).

- $\mathcal{O}_{rs}$: On input two verification keys pk_i, pk_j ($pk_i \neq pk_j$), a message m_i, and a signature σ_i at level ℓ, the challenger responds with

$$\texttt{ReSign}(\texttt{ReKeyGen}(pk_i, pk_j, sk_j), pk_i, m_i, \sigma_i, \ell).$$

Forgery: The adversary outputs a message m^*, a verification key pk^*, and a signature σ^* at level ℓ^*. The adversary *wins* if all the following requirements are satisfied:

- $\texttt{Verify}(pk^*, m^*, \sigma^*, \ell^*) = 1$.

- pk^* is uncorrupted.

- The adversary has not made a signature query on (pk^*, m^*).

- The adversary has not made a signature query on (pk', m^*), where pk' is uncorrupted, and there exists such a delegation subchain from pk' to pk^* that does not contain any uncorrupted delegation pair;

- The adversary has not made a re-signature key query on (pk_i, pk_j), which satisfies all the following conditions:

  - pk_i is corrupted,

  - pk_j is uncorrupted,

  - There exists such a delegation subchain from pk_j to pk^* that does not contain any uncorrupted delegation pair.

- The adversary has not made a re-signature query on $(pk_i, pk_j, m^*, \sigma_i, *)$, where pk_j is uncorrupted, and there exists such a delegation subchain from pk_j to pk^* that does not contain any uncorrupted delegation pair.

We define $\mathbf{Adv}_{\texttt{PRS}}^{\texttt{EU}}(\lambda)$ to be the probability that adversary $\mathcal{A}$ wins in the above game for the security parameter λ.

Definition 2.2.7. A private proxy and unidirectional scheme is existentially unforge-able with respect to adaptive chosen message attacks if for all p.p.t. adversaries $\mathcal{A}$,

$\mathbf{Adv}^{EU}_{PRS}(\lambda)$ is negligible for the security parameter λ.

Remark 2.2.8 (Requirement). The requirements for the adversary's success guarantee that $(pk^*, m^*, \sigma^*, \ell^*)$ is a valid signature; the adversary cannot trivially obtain a valid owner-type signature by obtaining the signing key; the adversary cannot trivially obtained a valid owner-type signature by the signature oracle; and the adversary cannot trivially obtain a valid nonowner-type signature by the re-signature key oracle and re-signature oracle.

Remark 2.2.9 (Chosen-Key Model). Following the spirit in [193], we can easily extend our security model into the chosen-key model [23], where the certificate authority does not need to verify that the owner of one verification key indeed knows the corresponding signing key. In particular, the challenger is no longer responsible for generating the key pairs for corrupted users, and the adversary generates all the corrupted verification keys involved in other oracles.

The static mode can be modified to adaptive mode by removing the restriction that the adversary should decide which users and proxies are to be corrupted before the game starts.

2.3   Proxy Re-Encryption

Proxy re-encryption (PRE) is also proposed by Blaze et al. [29] at Eurocrypt 1998. In such an encryption scheme, a semi-trusted proxy, holding some information (a.k.a. re-encryption key), can transform a ciphertext under one public key (delegator Alice) to another ciphertext under another public key (delegatee Bob). These two ciphertexts are corresponding to the same plaintext. However, the proxy cannot obtain the plaintext.

2.3.1   Properties and Definition

PRE has the similar desired properties with PRS.

- *Unidirectional:* In this scheme, a re-encryption key allows the proxy to transform Alice's ciphertext to Bob's but not vice versa. In a bidirectional scheme, on the other hand, the re-encryption key allows the proxy to transform Alice's ciphertext to Bob's as well as B's ciphertext to Alice's.

- *Multiuse:* In a multiuse scheme, a transformed ciphertext can be re-transformed again by a proxy. In a single-use scheme, the proxy can transform only the ciphertexts that have not been transformed.

- *Private proxy:* The proxy can keep the re-encryption key as a secret in a private proxy scheme, but anyone can recompute the re-encryption key by observing the process of re-enncryption passively in a public proxy scheme.

- *Transparent:* In a transparent scheme, users may not even know the existence of a proxy.

- *Key-optimal:* In a key-optimal scheme, a user is required to protect and store only a small constant amount of secrets no matter how many decryption delegations the user gives or accepts.

- *Noninteractive:* The delegatee is not required to participate in delegation process.

- *Non-transitive:* A re-encryption key cannot be generated from two other re-encryption keys. For example, the re-encryption key from Alice to Bob cannot be generated from the re-encryption keys from Alice to Tina and Tina to Bob.

- *Temporary:* A re-encryption right is temporary. This can be done by either revoking the right [11] or expiring the right.

- *Collusion-resistant:* The delegator can delegate the decryption rights to the delegatee via the proxy, while keeping signing rights for the same public key.

Definition 2.3.1 (Proxy Re-encryption). A PRE scheme is a tuple of p.p.t. algorithms KeyGen, ReKeyGen, Enc, ReEnc, Dec.

- KeyGen: It takes as input the security parameter λ, and returns a public key pk and a private key sk.

- ReKeyGen: It takes as input delegator Alice's key pair (pk_A, sk_A) and delegatee Bob's key pair (pk_B, sk_B), and returns a re-encryption key $rk_{A \to B}$ for the proxy. If the PRE scheme is unidirectional, the delegatee's private key is not included in the input. If the PRE scheme is bidirectional, then the proxy can easily obtain $rk_{B \to A}$ from $rk_{A \to B}$. In many bidirectional schemes [65], $rk_{A \to B} = 1/rk_{B \to A}$.

- Enc: It takes as input a public key pk, a positive integer ℓ, and a message m from the message space, and returns a ciphertext at level ℓ. If the PRE scheme is single-use, then $\ell \in \{1, 2\}$.

- ReEnc: It takes as a re-encryption key $rk_{A \to B}$ and a ciphertext C_A under pk_A at level ℓ, and returns the ciphertext C_B under pk_B at level $\ell + 1$. If the PRE scheme is single-use, then $\ell = 1$ and it can be omitted.

- Dec: It takes as input a private key sk and a ciphertext C at level ℓ, and returns m in the message space or a special symbol reject.

Correctness. The correctness property has two requirements. For any message m in the message space and any key pairs $(pk, sk), (pk', sk') \leftarrow$ KeyGen(1^λ). Then the following two conditions must hold:

$$\text{Dec}(sk, \text{Enc}(pk, m, \ell), \ell) = m \text{ and}$$
$$\text{Dec}(sk', \text{ReEnc}(\text{ReKeyGen}(sk, pk'), C, \ell), \ell + 1) = m,$$

where C is the ciphertext for message m under pk from algorithm Enc or ReEnc.

2.3.2 Related Work

Since the introduction of PRE by Blaze *et al.* [29], there have been many papers
[7, 9, 10, 29, 65, 86, 96, 130, 145, 156, 246] that have proposed different PRE schemes
with different security properties.

The first bidirectional, multiuse PRE scheme secure against chosen-plaintext attack
(CPA) was proposed by Blaze *et al.* [29]; however, it is not collusion resistant. Based
on public key encryption with double trapdoors, Ateniese *et al.* [9, 10] proposed the
first collusion resistant, unidirectional, single-use PRE schemes with CPA security. At
TCC 2007, Hohenberger *et al.* [145] proposed a new collusion resistant, CPA-secure,
unidirectional PRE scheme, where the re-encryption key together with the re-encryption
algorithm can be treated as an obfuscated re-encryption program.

However, many PRE's applications, such as the distributed file system, demand that
the underlying PRE scheme is secure against chosen-ciphertext attack (CCA). To solve
the problem, Canetti and Hohenberger [65] and Ateniese and Green [130] proposed the
definition of CCA security independently. There are two kinds of CCA-secure PRE
schemes: with pairings and without pairings.

By using pairings and the CHK paradigm [36, 63], Canetti and Hohenberger [65]
proposed the first CCA-secure (bidirectional) PRE scheme in the standard model. However, their scheme suffers from collusion attacks. Furthermore, they didn't propose any
CCA-secure unidirectional PRE scheme, and left it as an open problem. Based on
Canetti-Hohenberger technique, Libert and Vergnaud [194, 195] proposed a new unidirectional PRE scheme that is CCA-secure and collusion resistant in the standard model.
Recently, Weng *et al.* [293] and we [248] proposed CCA-secure and collusion-resistant
unidirectional PRE schemes by improving Libert and Vergnaud's method.

Due to the heavy cost of pairing computation, it is desired to design CCA-secure
PRE schemes without pairings. We [246] and Chow et al. [86] proposed CCA-secure
and collusion-resistant unidirectional PRE schemes without pairings. However, the pro-

posed schemes are only proven secure in the random oracle model. Most recently, Matsuda *et al.* [210] proposed a new pairing-free, CCA-secure (bidirectional) PRE scheme, which is proven secure in the standard model. However, it suffers from the collusion attack. Furthermore, Weng et al. [295] pointed out that the scheme of Matsuda et al. is not CCA-secure, but they did not suggest any improvement.

In 2009[1], we [249] proposed a generic construction for CCA-secure, single-use, unidirectional PRE from CCA-secure (2,2) threshold public key encryption. With the generic construction, we can obtain two kinds of single-use, unidirectional PRE schemes. One is secure in the random oracle model without pairings, and the other is secure in the standard model with pairings. At CT-RSA 2012, Hanaoka et al. [136] obtained a similar result by using similar methods.

There are also some PRE schemes with some special properties, such as threshold [249], conditional [255, 272, 294], invisible proxy [160, 272], identity-based [130, 255, 283], anonymous [7, 253, 254], and searchable [247].

In the rest of this section, we first introduce the security models of (unidirectional) PRE. In what follows, we introduce the PRE scheme [246] proposed at PKC 2009.

2.3.3   Security Models

Usually, unidirectional PRE is better than bidirectional PRE. For example, the delegator delegates his decryption rights to the delegatee, the delegatee does not always want to do the reverse delegation. Furthermore, any unidirectional scheme can be easily transformed to a bidirectional one by running the former in both directions, while whether the reverse holds is unknown. Hence, we only give the security models for single-use unidirectional PRE and multiuse unidirectional PRE in this section.

The security models below are defined by games played between a challenger C and an adversary $\mathcal{A}$. We assume that the public keys input into the oracles by the adversary

---

[1]The paper [249] was published online in May, 2009.

are all from public key generation oracle $\mathcal{O}_{pk}$. We say that a public key is corrupted if it has been queried to private key generation oracle $\mathcal{O}_{sk}$ by the adversary.

Chosen Ciphertext Security for Single-use, Unidirectional Proxy Re-encryption (SUPRE)

In almost all existing unidirectional PRE schemes except one [119], there are two types of ciphertexts. One is original ciphertext ($\ell - 1$) that can only be generated by Enc, and the other is re-encrypted ciphertext ($\ell = 2$) that can be generated by not only Enc but also ReEnc. Hence, we have two security models for SUPRE, one aims to protect the plaintext corresponding to the original ciphertext, and the other aims to protect the plaintext corresponding to the re-encrypted ciphertext.

The challenge ciphertext is an original ciphertext.

Setup: The challenger sets up the system parameters.

Phase 1: The adversary $\mathcal{A}$ can issue the following queries adaptively.

- *Public key generation oracle* $\mathcal{O}_{pk}$: The challenger takes a security parameter k, runs $\text{KeyGen}(1^\lambda)$ to generate a key pair (pk_i, sk_i), gives pk_i to $\mathcal{A}$ and records (pk_i, sk_i) in Table T_k. In the following, sk_i is the corresponding private key to pk_i.

- *Private key generation oracle* $\mathcal{O}_{sk}$: On input pk_i, the challenger searches for pk_i in Table T_k and returns sk_i.

- *Re-encryption key generation oracle* $\mathcal{O}_{rk}$: On input (pk_i, pk_j), the challenger returns the re-encryption key $rk_{i,j} = \text{ReKeyGen}(sk_i, pk_j)$.

- *Re-encryption oracle* $\mathcal{O}_{re}$: On input (pk_i, pk_j, C), the challenger returns the re-encrypted ciphertext $C' = \text{ReEnc}(\text{ReKeyGen}(sk_i, pk_j), C)$.

- *Decryption oracle* $\mathcal{O}_{dec}$: On input (pk_i, C_i), the challenger returns $\mathtt{Dec}(sk_i, C_i)$.

Challenge: Once Phase 1 is over, $\mathcal{A}$ outputs two plaintexts m_0, m_1 with equal length from the message space, and a public key pk^* on which $\mathcal{A}$ wishes to challenge. There are two restrictions on the public key pk^*: (i) pk^* has not been queried to $\mathcal{O}_{sk}$; (ii) if $(pk^*, \star)$ has appeared in any query to $\mathcal{O}_{rk}$, $\star$ should not be queried to $\mathcal{O}_{sk}$. The challenger picks a random bit $\mathbf{b} \in \{0, 1\}$ and sets $C^* = \mathtt{Enc}(pk^*, m_\mathbf{b})$. It sends C^* as the challenge to $\mathcal{A}$.

Phase 2: This phase is almost the same as Phase 1 but with the following constraints:

- $\mathcal{O}_{sk}$: On input pk_i, if $pk_i = pk^*$, the challenger outputs $\mathtt{reject}$.

- $\mathcal{O}_{rk}$: On input (pk_i, pk_j), if $pk_i = pk^*$, and pk_j is corrupted, the challenger outputs $\mathtt{reject}$.

- $\mathcal{O}_{re}$: On input (pk_i, pk_j, C_i), if $(pk_i, C_i) = (pk^*, C^*)$ and pk_j is corrupted, the challenger outputs $\mathtt{reject}$.

- $\mathcal{O}_{dec}$: On input (pk_i, C_i), if (pk_i, C_i) is a $\mathtt{derivative}$ (see the part after Definition 2.3.2 below) of (pk^*, C^*), the challenger outputs $\mathtt{reject}$.

Guess: Finally, the adversary $\mathcal{A}$ outputs a guess $\mathbf{b}' \in \{0, 1\}$ and wins the game if $\mathbf{b} = \mathbf{b}'$.

We refer to such an adversary $\mathcal{A}$ as a CCA-O adversary. We define adversary $\mathcal{A}$'s advantage in attacking SUPRE as the following function:

$$\mathbf{Adv}_{\mathtt{SUPRE}}^{\mathtt{CCA-O}}(\lambda) = |\mathrm{Pr}[\mathbf{b} = \mathbf{b}'] - 1/2|.$$

Using the CCA-O game, we can define CCA-O security of SUPRE.

Definition 2.3.2. [CCA-O Security] We say that a single-use, unidirectional PRE scheme SUPRE is CCA-O-secure, if for any polynomial time CCA-O adversary $\mathcal{A}$, the function $\mathbf{Adv}_{\mathtt{SUPRE}}^{\mathtt{CCA-O}}(\lambda)$ is negligible.

The core of PRE's security is the definition of derivatives of ciphertext. Derivatives of (pk^*, C^*) could be obtained by combining some of the following rules:

R-1 (pk^*, C^*) is a derivative of itself.

R-2 If $\mathcal{A}$ has queried $\mathcal{O}_{re}$ on input (pk^*, pk, C^*) and obtained C, (pk, C) is a derivative of (pk^*, C^*).

R-3 If $\mathcal{A}$ has queried $\mathcal{O}_{rk}$ on input (pk^*, pk) and $C \leftarrow \texttt{ReEnc}(\mathcal{O}_{rk}(pk^*, pk), C^*)$, (pk, C) is a derivative of (pk^*, C^*).

R-4 If $\mathcal{A}$ has queried $\mathcal{O}_{rk}$ on input (pk^*, pk) and $\texttt{Dec}(pk, C) \in \{m_0, m_1\}$, (pk, C) is a derivative of (pk^*, C^*).

R-5 If (pk, C) is a re-encrypted ciphertext and $\texttt{Dec}(pk, C) \in \{m_0, m_1\}$, (pk, C) is a derivative of (pk^*, C^*).

The definition of derivatives we use in this section consists of relations R-1, R-2, and R-3, named $\texttt{Def}_{\texttt{der}}$. The single-use unidirectional case extended from the Canetti-Hohenberger definition [65] corresponds to relations R-1, R-2, and R-4, and the Libert-Vergnuad definition [194, 195] consists of items R-1 and R-5.

R-1 is the trivial reflexive relation. R-2 models the direct consequence of the re-encryption process. R-5 follows the spirit of the definition of re-randomizable CCA security for the re-encrypted ciphertext. One possible stronger rule (as R-4) is that only when $\mathcal{A}$ has queried the required re-encryption key, then all re-encrypted ciphertexts that are generated by such a key and have either one of the challenge messages as the plaintext are disallowed to be queried to $\mathcal{O}_{dec}$. R-3, in a strict sense, is still deviated from the regular CCA security since a class of ciphertexts instead of a single ciphertext is disallowed.

Most importantly, $\texttt{Def}_{\texttt{der}}$ disallows a *smaller* class of ciphertexts when compared with other definitions. Consider an adversary which has queried the re-encryption oracle with (pk^*, pk, C^*) and obtained (pk, C), where pk is an uncorrupted user. It is possible that this adversary can change C into C' while the adversary does not require to run ReEnc algorithm, yet the output of $\text{Dec}(C, pk)$ equals to the one of $\text{Dec}(C', pk)$, or it can be computed easily from the one of $\text{Dec}(C', pk)$. According to the extension of the Canetti-Hohenberger definition and the Libert-Vergnaud definition, (pk, C') *is a* derivative of (pk^*, C^*). However, (pk, C') *is not* a derivative of (pk^*, C^*) in Def_{der}.

Essentially, Def_{der} *only* takes the re-encrypted ciphertext obtained *directly* either from the algorithm ReEnc or the oracle $\mathcal{O}_{re}$ as the derivative. This feature makes Def_{der} closer to the CCA security definition for the typical public key encryption.

The challenge ciphertext is a re-encrypted level ciphertext. The CCA-R security of SUPRE is defined by the same approach of the CCA-O security.

Phase 1, Guess: Identical to that in CCA-O game.

Challenge: Once Phase 1 is over, $\mathcal{A}$ outputs two plaintexts m_0, m_1 with equal length from the message space, and two uncorrupted public keys pk, pk^* on which $\mathcal{A}$ wishes to challenge. The challenger picks a random bit $\text{b} \in \{0, 1\}$ and sets $C^* = \text{ReEnc}(rk, \text{Enc}(pk, m_{\text{b}}))$, where rk is a re-encryption key from pk to pk^*. It sends C^* as the challenge to $\mathcal{A}$.

Phase 2: Almost the same as in Phase 1 but with the following constraints:

- $\mathcal{O}_{sk}$: On input pk_i, if $pk_i = pk$ or $pk_i = pk^*$, the challenger outputs reject.

- $\mathcal{O}_{dec}$: On input (pk_i, C_i), if $(pk_i, C_i) = (pk^*, C^*)$, the challenger outputs reject.

We also define $\mathbf{Adv}_{\text{SUPRE}}^{\text{CCA-R}}(\lambda) = |\Pr[\text{b} = \text{b}'] - 1/2|$ for the security parameter λ as in CCA-O security.

Definition 2.3.3 (CCA-R Security). We say that a single-use, unidirectional PRE scheme SUPRE is CCA-R-secure, if for any polynomial time CCA-R adversary $\mathcal{A}$ the function $\mathbf{Adv}_{\mathrm{SUPRE}}^{\mathrm{CCA\text{-}R}}(\lambda)$ is negligible.

Remark 2.3.4. There is another security, named collusion resistance, guaranteeing that any collusion of delegatee and proxy cannot lead to the reveal of the delegator's private key. It is easy to see that the CCA-R security implies collusion resistance, since anyone obtaining the delegator's private key can definitely decrypt the re-encrypted ciphertext for the delegator.

Chosen-Ciphertext Security for Multiuse, Unidirectional Proxy Re-Encryption

Before giving our security model, we will first define several terms for PRE.

- If user A delegates his decryption rights to user B via a proxy P, then both users A and B are said to be in a *delegation chain*, denoted as (A, B). If user B delegates his decryption rights to user C via a proxy P', then users A and C are said to be in a *delegation chain* too, the resulting delegation chain is denoted as (A, B, C). If user C delegates his decryption rights to other people, the delegation chain is further extended.

- One user is the smallest delegation chain.

- If the first and last user in a delegation chain are users A and B, respectively, then the delegation chain can be denoted as $(A, \ldots, B)$.

- For a delegation chain $(A, \ldots, B)$, if the ciphertext for user A can be transformed to the ciphertext for user B by the outputs of the re-encryption key generation oracle and the private key generation oracle, the delegation chain $(A, \ldots, B)$ is *corrupted*; otherwise, it is *uncorrupted*.

Now, we give the CCA security for multiuse, unidirectional PRE MUPRE by the following CCA game.

Setup: $\mathcal{C}$ sets up the system parameters, and sends the system parameters to $\mathcal{A}$.

Phase 1: $\mathcal{A}$ adaptively issues the following queries:

- *Public key generation oracle $\mathcal{O}_{pk}$:* On input an index i, $\mathcal{C}$ first takes a security parameter 1^λ, and then runs algorithm $\texttt{KeyGen}(1^\lambda)$ to generate a key pair (pk_i, sk_i), gives pk_i to $\mathcal{A}$ and records (pk_i, sk_i) in Table T_k. In the following oracles, sk_i is the private key corresponding to pk_i.

- *Private key generation oracle $\mathcal{O}_{sk}$:* On input pk_i, $\mathcal{C}$ searches pk_i in Table T_k and returns the associated sk_i.

- *Re-encryption key generation oracle $\mathcal{O}_{rk}$:* On input (pk_i, pk_j), $\mathcal{C}$ returns the re-encryption key $rk_{i,j} = \texttt{ReKeyGen}(sk_i, pk_j)$.

- *Re-encryption oracle $\mathcal{O}_{re}$:* On input $(pk_i, pk_j, (\ell, C))$, $\mathcal{C}$ returns the re-encrypted ciphertext

$$(\ell + 1, C') \leftarrow \texttt{ReEnc}(\texttt{ReKeyGen}(sk_i, pk_j), (\ell, C)).$$

- *Decryption oracle $\mathcal{O}_{dec}$:* On input $(pk, (\ell, C))$, $\mathcal{C}$ returns $\texttt{Dec}(sk, (\ell, C))$.

Challenge: Once Phase 1 is over, $\mathcal{A}$ outputs two plaintexts m_0, m_1 with equal length from the message space, a positive integer ℓ^*, a set of public keys $\{pk_{i_j}\}_{j=1}^{j=\ell^*-1}$, and an uncorrupted public key pk^* on which $\mathcal{A}$ wishes to challenge. There is one constraint on the public key pk^*, i.e., there does not exist a corrupted delegation chain from pk^* to a corrupted public key. $\mathcal{C}$ picks a random bit $\mathbf{b} \in \{0,1\}$ and sets

$C^* = \texttt{ReEnc}(\texttt{ReKeyGen}(sk^*, pk_{i_{\ell^*-1}}), \texttt{ReEnc}(\texttt{ReKeyGen}(sk_{i_{\ell^*-1}}, pk_{i_{\ell^*-2}}), \ldots,$
$\texttt{ReEnc}(\texttt{ReKeyGen}(sk_{i_1}, pk_{i_2}), \texttt{Enc}(pk_1, m_\mathbf{b})))).$

It sends (ℓ^*, C^*) as the challenge to $\mathcal{A}$.

Note that when $\ell^* = 1$, then $\{pk_{i_j}\}_{j=1}^{j=\ell^*-1}$ is an empty set, and $C^* = \text{Enc}(pk^*, m_b)$.

Phase 2: $\mathcal{A}$ adaptively issues more queries:

- $\mathcal{O}_{pk}$: Identical to that in Phase 1.

- $\mathcal{O}_{sk}$: On input pk_i, if there exists a corrupted delegation chain from pk^* to pk_i, $\mathcal{C}$ outputs reject; otherwise, $\mathcal{C}$ executes the same steps as in Phase 1.

- $\mathcal{O}_{rk}$: On input (pk_i, pk_j), if one of the following conditions holds, $\mathcal{C}$ outputs reject; otherwise, $\mathcal{C}$ executes the same steps as in Phase 1.

    - pk_i is uncorrupted and pk_j is corrupted, and there is a corrupted delegation chain from pk^* to pk_i.

    - pk_i is uncorrupted and pk_j is corrupted, and there is a corrupted delegation chain from pk to pk_i, where the adversary has a derivative (see below) of $(pk^*, (\ell^*, C^*))$: $(pk, (\ell, C))$.

The definition of derivatives of $(pk^*, (\ell^*, C^*))$ is different from that in single-use case.

1. $(pk^*, (\ell^*, C^*))$ is a derivative of itself.

2. If $(pk, (\ell, C))$ is a derivative of $(pk^*, (\ell^*, C^*))$ and $(pk', (\ell', C'))$ is a derivative of $(pk, (\ell, C))$, $(pk', (\ell', C'))$ is a derivative of (pk^*, C^*).

3. If $(pk', (\ell', C')) \leftarrow \mathcal{O}_{re}(pk, pk', (\ell, C))$, $(pk', (\ell', C'))$ is a derivative of $(pk, (\ell, C))$.

4. If $(\ell', C') \leftarrow \text{ReEnc}(\mathcal{O}_{rk}(pk, pk'), (\ell, C))$, $(pk', (\ell', C'))$ is a derivative of $(pk, (\ell, C))$.

5. If $(\ell', C') \leftarrow \text{ReEnc}(\text{ReKeyGen}(sk, pk'), (\ell, C))$, $(pk', (\ell', C'))$ is a derivative of $(pk, (\ell, C))$, where sk is the private key of pk.

- $\mathcal{O}_{re}$: On input $(pk_i, pk_j, (\ell, C))$, if $(pk_i, (\ell, C))$ is a derivative of $(pk^*, (\ell^*, C^*))$, and there is a corrupted delegation chain from pk_j to a corrupted public key or pk_j is corrupted, $\mathcal{C}$ outputs reject; otherwise, $\mathcal{C}$ responds the same as in Phase 1.

- $\mathcal{O}_{dec}$: On input $(pk_i, (\ell, C))$, if $(pk_i, (\ell, C))$ is a derivative of $(pk^*, (\ell^*, C^*))$, $\mathcal{C}$ outputs reject; otherwise, $\mathcal{C}$ responds the same as in Phase 1.

Guess: Finally, $\mathcal{A}$ outputs a guess $\mathbf{b}' \in \{0, 1\}$ and wins the game if $\mathbf{b} = \mathbf{b}'$.

We also define $\mathbf{Adv}_{\mathrm{MUPRE}}^{\mathrm{CCA}}(\lambda) = |\Pr[\mathbf{b} = \mathbf{b}'] - 1/2|$ for the security parameter λ as that in CCA-O security for SUPRE.

Definition 2.3.5 (CCA Security). We say that a multiuse, unidirectional PRE scheme MUPRE is CCA-secure, if for any polynomial time CCA adversary $\mathcal{A}$ the function $\mathbf{Adv}_{\mathrm{MUPRE}}^{\mathrm{CCA}}(\lambda)$ is negligible.

Collusion Resistance for Multiuse, Unidirectional Proxy Re-encryption

We still use the above method to define collusion resistance. In the CR game, there exist the same oracles as that in Phase 1 of the CCA game for multiuse, unidirectional PRE. In the end, the adversary wins the game if it outputs the private key of an uncorrupted user. The adversary is named as a CR adversary.

Definition 2.3.6. We say that a multiuse, unidirectional PRE scheme MUPRE is collusion resistant, if for any polynomial time CR adversary $\mathcal{A}$ the function $\mathbf{Adv}_{\mathrm{MUPRE}}^{\mathrm{CCA-O}}(\lambda)$ is negligible.

Remark 2.3.7 (CCA Security vs. Collusion Resistance). Unlike the single-use unidirectional proxy, the CCA security (for ℓ-level ciphertext, $\ell > 1$) of multiuse, unidirectional PRE cannot guarantee the CR security, since in the CCA game, the adversary cannot query $\mathcal{O}_{rk}$ to get the re-encryption key from the target user to any corrupted users, but it is allowed in the CR game.

2.3.4 Single-Use, Unidirectional Scheme

From the security model of SUPRE, we see that there are three sufficient conditions for CCA-secure SUPRE:

1) Each of the proxy and delegator can verify the validity of the original ciphertext.

2) The delegatee can verify the validity of the re-encrypted ciphertext.

3) The delegatee colluding with the proxy cannot get the private key of the delegator.

The first condition can be easily met if the original ciphertext is publicly verifiable. There are many methods to obtain public verifiability, such as signature of knowledge. There are as well many methods to get the second condition such as Fujisaki-Okamoto conversion [115]. The last condition is usually the most difficult one to realize on designing CCA-secure unidirectional PRE. From the definition of PRE, the delegatee colluding with the proxy can definitely get a value that can be used to do decryption as the private key of the delegator. Hence, the value, obtained by the collusion of the delegatee and proxy, should not be the real private key of the delegator, instead a sub-private key. The similar concept of sub-private key has appeared in other cryptographic primitives, such as public key encryption with double trapdoors (both trapdoors can be used to do decryption, and one trapdoor can be computed from another one but not vice versa).

Following the above method, we [246] proposed the first CCA-secure, single-use, unidirectional PRE scheme S_{SUPRE} in the random oracle model by using Fujisaki-Okamoto conversion, signature of knowledge and the public key encryption with double trapdoors proposed by Bresson et al. [54] (named BCP03).

Signature of Knowledge

The following noninteractive zero-knowledge proof of knowledge, named signature of knowledge of equality of two discrete logarithms [8, 61, 258], will be used in followings:

Definition 2.3.8. Let $y_1, y_2, g, h \in \mathbb{G}$, $\mathbb{G}$ be a cyclic group of quadratic residues modulo N^2 (N is a safe-prime modulus), and $H(\cdot) : \{0,1\}^* \rightarrow \{0,1\}^\lambda$ (λ is the security parameter). A pair (c, s), verifying $c = H(y_1||y_2||g||h||g^s y_1^c||h^s y_2^c||m)$ is a signature of knowledge of the discrete logarithm of both $y_1 = g^x$ w.r.t. base g and $y_2 = h^x$ w.r.t. base h, on a message $m \in \{0,1\}^*$.

The party in possession of the secret x is able to compute the signature, provided that $x = \log_g y_1 = \log_h y_2$, by choosing a random $t \in \{0, \ldots, 2^{|N^2|+k} - 1\}$ ($|n|$ is the bit-length of n). And then computing c and s as:

$$c = H(y_1||y_2||g||h||g^t||h^t||m) \text{ and } s = t - cx.$$

We denote $\mathrm{SoK.Gen}(y_1, y_2, g, h, m)$ as the generation of the proof.

The Public Key Encryption with Double Trapdoors: BCP03

The following description is from [54]. Let $N = pq$ be a safe prime modulus like $p = 2p' + 1$, $q = 2q' + 1$, and p, p', q, q' are primes. Assume $\mathbb{G}$ is the cyclic group of quadratic residues modulo N^2, then the order of $\mathbb{G}$ is $Np'q'$.

- KeyGen: Choose a random element $\alpha \in \mathbb{Z}_{N^2}^*$, a random value $a \in [1, Np'q']$, and set $g = \alpha^2 \mod N^2$ and $h = g^a \mod N^2$. The public key is (N, g, h) and the private key is a.

- Enc: On input a public key pk and a message $m \in \mathbb{Z}_N$, the ciphertext (A, B) is computed as

$$A = g^r \mod N^2, \quad B = h^r(1 + mN) \mod N^2,$$

where r is a random number from $\mathbb{Z}_{N^2}$.

- Dec: There are two methods to decrypt:

    - Knowing a, one can compute m by

    $$m = \frac{B/(A^a) - 1 \bmod N^2}{N}.$$

    - Knowing p', q', one can compute m by

    $$m = \frac{D - 1 \bmod N^2}{N} \cdot \pi \bmod N,$$

    where $D = \left(\frac{B}{g^{w_1}}\right)^{2p'q'}$, $w_1 = ar \bmod N$, $ar \bmod pqp'q' = w_1 + w_2 N$, π is the inverse of $2p'q' \bmod N$.

    Note that the values of $a \bmod N$ and $r \bmod N$ can be computed given $h = g^a \bmod N^2$, $A = g^r \bmod N^2$, and p', q', by the method in [225] (Theorem 1 in [225]).

The Description of Scheme S_{SUPRE}

Scheme S_{SUPRE} contains three cryptographic hash functions: $H_1(\cdot) : \{0, 1\}^* \rightarrow \{0, 1\}^{\lambda_1}$, $H_2(\cdot) : \{0, 1\}^* \rightarrow \{0, 1\}^n$, and $H_3(\cdot) : \{0, 1\}^* \rightarrow \{0, 1\}^{\lambda_2}$, where λ_1 and λ_2 are the security parameters, n is the bit-length of messages to be encrypted. The details are as follows:

- KeyGen: Choose a safe-prime modulus $N = pq$, three random numbers $\alpha \in \mathbb{Z}_{N^2}^*$, $a, b \in [1, pp'qq']$, a hash function $H(\cdot)$, where $p = 2p' + 1$, $q = 2q' + 1$, p, p', q, q' are primes, and $H(\cdot) : \{0, 1\}^* \rightarrow \mathbb{Z}_{N^2}$. Furthermore, set $g_0 = \alpha^2 \bmod N^2$, $g_1 = g_0^a \bmod N^2$, and $g_2 = g_0^b \bmod N^2$. The public key is $pk = (H(\cdot), N, g_0, g_1, g_2)$, the sub-private key is $wsk = (a, b)$, and the private key is $sk = (p, q, p', q')$.

- ReKeyGen: On input a public key $pk_Y = (H_Y(\cdot), N_Y, g_{Y0}, g_{Y1}, g_{Y2})$, a sub-private key $wsk_X = a_X$, and a private key $sk_X = (p_X, q_X, p'_X, q'_X)$, it outputs the unidirectional re-encryption key $rk_{X \to Y} = (rk_{X \to Y}^{(1)}, rk_{X \to Y}^{(2)})$, where $rk_{X \to Y}^{(1)} = (\dot{A}, \dot{B}, \dot{C})$, and computed as follows:

  - Choose two random numbers $\dot{\sigma} \in \mathbb{Z}_N$, $\dot{\beta} \in \{0,1\}^{\lambda_1}$.

  - Compute $rk_{X \to Y}^{(2)} = a_X - \dot{\beta} \mod (p_X q_X p'_X q'_X)$.

  - Compute $r_{X \to Y} = H_Y(\dot{\sigma} || \dot{\beta})$, $\dot{A} = (g_{Y0})^{r_{X \to Y}} \mod (N_Y)^2$, $\dot{C} = H_1(\dot{\sigma}) \oplus \dot{\beta}$,

  $$\dot{B} = (g_{Y2})^{r_{X \to Y}} \cdot (1 + \dot{\sigma} N_Y) \mod (N_Y)^2 \tag{2.1}$$

- Enc: On input a public key $pk = (H(\cdot), N, g_0, g_1, g_2)$ and a message $m \in \{0,1\}^n$, it proceeds as follows:

  - Choose a random number $\sigma \in \mathbb{Z}_N$.

  - Compute $r = H(\sigma || m)$, $A = (g_0)^r \mod N^2$, $C = H_2(\sigma) \oplus m$, $D = (g_2)^r \mod N^2$,

  $$B = (g_1)^r \cdot (1 + \sigma N) \mod N^2. \tag{2.2}$$

  - Run $(c, s) \leftarrow \text{SoK.Gen}(A, D, g_0, g_2, (B, C))$, where the underlying hash function is H_3.

  - Output the ciphertext $K = (A, B, C, D, c, s)$.

- ReEnc: On input a re-encryption key $rk_{X \to Y} = (rk_{X \to Y}^{(1)}, rk_{X \to Y}^{(2)})$ and a ciphertext $K = (A, B, C, D, c, s)$ under key $pk_X = (H_X(\cdot), N_X, g_{X0}, g_{X1}, g_{X2})$, check whether $c = H_3(A || D || g_{X0} || g_{X2} || (g_{X0})^s A^c || (g_{X2})^s D^c || (B || C))$. If not hold, output reject and terminate; otherwise, re-encrypt the ciphertext to be one under key pk_Y as:

- Compute $A' = A^{rk^{(2)}_{X \to Y}} = (g_{X0})^{r(a_X - \dot{\beta})} \bmod (N_X)^2$.

- Output the new ciphertext $(A, A', B, C, rk^{(1)}_{X \to Y}) = (A, A', B, C, \dot{A}, \dot{B}, \dot{C})$.

- Dec: On input a private key and any ciphertext K, parse $K = (A, B, C, D, c, s)$, or $K = (A, A', B, C, \dot{A}, \dot{B}, \dot{C})$.

  - *Case $K = (A, B, C, D, c, s)$:* Check whether

  $$c = H_3(A||D||g_0||g_2||(g_0)^s A^c||(g_2)^s D^c||(B||C)),$$

  if not, output `reject` and terminate; otherwise,

    * compute $\sigma = \frac{B/(A^a) - 1 \bmod N^2}{N}$, if the input private key is the sub-private key a;

    * compute $\sigma = \frac{(B/g_0^{w_1})^{2p'q'} - 1 \bmod N^2}{N} \cdot \pi (\bmod N)$, where w_1 is computed as that in scheme BCP03, and π is the inverse of $2p'q' \bmod N$, if the private key is the private key (p, q, p', q').

  Compute $m = C \oplus H_2(\sigma)$, if $A = (g_0)^{H(\sigma||m)}$ and $B = (g_1)^{H(\sigma||m)} \cdot (1 + \sigma N) \bmod N^2$ both hold, output m; otherwise, output `reject` and terminate.

  - *Case $K = (A, A', B, C, \dot{A}, \dot{B}, \dot{C})$:* In this case, the decryptor should know the delegator's (Alice's) public key $(H'(\cdot), N', g_0', g_1', g_2')$.

    * If the input private key is the sub-private key b, compute

    $$\dot{\sigma} = \frac{\dot{B}/(\dot{A}^b) - 1 \bmod N^2}{N}.$$

    * If the input private key is the private key (p, q, p', q'), computes

    $$\dot{\sigma} = \frac{(\dot{B}/g_0^{w_1})^{2p'q'} - 1 \bmod N^2}{N} \cdot \pi (\bmod N),$$

where w_1 is computed as that in scheme BCP03, and π is the inverse of $2p'q' \bmod N$.

Compute $\dot{\beta} = \dot{C} \oplus H_1(\dot{\sigma})$, if $\dot{A} = (g_0)^{H(\dot{\sigma}||\dot{\beta})}$ and $\dot{B} = (g_2)^{H(\dot{\sigma}||\dot{\beta})} \cdot (1 + \dot{\sigma}N) \bmod N^2$ both hold, then compute

$$\sigma = \frac{B/(A' \cdot A^{\dot{\beta}}) - 1 \bmod N'^2}{N'}, \quad m = C \oplus H_2(\sigma);$$

otherwise, output reject and terminate. If $A = (g_0')^{H'(\sigma||m)} \bmod N'^2$ and $B = (g_1')^{H'(\sigma||m)} \cdot (1 + \sigma N') \bmod N'^2$ both hold, output m; otherwise, output reject and terminate.

Note that $(H(\cdot), N, g_0, g_1, g_2)$ is the public key of the decryptor.

Theorem 2.3.9 (CCA-O security). *In the random oracle model, scheme S_{SUPRE} is CCA-O-secure under the assumptions that the DDH problem over $\mathbb{Z}_{N^2}^*$ is hard, and that the signature of knowledge is secure.*

Proof. . We show that if an algorithm $\mathcal{A}$ exists and can break the CCA-O security of S_{SUPRE} with probability ϵ in time t, there is another algorithm $\mathcal{B}$ that uses $\mathcal{A}$ to solve the DDH problem over $\mathbb{Z}_{N^2}^*$, i.e., on DDH input $(\mathbf{N}, \mathbf{g}, \mathbf{g^u}, \mathbf{g^v}, \mathbf{T})$, $\mathcal{B}$ decides if $\mathbf{T} = \mathbf{g^{uv}}$ or not.

$\mathcal{B}$ interacts with $\mathcal{A}$ in a CCA-O game as follows ($\mathcal{B}$ simulates the challenger for $\mathcal{A}$). In the following, we use starred letters $(A^*, B^*, C^*, D^*, c^*, s^*)$ to refer to the challenge ciphertext corresponding to the target public key pk^*.

Hash Oracles:

$\mathcal{O}_H$: One oracle $\mathcal{O}_H$ is corresponding to one hash function $H(\cdot)$ which is a part of user's public key. As a result, there are many such oracles, and they are all constructed in the following method. On input (σ_i, m_i), $\mathcal{B}$ first checks whether triple $(\sigma_i, m_i, \alpha_i)$ is in Table T_H. If yes, $\mathcal{B}$ responds $\mathcal{A}$ with α_i; otherwise $\mathcal{B}$ chooses a random number $\alpha_i \in \mathbb{Z}_{N^2}$, responds $\mathcal{A}$ with α_i, and records $(\sigma_i, m_i, \alpha_i)$ in Table T_H, where N is the corresponding safe-prime modulus and a part of user's public key.

$\mathcal{O}_{H_1}$: On input σ_i, $\mathcal{B}$ first checks whether pair (σ_i, β_i) is in Table T_{H_1}. If yes, $\mathcal{B}$ responds $\mathcal{A}$ with β_i; otherwise, $\mathcal{B}$ chooses a random number $\beta_i \in \{0,1\}^{\lambda_1}$, responds $\mathcal{A}$ with β_i, and records (σ_i, β_i) in Table T_{H_1}.

$\mathcal{O}_{H_2}$: On input σ_i, $\mathcal{B}$ first checks whether pair (σ_i, γ_i) is in Table T_{H_2}. If yes, $\mathcal{B}$ responds $\mathcal{A}$ with γ_i; otherwise, $\mathcal{B}$ chooses a random number $\gamma_i \in \{0,1\}^n$, responds $\mathcal{A}$ with γ_i, and records (σ_i, γ_i) in Table T_{H_2}.

$\mathcal{O}_{H_3}$: On input $(A_i, D_i, g_{i0}, g_{i2}, E_i, F_i, B_i, C_i)$, $\mathcal{B}$ first checks whether tuple $(A_i, D_i, g_{i0}, g_{i2}, E_i, F_i, B_i, C_i, \delta_i)$ is in Table T_{H_2}. If yes, $\mathcal{B}$ responds $\mathcal{A}$ with δ_i; otherwise, $\mathcal{B}$ chooses a random number $\delta_i \in \{0,1\}^{\lambda_2}$, responds $\mathcal{A}$ with δ_i, and records $(A_i, D_i, g_{i0}, g_{i1}, E_i, F_i, B_i, C_i, \delta_i)$ in Table T_{H_3}.

Phase 1:

$\mathcal{O}_{pk}$: On input an index i, $\mathcal{B}$ decides whether pk_i is the target public key pk^*.

- If yes, $\mathcal{B}$ sets $N = \mathbf{N}$, $H(\cdot) : \{0,1\}^* \rightarrow \mathbb{Z}_{\mathbf{N}^2}$, $g_0 = \mathbf{g}$, $g_1 = \mathbf{g}^\mathbf{u}$, and $g_2 = \mathbf{g}^w$, where $w \in \mathbb{Z}_{\mathbf{N}^2}$. And then, $\mathcal{B}$ records

$$(H(\cdot), N, g_0, g_1, g_2, \perp, \perp, \perp, \perp)$$

in Table T_k.

- Otherwise, $\mathcal{B}$ runs KeyGen to get the public key $(H(\cdot), N, g_0, g_1, g_2)$, the private key (p', q'), and weak private key (a, b); and records

$$(H(\cdot), N, g_0, g_1, g_2, a, b, p', q')$$

in Table T_k.

Finally, $\mathcal{B}$ returns $(H(\cdot), N, g_0, g_1, g_2)$ to $\mathcal{A}$ as pk_i.

$\mathcal{O}_{sk}$: On input $pk_X = (H_X(\cdot), N_X, g_{X0}, g_{X1}, g_{X2})$, $\mathcal{B}$ checks whether pk_X is in T_k. If not, $\mathcal{B}$ terminates. Otherwise, if pk_X is the guessed attacked public key, $\mathcal{B}$ reports failure and aborts; otherwise, $\mathcal{B}$ responds $\mathcal{A}$ with corresponding (p'_X, q'_X), and records pk_X in Table T_{sk}.

$\mathcal{O}_{rk}$: On input (pk_X, pk_Y), $\mathcal{B}$ checks whether pk_X and pk_Y are both in T_k. If not, $\mathcal{B}$ terminates. Otherwise, $\mathcal{B}$ checks whether $(pk_X, pk_Y, rk^{(1)}_{X \to Y}, rk^{(2)}_{X \to Y})$ is in Table T_{rk}, or $(pk_X, pk_Y, \beta_{X \to Y}, rk^{(1)}_{X \to Y}, rk^{(2)}_{X \to Y})$ is in Table T_{urk}, if it exists, $\mathcal{B}$ returns $(rk^{(1)}_{X \to Y}, rk^{(2)}_{X \to Y})$ to $\mathcal{A}$; otherwise,

- if pk_X is in Table T_{sk} or pk_X is not the guessed attacked public key, $\mathcal{B}$ responds $\mathcal{A}$ with $(rk^{(1)}_{X \to Y}, rk^{(2)}_{X \to Y}) \leftarrow \mathsf{ReKeyGen}(sk_X, pk_Y)$, and records $(pk_X, pk_Y, rk^{(1)}_{X \to Y}, rk^{(2)}_{X \to Y})$ in Table T_{rk};

- if pk_X is the guessed attacked public key, and pk_Y is not in Table T_{sk}, $\mathcal{B}$ proceeds as follows.

  - Choose three random numbers $\dot{\beta} \in \{0,1\}^{\lambda_1}, rk^{(2)}_{X \to Y} \in \mathbb{Z}_{N^2}, \dot{\sigma} \in \mathbb{Z}_N$.
  - Compute

$$r_{X \to Y} = H_Y(\dot{\sigma} || \dot{\beta}),$$
$$\dot{A} = (g_{Y0})^{r_{X \to Y}} \bmod (N_Y)^2,$$
$$\dot{B} = (g_{Y1})^{r_{X \to Y}} \cdot (1 + \dot{\sigma} N_Y) \bmod (N_Y)^2,$$
$$\dot{C} = H_1(\dot{\sigma}) \oplus \dot{\beta},$$

  where $(H_Y(\cdot), N_Y, g_{Y0}, g_{Y1}, g_{Y2}) = pk_Y$.

  - Set $rk^{(1)}_{X \to Y} = (\dot{A}, \dot{B}, \dot{C})$.
  - Return $(rk^{(1)}_{X \to Y}, rk^{(2)}_{X \to Y})$ to $\mathcal{A}$.

- Record $(pk_X, pk_Y, \dot{\beta}, rk^{(1)}_{X \to Y}, rk^{(2)}_{X \to Y})$ in Table T_{urk}.

If $\mathcal{B}$ successfully responds the decryption queries with the re-encrypted ciphertexts that re-encrypted by the above re-encryption keys, these keys are indistinguishable from that in the real execution from the viewpoint of $\mathcal{A}$, due to the security property of scheme BCP03 with Fujisaki-Okamoto conversion.

- If pk_X is the guessed attacked public key, and pk_Y is in Table T_{sk}, $\mathcal{B}$ reports `failure` and aborts.

$\mathcal{O}_{rc}$: On input (pk_X, pk_Y, K), $\mathcal{B}$ checks whether pk_X and pk_Y are both in Table T_k. If not, $\mathcal{B}$ terminates. Otherwise, $\mathcal{B}$ parses $K = (A, B, C, D, c, s)$, and checks whether $c = H_3(A||D||g_{X0}||g_{X2}||(g_{X0})^s A^c||(g_{X2})^s D^c||(B||C))$, where $pk_X = (H_X(\cdot), N_X, g_{X0}, g_{X1}, g_{X2})$, if not, $\mathcal{B}$ outputs `reject` and terminates; otherwise, $\mathcal{B}$ proceeds as follows.

- If pk_X is the guessed attacked public key, and pk_Y is in Table T_{sk}, $\mathcal{B}$ executes the following steps:

  1. Set two empty sets, S_1 and S_2.
  2. Find all elements $(\sigma_i, m_i, \alpha_i)$ in Table T_{H_X} such that $A = (g_{X0})^{\alpha_i} \bmod (N_X)^2$, and put them into Set S_1. If $S_1 = \varnothing$, output `reject` and terminate.
     This step makes this oracle distinguishable from the real execution when the adversary can guess the correct value of $H_X(\sigma_i||m_i)$ without querying $\mathcal{O}_{H_X}$. The probability of this event is $q_{H_X}/|\mathbb{Z}_{(N_X)^2}|$, where q_{H_X} is the number of queries to $\mathcal{O}_{H_X}$.
  3. For every $(\sigma_i, m_i, \alpha_i)$ in Set S_1, find all elements in Table T_{H_2} such that $\sigma_j = \sigma_i$ and put them (i.e., $(\sigma_i, m_i, \alpha_i)||(\sigma_j, \gamma_j)$'s) into Set S_2. If $S_2 = \varnothing$, output `reject` and terminate.

This step makes this oracle distinguishable from the real execution when the adversary can guess the correct value of $H_2(\sigma_i)$ without querying $\mathcal{O}_{H_2}$. The probability of this event is $q_{H_2}/2^n$, where q_{H_2} is the number of queries to $\mathcal{O}_{H_2}$.

4. Check if $(\sigma_i, m_i, \alpha_i)||(\sigma_j, \gamma_j)$ is in Set S_2, such that $(g_{X1})^{\alpha_i} \cdot (1 + \sigma_i N_X) \bmod (N_X)^2 = B$ and $\gamma_j \oplus m_i = C$. If it does not exist or more than one exist, output reject and terminate.

5. Search $(pk_X, pk_Y, \dot{\beta})$ in Table T_{frk}, if not, choose a random number from $\{0,1\}^{\lambda_1}$ as $\dot{\beta}$, and record $(pk_X, pk_Y, \dot{\beta})$ in Table T_{frk}.

6. Choose a random number $\dot{\sigma} \in \mathbb{Z}_N$.

7. Compute

$$r_{X \to Y} = H_Y(\dot{\sigma}||\dot{\beta}),$$

$$\dot{A} = (g_{Y0})^{r_{X \to Y}},$$

$$\dot{B} = (g_{Y1})^{r_{X \to Y}} \cdot (1 + \dot{\sigma} N_Y) \bmod (N_Y)^2,$$

$$\dot{C} = H_1(\dot{\sigma}) \oplus \beta_{X \to Y},$$

where $pk_Y = (H_Y(\cdot), N_Y, g_{Y0}, g_{Y1}, g_{Y2})$.

8. Set $rk_{X \to Y}^{(1)} = (\dot{A}, \dot{B}, \dot{C})$.

9. Return $(A, (g_{X1})^{\alpha_i} \cdot (g_{X0})^{-\dot{\beta}}, B, C, rk_{X \to Y}^{(1)})$ to $\mathcal{A}$.

- Otherwise, $\mathcal{B}$ calls oracle $\mathcal{O}_{rk}$ to get the re-encryption key $rk_{X \to Y}$, and returns $\text{ReEnc}(rk_{X \to Y}, K)$.

$\mathcal{O}_{dec}$: On input (pk_X, K), $\mathcal{B}$ checks whether pk_X is in Table T_K, if not, $\mathcal{B}$ terminates. Otherwise, $\mathcal{B}$ proceeds as follows:

- If pk_X is not the guessed attacked public key, then sk_X is known to $\mathcal{B}$, who responds $\mathcal{A}$ with $\text{Dec}(sk_X, K)$.

- If pk_X is the guessed attacked public key and $K = (A, B, C, D, c, s)$, $\mathcal{B}$ checks whether $c = H_3(A||D||g_{X0}||g_{X2}||(g_{X0})^s A^c||(g_{X2})^s D^c||(B||C))$, if not, $\mathcal{B}$ outputs reject and terminates; otherwise, $\mathcal{B}$ does:

  1. Set two empty sets S_1 and S_2.

  2. Find all elements $(\sigma_i, m_i, \alpha_i)$ in Table T_{H_X} such that $A = (g_{X0})^{\alpha_i} \bmod (N_X)^2$, and put them into Set S_1. If $S_1 = \varnothing$, output reject and terminate.

     This step makes this oracle distinguishable from the real execution when the adversary can guess the correct value of $H_X(\sigma_i||m_i)$ without querying $\mathcal{O}_{H_X}$. The probability of this event is $q_{H_X}/|\mathbb{Z}_{(N_X)^2}|$.

  3. For every $(\sigma_i, m_i, \alpha_i)$ in Set S_1, find all elements in Table T_{H_2} such that $\sigma_j = \sigma_i$ and put them (i.e., $(\sigma_i, m_i, \alpha_i)||(\sigma_j, \gamma_j)$'s) into Set S_2. If $S_2 = \varnothing$, output reject and terminate.

     This step makes this oracle distinguishable from the real execution when the adversary can guess the correct value of $H_2(\sigma_i)$ without querying $\mathcal{O}_{H_2}$. The probability of this event is $q_{H_2}/2^n$.

  4. Check if $(\sigma_i, m_i, \alpha_i)||(\sigma_j, \gamma_j)$ is in Set S_2, such that $(g_{X1})^{\alpha_i} \cdot (1 + \sigma_i N_X) \bmod (N_X)^2 = B$ and $\gamma_j \oplus m_i = C$. If none exists or more than one exist, output reject and terminate; otherwise, output m_i.

- If pk_X is the guessed attacked public key and $K = (A, A', B, C, \dot{A}, \dot{B}, \dot{C})$, $\mathcal{B}$ searches $(pk_Y, pk_X, \bigstar_1, \bigstar_2, \bigstar_3)$ in Table T_{urk}, such that $\bigstar_2 = (\dot{A}, \dot{B}, \dot{C})$.

  If it does not exist, $\mathcal{B}$ proceeds as follows:

  1. Set two empty sets S_1 and S_2.

  2. Find all elements $(\sigma_i, m_i, \alpha_i)$ in Table T_{H_X} such that $\dot{A} = (g_{X0})^{\alpha_i} \bmod (N_X)^2$, and put them into Set S_1. If $S_1 = \varnothing$, output reject and terminate.

This step makes this oracle distinguishable from the real execution when the adversary can guess the correct value of $H_X(\sigma_i \| m_i)$ without querying $\mathcal{O}_{H_X}$. The probability of this event is $q_{H_X} / |\mathbb{Z}_{(N_X)^2}|$.

3. For every $(\sigma_i, m_i, \alpha_i)$ in Set S_1, find all elements in Table T_{H_1} such that $\sigma_j = \sigma_i$ and put them (i.e., $(\sigma_i, m_i, \alpha_i) \| (\sigma_j, \beta_j)$'s) into Set S_2. If $S_2 = \varnothing$, output reject and terminate.

   This step makes this oracle distinguishable from the real execution when the adversary can guess the correct value of $H_1(\sigma_i)$ without querying $\mathcal{O}_{H_1}$. The probability of this event is $q_{H_1} / 2^{\lambda_1}$, where q_{H_1} is the number of queries to $\mathcal{O}_{H_1}$.

4. Check if $(\sigma_i, m_i, \alpha_i) \| (\sigma_j, \beta_j)$ is in Set S_2, such that $\dot{B} = (g_{X1})^{\alpha_i} \cdot (1 + \sigma_i N_X) \bmod (N_X)^2$ and $\beta_j \oplus m_i = \dot{C}$. If none exists or more than one exist, output reject and terminate.

5. Compute $\sigma = \frac{(B/(A' \cdot A^{m_i}) - 1 \bmod (N_Y)^2)}{N_Y}$, $m = C \oplus H_2(\sigma)$. If $A = (g_{Y0})^{H_Y(\sigma \| m)} \bmod (N_Y)^2$, output m, where $pk_Y = (H_Y(\cdot), N_Y, g_{Y0},$
   $g_{Y1}, g_{Y2})$ is the corresponding delegator's public key; otherwise, output reject and terminate.

If it does exist, $\mathcal{B}$ checks $A' \stackrel{?}{=} A^{\star_3}$. If not, output reject and terminate; otherwise, $\mathcal{B}$ proceeds as follows:

1. Set two empty sets S_1 and S_2.

2. Find all elements $(\sigma_i, m_i, \alpha_i)$ in Table T_{H_Y} such that $A = (g_{Y0})^{\alpha_i} \bmod (N_Y)^2$, and put them into Set S_1. If $S_1 = \varnothing$, output reject and terminate.

   This step makes this oracle distinguishable from the real execution when the adversary can guess the correct value of $H_Y(\sigma_i \| m_i)$ without querying $\mathcal{O}_{H_Y}$. The probability of this event is $q_{H_Y} / |\mathbb{Z}_{(N_Y)^2}|$, where

q_{H_Y} is the number of queries to $\mathcal{O}_{H_Y}$.

3. For every $(\sigma_i, m_i, \alpha_i)$ in Set S_1, find all elements in Table T_{H_2} like $\sigma_j = \sigma_i$ and put them (i.e., $(\sigma_i, m_i, \alpha_i)||(\sigma_j, \gamma_j)$'s) into Set S_2. If $S_2 = \varnothing$, output reject and terminate.

   This step makes this oracle distinguishable from the real execution when the adversary can guess the correct value of $H_2(\sigma_i)$ without querying $\mathcal{O}_{H_2}$. The probability of this event is $q_{H_2}/2^n$.

4. Check if $(\sigma_i, m_i, \alpha_i)||(\sigma_j, \gamma_j)$ is in Set S_2, such that $(g_{Y1})^{\alpha_i} \cdot (1 + \sigma_i N_Y) \bmod (N_Y)^2 = B$ and $\gamma_j \oplus m_i = C$. If none exists or more than one exist, output reject and terminate; otherwise, output m_i.

Challenge: At some point, $\mathcal{A}$ outputs a challenge tuple (pk^*, m_0, m_1). If pk^* is not the public key $\mathcal{B}$ guessed in oracle $\mathcal{O}_{pk}$, $\mathcal{B}$ reports failure and aborts. Otherwise, $\mathcal{B}$ chooses random $d \in \{0,1\}, \sigma \in \mathbb{Z}_N$ and sets:

$$A^* = \mathbf{g}^\mathbf{v} \bmod \mathbf{N}^2, \ B^* = \mathbf{T}(1 + m_d \mathbf{N}) \bmod \mathbf{N}^2,$$

$$C^* = H_2(\sigma) \oplus m_d, \ D^* = (\mathbf{g}^\mathbf{v})^w \bmod \mathbf{N}^2.$$

And then $\mathcal{B}$ chooses two random numbers $c^* \in \{0,1\}^{\lambda_2}, s^* \in \{0, \dots, 2^{L(\mathbf{N}^2)+\lambda_2} - 1\}$, computes $E^* = (\mathbf{g})^{s^*} A^{*c^*} \bmod \mathbf{N}^2$ and $F^* = (\mathbf{g}^\mathbf{u})^{s^*} D^{*c^*} \bmod \mathbf{N}^2$, and checks whether $(A^*, D^*, \mathbf{g}^\mathbf{u}, \mathbf{g}^w, E^*, F^*, B^*, C^*, \bigstar)$ is in Table T_{H_3}. If yes, $\mathcal{B}$ reports failure and aborts; otherwise, $\mathcal{B}$ outputs $(A^*, B^*, C^*, D^*, c^*, s^*)$, and records

$$(A^*, D^*, \mathbf{g}^\mathbf{u}, \mathbf{g}^w, E^*, F^*, B^*, C^*, c^*)$$

in Table T_{H_3}.

Phase 2:

$\mathcal{O}_{pk}$: $\mathcal{B}$ responds as in Phase 1.

$\mathcal{O}_{sk}$: On input pk_i, if $pk_i = pk^*$, or (pk^*, pk_i) is in Table T_{rk}, then $\mathcal{B}$ terminates. Otherwise, $\mathcal{B}$ responds as in Phase 1.

$\mathcal{O}_{rk}$: On input (pk_i, pk_j), if $pk_i = pk^*$, and pk_j is in Table T_{sk}, $\mathcal{B}$ terminates. Otherwise, $\mathcal{B}$ responds as in Phase 1.

$\mathcal{O}_{re}$: On input (pk_i, pk_j, K), if $(pk_i, K) = (pk^*, K^*)$ and pk_j is in Table T_{sk}, $\mathcal{B}$ terminates. Otherwise, $\mathcal{B}$ responds as in Phase 1, except when $pk_i = pk^*$ and $(A, B, C, D, c, s) = (A^*, B^*, C^*, D^*, c^*, s^*)$, $\mathcal{B}$ should record the result $(pk_j, A', C, D, \dot{A}, \dot{C}, \dot{D})$ in Table T_{der}, where the derivatives of the challenge ciphertext are recorded.

$\mathcal{O}_{dec}$: On input (pk_i, K), if $(pk_i, K) = (pk^*, K^*)$, or (pk_i, K) is in T_{der}, or $K = \texttt{ReEnc}(\mathcal{O}_{rk}(pk^*, pk_i), K^*)$, then $\mathcal{B}$ terminates. Otherwise, $\mathcal{B}$ responds as in Phase 1.

Guess: Finally, the adversary $\mathcal{A}$ outputs a guess $d' \in \{0, 1\}$. If $d = d'$, then $\mathcal{B}$ outputs 1 (i.e., DDH instance), otherwise, $\mathcal{B}$ outputs 0 (i.e., not a DDH instance).

Firstly, we analyze the probability of that $\mathcal{B}$ does not output `failure`, which contains the following two events:

1. $\mathcal{B}$ guesses the right attacked public key.

2. The record $(A^*, D^*, \mathbf{g^u}, \mathbf{g}^w, E^*, F^*, B^*, C^*, \bigstar)$ is not in table T_{H_3} before challenge phase.

Suppose $\mathcal{A}$ makes a total of q_{pk} queries to public key generation oracle, q_{rk} queries to re-encryption key generation orale, q_{de} queries to decryption orale, q_H queries to H hash function oracle, q_{H_1} queries to H_1 hash function oracle, q_{H_2} queries to H_2 hash function oracle, and q_{H_3} queries to H_3 hash function oracle.

The probabilities of the above two events are $1/q_{pk}$ and $1 - (q_{H_3} + 1)/2^{\lambda_2}$, respectively. Therefore, the probability that $\mathcal{B}$ does not output `failure` during the simulation is $(1 - (q_{H_3} + 1)/2^{\lambda_2})/q_{pk}$.

Secondly, oracles $\mathcal{O}_{re}$ and $\mathcal{O}_{dec}$ are indistinguishable from the corresponding real executions with probabilities at least

$$\left(\frac{1 + q_{max} + (q_{max})^2}{(1 + q_{max})^2} + \frac{q_{max}}{(1 + q_{max})^2} \left(1 - \frac{q_{H_X}}{|\mathbb{Z}_{N_{mX}}|} \right) \left(1 - \frac{q_{H_1}}{2^{\lambda_1}} \right) \right)^{q_{re}}$$

$$> \left(\frac{1 + q_{max} + (q_{max})^2}{(1 + q_{max})^2} \right)^{q_{re}}$$

and

$$\left(\frac{q_{max}}{1 + q_{max}} + \frac{1}{1 + q_{max}} \left(1 - \frac{q_{H_X}}{|\mathbb{Z}_{N_{mX}}|} \right) (1 - q_2) \right)^{q_{de}} > \left(\frac{q_{max}}{1 + q_{max}} \right)^{q_{de}},$$

respectively, where q_{H_X} is the amount of queries to the same kind of oracle $\mathcal{O}_H$, N_{mX} is the largest number among users' public key N's, $|\mathbb{Z}_{N_{mX}}|$ is the size of $\mathbb{Z}_{N_{mX}}$, and $q_2 = \max(\frac{q_{H_1}}{2^{\lambda_1}}, \frac{q_{H_2}}{2^n})$.

Finally, in the re-encryption oracle, we assume that the signature of knowledge is secure, hence, we should minus the probability of breaking the signature of knowledge, ξ.

As a result, $\mathcal{B}$ has an advantage that is at least

$$\epsilon \cdot \frac{(1 - (q_{H_3} + 1)/2^{\lambda_2}) \cdot (1 + q_{max} + (q_{max})^2)^{q_{re}} \cdot (q_{max})^{q_{de}}}{q_{pk} \cdot (1 + q_{max})^{2q_{re} + q_{de}}} - \xi$$

and its running time is at most

$$t + O(3q_{pk} + (7 + q_H)q_{rk} + (5 + q_H)q_{de})t_e,$$

where t_e is the time of computing one exponentiation in a cyclic group of quadratic residues modulo. We here only consider the exponentiation computation.

This completes the proof. □

Theorem 2.3.10 (CCA-R security). *In the random oracle model, scheme S_{SUPRE} is CCA-*

R-secure under the assumptions that the DDH problem over $\mathbb{Z}^*_{N^2}$ is hard, and that the signature of knowledge is secure.

We can use a similar method in the proof of Theorem 2.3.9 to prove the above theorem. The main difference is that the public key of the target user is $(N = \mathbf{N}, g_0 = g, g_1 = \mathbf{g}^w, g_2 = \mathbf{g}^\mathbf{u})$ not $(N = \mathbf{N}, g_0 = g, g_1 = \mathbf{g}^\mathbf{u}, g_2 = \mathbf{g}^w)$. This change could help us to generate any re-encryption key from the target user to any other user.

We leave this proof to the reader for practice.

2.4   Notes

In this chapter, we give a comprehensive introduction of security definitions of proxy re-cryptography. We also introduce one of our PRS schemes and one of PRE schemes, which are published at Indocrypt 2007 and PKC 2009, respectively. We refer to [11, 193, 252] and [9, 10, 130, 194, 195, 249, 255, 283] for more PRS schemes and PRE schemes, respectively.

There are some interesting problems in proxy re-cryptography still remaining open. For instance,

- How to design proxy re-signature (re-encryption) with multi-usability, unidirectionality, and constant size.

- How to design pairing-free proxy re-signature (re-encryption) with collusion resistance, multi-usability, bidirectionality, and constant size.

- How to design generic constructions of PRE with some properties from the public key encryption with the specific properties, like the generic construction from ID-based (attribute-based or certificateless) encryption to ID-based (attribute-based or certificateless) PRE.

Chapter 3

Attribute-Based Cryptography

3.1 Introduction

With the development of communication networks, there is a trend for users to store their sensitive data on the Internet. To distribute a message to a specific set of users, a trivial method is to encrypt it under each user's public key or identity in traditional cryptosystem [34, 37, 87, 89, 90]. As expected, ciphertext size and computational cost of encryption/decryption algorithms are linear with the number of receivers. Therefore, it is less attractive or even intolerable when the number of receivers is large. Indeed, in most cases, the qualified receivers share some common attributes, such as working location, gender, and age range.

For this reason, Sahai and Waters took the first step to solve the problem and introduced the concept of attribute-based encryption. In an attribute-based encryption mechanism (ABE), user's keys and ciphertexts are labeled with sets of descriptive attributes and only when the attributes of the ciphertext match those of the user's key, the corresponding ciphertext can be decrypted. The proposed scheme allows for decryption when a threshold number k is less than the size of the overlap between a ciphertext and a private key. The attribute-based cryptosystems that each ciphertext is labeled by the encryptor with a set of descriptive attributes and each private key is associated with an access structure that specifies the types of ciphertexts the key can decrypt, are named key-policy attribute-based encryption (KPABE). While in a ciphertext-policy attribute-based encryption scheme (CPABE), each user is identified by an attribute set and receives the private keys corresponding to those attributes from the authority. The sender who aims to distribute a message will construct an access policy associated to the ciphertexts by connecting the attributes with OR, AND, and threshold gates.

The remainder of this chapter is organized as follows. Some universal definitions are introduced in Section 3.2. Then, a bounded ciphertext-policy encryption scheme, a multi-authority encryption scheme, an interval encryption scheme, and a fuzzy identity-based signature scheme are proposed in Section 3.3, 3.4, 3.5, and 3.6, respectively. Finally, a brief conclusion is given and some future works are suggested in Section 3.7.

3.2   Universal Definitions

Definition 3.2.1 (Access Structure [20]). Let $\{P_1, P_2, \ldots, P_n\}$ be a set of parties. A collection $\mathbb{A} \subseteq 2^{\{P_1, P_2, \ldots, P_n\}}$ is monotone, for $\forall\ B, C$, if $B \in \mathbb{A}$ and $B \subseteq C$, then $C \in \mathbb{A}$. An access structure (monotone access structure) is a collection (monotone collection) $\mathbb{A}$ of nonempty subsets of $\{P_1, P_2, \ldots, P_n\}$, i.e., $\mathbb{A} \subseteq 2^{\{P_1, P_2, \ldots, P_n\}} \setminus \{\emptyset\}$. The sets in $\mathbb{A}$ are called the authorized sets, and the sets not in $\mathbb{A}$ are called the unauthorized sets.

In ABE, the access structure $\mathbb{A}$ contains the authorized sets of attributes. It is shown [20] that any monotone access structure can be realized by a linear secret sharing scheme.

Definition 3.2.2 (Linear Secret-Sharing Schemes (LSSS) [291]). A secret sharing scheme Π over a set of parties $\mathbb{P}$ is called linear (over $\mathbb{Z}_p$) if

1. The shares for each party form a vector over $\mathbb{Z}_p$.

2. There exists a matrix A called the share-generating matrix for Π. The matrix A has l rows and n columns. For $i = 1, \ldots, l$, the ith row of A is labeled by a party $\rho(i)(\rho$ is a function from $\{1, \ldots, l\}$ to $\mathbb{P})$. When we consider the column vector $\vec{v} = (s, r_2, \ldots, r_n)$, where $s \in \mathbb{Z}_p$ is the secret to be shared and $r_2, \ldots, r_n \in \mathbb{Z}_p$ are randomly chosen, $A\vec{v}$ is the vector of l shares of the secret s according to Π. The share $(A\vec{v})_i$ belongs to party $\rho(l)$.

It is shown that each linear secret-sharing scheme according to the above definition also enjoys the linear reconstruction property [20], defined as follows: Suppose that Π is an LSSS for access structure $\mathbb{A}$. Let $S \in \mathbb{A}$ be an authorized set, and let $I \subset \{1, 2, \ldots, l\}$ be defined as $I = \{i : \rho(i) \in S\}$. There exist constants $\{\omega_i \in \mathbb{Z}_p\}_{i \in I}$ such that if $\{\lambda_i\}$ are valid shares of any secret s according to Π, then $\sum_{i \in I} \omega_i \lambda_i = s$. Furthermore, these constants $\{\omega_i\}$ can be found in time polynomial to the size of the share-generating matrix A. For any unauthorized set, no such constants exist. In expressive CPABE systems, LSSS matrix (A, ρ) is always used to express an access policy associated to a ciphertext.

3.3 Bounded Ciphertext-Policy Encryption Schemes

The concept of attribute-based encryption (ABE) was introduced by Sahai and Waters [241]. In their scheme, the secret key is associated with an attribute set, and the ciphertext is also associated with another set of attributes. The decryption is only successful while these two sets overlap more than a preset threshold. Later, Goyal *et al.* [129] further classified ABE into two categories: ciphertext-policy (CP) ABE and key-policy (KP) ABE. While KPABE (the user's secret key represents an access policy of attributes) is well developed by the subsequent research [129, 223], how to design an efficient and secure CPABE (the ciphertext represents an access policy chosen by the sender) remains open [27, 84, 127].

The first CPABE scheme was proposed by Bethencourt et al. [27] It allows users to encrypt a message under an expression consisting of threshold gates between attributes (called a fine-grained access structure). However, it only has a security argument in the generic group model and the random oracle model. Later, Cheung and Newport [84] gave a provably secure CPABE in the standard model. Their scheme supports an access policy with "AND" gate on positive and negative attributes but can not resist collusion attack while extending to the threshold gate. Recently, a bounded CPABE scheme, supporting fine-grained access policy, was proposed in [127].

Waters [291] presented several CPABE schemes. The construction of the first scheme is very elegant, and the security can be reduced to decisional q-Bilinear Diffie–Hellman-Exponent(BDHE) problem. The ciphertext size only linearly increases with the number of the attributes presented in the access structure. Another scheme, based on DBDH assumption, is less efficient, since the ciphertext size is restricted by the length and the width of the matrix, which is dependent on the size of the access structure.

Overview of Schemes

Ciphertext-policy attribute-based encryption (CPABE) cryptosystems are widely used to realize the sensitive user data sharing and access control on the Internet. For example,"(AGE>25) AND (CS)" represents the restrictions for a qualified user who at least holds the secret keys representing age more than 25 and the ownership of a computer science degree; three attributes (DB,OS,DM) connecting by a "two out of three threshold gate" restrict the decryption only to be successful when a user at least registers two courses from the list (Database, Operating System, and Discrete Mathematics). As in Figure 3.1, a sender will store the ciphertext encrypted under the access policy in the server S_1. The users U_2 and U_3 have the access right to the message since they both have two attributes from the attribute set, while U_1 does not, since he/she only has one attribute. This is an efficient and convenient approach for a user to broadcast his message to the others in practice. Therefore, we mainly focus on developing a more efficient CPABE scheme.

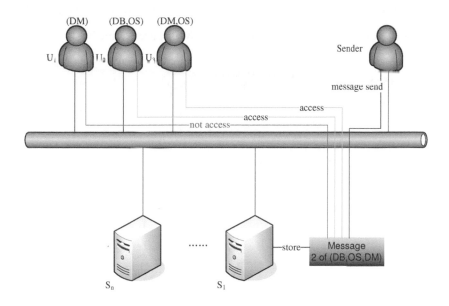

Figure 3.1: A sample for CPABE system

Considering the security proof of an ABE scheme, more flexible access policy adopted on the sender's side makes the simulation more difficult. Thus, restricting the size of the access policy is necessary for designing a CPABE which reduces the security to an assumption in number theory. Goyal *et al.* introduced a new bounded CPABE (BCPABE), in which the encryption access tree must be limited by two properties: the maximum height and the maximum cardinality of non-leaf nodes. For example, the access tree (to the left of Figure 3.2) with height 2 has the maximum cardinality 3.

In [127], a bounded CPABE scheme, secure in the standard model, was proposed to support a bounded size access tree with threshold gates as its non-leaf nodes. At the beginning, the system manager pre-sets two bounds (d, c) and a unique (d, c)-universal access tree $\mathcal{T}_u$ (Figure 3.3). Later, the system manager will publish the public parameter and generate users' secret keys according to this universal access tree. If a sender wants to distribute his message, it requires him to first convert a (d, c)-bounded access tree $\mathcal{T}$ (the left in Figure 3.2) into its (d, c)-normal form[1] $\mathcal{T}_n$ (the right in Figure 3.2), and then to complete the encryption procedure according to a map constructed from a (d, c)-normal form access tree $\mathcal{T}_n$ to the (d, c)-universal access tree $\mathcal{T}_u$.

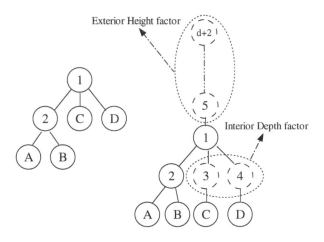

Figure 3.2: A conversion from (d, c)-bounded access tree $\mathcal{T}$ to its (d, c)-normal form $\mathcal{T}_n$

In the (d, c)-universal access tree $\mathcal{T}_u$, only leaf nodes of depth d are associated with real attributes. This setting leads to the fact that users have to construct a map from $\mathcal{T}_n$ to $\mathcal{T}_u$ in order to ensure that the leaf nodes in $\mathcal{T}_n$ share same real attributes with their corresponding leaf nodes in $\mathcal{T}_u$. However, this conversion of normal form actually expands the original tree $\mathcal{T}$ with a great many non-leaf nodes which leads to a boost on computational cost. We conclude that the expanded size is mainly due to two factors called exterior height factor and interior depth factor (refer to Figure 3.2). Exterior height factor correlates with the height h of a (d, c)-bounded access tree $\mathcal{T}$. $d - h$ nodes must be added in order to expand the height of its normal form to d. Goyal *et*

---

[1] The definition of normal form can be looked up in Section 2.4

al. provide a method to eliminate this factor by constructing multiple parallel schemes with different-sized universal access trees, though it is very inefficient. Interior depth factor is the relative depth between leaf nodes in $\mathcal{T}$. Non-leaf nodes must be added in order to make the leaf nodes all at the same depth (which is the deepest level). In order to eliminate both factors, we neglect the interim step, i.e., converting the access tree to normal form, and directly map the (d, c)-bounded access tree selected by the sender to the (d, c)-universal tree. In other words, the redundant steps which pull all the leaf nodes to the deepest level by adding non-leaf nodes are skipped and thus the computational cost is reduced.

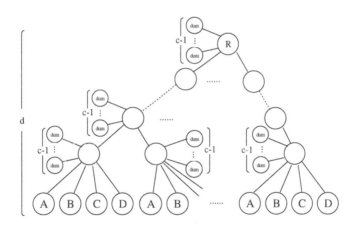

Figure 3.3: A (d, c)-universal access tree $\mathcal{T}_u$ in GJPS

As mentioned in GJPS, to construct a more efficient BCPABE scheme based on some assumptions in number theory is an important open problem. We will provide an affirmative answer to this problem.

This section will present a bounded CPABE scheme BCP_1 which is more efficient than the previous works in GJPS. The security of BCP_1 can be reduced to the Decisional Bilinear Diffie–Hellman assumption in the standard model. Different from GJPS, the computational cost of encryption and decryption in our scheme are largely reduced since we skip redundant steps, meanwhile the ciphertext size is shorter. Nevertheless, as a tradeoff, we demonstrate that the spacial cost, like the size of public parameter and secret keys, increases but be less than twice of the counterpart of GJPS. Furthermore, we propose a provably secure BCPABE scheme BCP_2 in the standard model under a chosen ciphertext secure notion by adopting a one-time signature technique. There are two approaches which could be used for extending BCP_1 to BCP_2. These two methods make a tradeoff between ciphertext size and the size of public/secret parameters.

For illustration, we take a concrete example to explain the primitive idea used to reduce the computational cost of encryption/decryption algorithms.

Consider a BCPABE scheme setup with bounds $(d, c)^2$ and an encryption under a (d, c)-bounded access tree $\mathcal{T}$ shown in the left of Figure 3.2. The threshold values of nodes 1 and 2 are both set to be 2, and nodes A, B, C, D represent four different real attributes. Now, we will show how the encryption algorithm of GJPS differs from ours.

In GJPS, to encrypt under $\mathcal{T}$, a user will first convert $\mathcal{T}$ to its normal form access tree $\mathcal{T}_n$ (shown in the right of Figure 3.2). The threshold values of nodes 3, 4, 5, $\cdots$, $d + 2$ in $\mathcal{T}_n$ are all set to be 1. Assume the computational cost of a single node is T, the total cost is $(4+c-2+c-2+d\times(c-1))\cdot T$ where exterior factor takes $(d-2)\cdot(c-1)\cdot T$ and interior factor takes $2 \cdot (c - 1) \cdot T$. Likewise, if a user has a secret key associated with attributes (C, D), he will at least expend $(1 + 2 + d - 2) \times c = (d+1) \times c$ paring computation. It is obvious that the larger the two initial parameters (d, c) are set to be, the more cost a user will spend on encryption and decryption.

In contrast, our scheme defines a map from $\mathcal{T}$ to $\mathcal{T}_u$ and the total cost on encryption takes $(4 + c - 2 + c - 2) \cdot T$. The user with attributes (C, D) will expend only c paring computation. Therefore, our scheme saves $d \times (c - 1) \cdot T$ on encryption and at least dc times paring computation on decryption. The cost is even irrelevant with the initial parameter d. The comparison of general case can be found in Table 3.1 in the following

---

[2]$d \geq 2, c \geq 3$

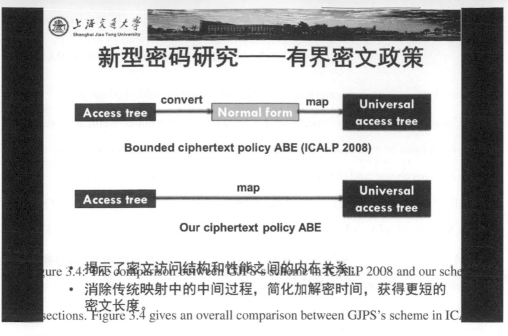

Figure 3.4: The comparison between GJPS's scheme in ICALP 2008 and our scheme.

sections. Figure 3.4 gives an overall comparison between GJPS's scheme in ICALP 2008 and our scheme.

3.3.1 Definitions

Decisional Bilinear Diffie–Hellman Problem. An algorithm S is an ε'-solver of the DBDH problem if it distinguishes with probability at least $1/2 + \varepsilon'$ between the two following probability distributions:

$\mathcal{D}_{bdh} = (g, g^a, g^b, g^c, e(g,g)^{abc})$, where a, b, c are chosen randomly in Z_p,

$\mathcal{D}_{rand} = (g, g^a, g^b, g^c, Z)$, where a, b, c are chosen randomly in Z_p and Z is chosen randomly in G_T.

Definition 3.3.1. The DBDH assumption holds in G and G_T if there is no probabilistic polynomial-time ε'-solver of the DBDH problem for non-negligible value ε'.

One-Time Signature. A one-time signature scheme $\mathtt{ots}$ [36, 63] consists of three algorithms ($\mathtt{ots}.\textbf{KGen}$, $\mathtt{ots}.\textbf{Sig}$, $\mathtt{ots}.\textbf{Ver}$). $\mathtt{ots}.\,\textbf{KGen}(1^k) \to (sk, vk)$ is the key generation algorithm, which outputs a secret key sk and a public verification key vk. $\mathtt{ots}.\textbf{Sig}(sk, m) \to \sigma$ is the sign algorithm which takes the secret key sk and a message m as its input, and outputs a signature σ. Finally, the verification algorithm $\mathtt{ots}.\textbf{Ver}(\sigma, m, vk) \to 0$ or 1 takes the signature σ, a message m, and a public verification key vk as its input, and outputs 1 if the signature is valid; 0 otherwise.

Concerning the security issue, an adversary first receives a public verification key vk generated from $\mathtt{ots}.\mathbf{KGen}(1^k)$. He then makes at most one signature query for message m of his choice, and obtains as an answer the valid signature $\mathtt{ots}.\mathbf{Sig}(sk, m) \to \sigma$. Finally, he outputs a pair (m', σ'). The adversary succeeds if $(m', \sigma') \neq (m, \sigma)$ and $\mathtt{ots}.\mathbf{Ver}(\sigma', m', vk) \to 1$.

A one-time signature scheme $\mathtt{ots}$ is ε_{ots}-secure if every polynomial-time adversary against $\mathtt{ots}$ has a success probability bounded by ε_{ots}.

Several definitions and notions are given below.

Attribute Set: n real attributes, indexed from 1 to n. Any attribute set $\gamma \subseteq \{1, \cdots, n\}$. $c - 1$ dummy attributes, indexed from $n + 1$ to $n + c - 1$.

Access Tree: Let $\mathcal{T}$ represent an access tree with its root r. Each non-leaf node x can be seen as a threshold gate with threshold value k_x. If x has c_x child nodes, it is required that $0 < k_x \leq c_x$. If x is a leaf node of the access tree, it is associated with a single attribute, denoted as $\mathrm{att}(x)$.

We fix the root of an access tree to be at depth 0. Let $\Sigma_{\mathcal{T}}$ denote the set of all the non-leaf nodes, and $\Theta_{\mathcal{T}}$ denote all the leaf nodes. Let $\mathrm{p}(x)$ denote the parent of node x. For each non-leaf node x, we define an order among x's child nodes, that is, every child node z is numbered from 1 to c_x. $\mathrm{index}(z)$ returns such a number associated to node z. For simplicity, if z is a leaf node, we let $\mathrm{att}(z) = \mathrm{index}(z)$.

Satisfying an Access Tree: Let $\mathcal{T}$ be an access tree with root r. $\mathcal{T}_x$ is a subtree rooted at a node x in $\mathcal{T}$. If an attribute set γ satisfies the access tree $\mathcal{T}_x$, we denote $\mathcal{T}_x(\gamma) = 1$. If x is a non-leaf node, evaluate $\mathcal{T}_z(\gamma)$ for all children z of node x. $\mathcal{T}_x(\gamma)$ returns 1 if and only if at least k_x children of x return 1. If x is a leaf node, then $\mathcal{T}_x(\gamma) = 1$ iff $\mathrm{att}(x) \in \gamma$. If an access tree rooted at x is a γ-satisfied tree, then we call the node x as a γ-satisfied node. Suppose $\mathcal{T}$ is a γ-satisfied tree, we call a subtree of $\mathcal{T}$ with root r as a γ-satisfied non-redundant tree, if the cardinality of each non-leaf node

x is equal to x's threshold value in $\mathcal{T}$ and every node is a qualified node. Let $\hat{\mathcal{T}}$ denote the γ-satisfied nonredundant tree with minimum non-leaf nodes.[3]

Universal Access Tree: Here, we describe a universal access tree (Figure 3.5) with two input parameters d and c. First, we define a complete c-ary tree $\mathcal{T}'$ of height $d - 1$, where each node has a threshold value c. Next, $c - 1$ new leaf nodes named "dummy nodes" representing $c - 1$ dummy attributes and n new leaf nodes named "real nodes" representing n real attributes are attached to each node in $\mathcal{T}'$. The resultant tree $\mathcal{T}_u$ is called a (d, c)-universal access tree. Let each node x except root has an index related with its parent node and $\text{att}(x) = \text{index}(x)$ if x is a leaf node of a universal access tree. Here, for all the child nodes of one non-leaf node x in $\mathcal{T}$, real nodes and dummy nodes will take indexes from $\{1, \cdots, n\}$ and $\{n + 1, \cdots, n + c - 1\}$, respectively, and other non-leaf nodes will take indexes from $\{n + c, \cdots, n + 2c - 1\}$.

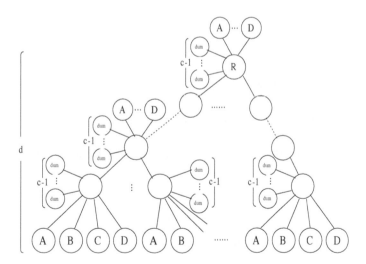

Figure 3.5: Modified universal tree

Bounded Access Tree: We call that $\mathcal{T}$ a (d, c)-bounded access tree if its height $d' \leq d$ and each non-leaf node x in $\mathcal{T}$ has at most c non-leaf child nodes.

---

[3]The decryption cost depends on the non-leaf nodes in the γ-satisfied nonredundant tree the decryptor chooses.

Normal Form: Consider a (d, c)-bounded access tree $\mathcal{T}_n$. $\mathcal{T}_n$ exhibits the (d, c)-normal form if (a) its height $d' = d$, and (b) all the leaf nodes are at depth d. Any (d, c)-bounded access tree $\mathcal{T}$ can be converted to its normal form without modifying its satisfying logic. (This is a special technique used in GJPS.)

Map between Access Trees: The map, constructed from a (d, c)-bounded access tree $\mathcal{T}$ to (d, c)-universal access tree $\mathcal{T}_u$, is defined in the following way in a top-down manner. First, the root of $\mathcal{T}$ is mapped to the root of $\mathcal{T}_u$. Now, suppose that x' in $\mathcal{T}$ is mapped to x in $\mathcal{T}_u$. Let $z'_1, \cdots, z'_{c_{x'}}$ be the child nodes of x'. For $i \in \{1, \cdots, c_{x'}\}$, if z'_i is a leaf node $\mathrm{att}(z'_i) \in \mathcal{N}$, set x's child node z like $z = \mathrm{map}(z'_i)$ where $\mathrm{index}(z) = \mathrm{index}(z'_i)$; if z'_i is a non-leaf node, set x's child node z such that $z = \mathrm{map}(z'_i)$ where $\mathrm{index}(z) = \mathrm{index}(z'_i) + n + c - 1$. This procedure is performed recursively, until each node in $\mathcal{T}$ is mapped to a corresponding node in $\mathcal{T}_u$. Notice that in a (d, c)-bounded access tree, each non-leaf node x in $\mathcal{T}$ has at most c non-leaf child nodes, and this recursive procedure can be terminable.

3.3.2 Security Models

Bounded Ciphertext-policy Encryption Model.

Definition 3.3.2. A BCPABE scheme includes a tuple of probabilistic polynomial-time algorithms as follows:

- **Setup**$(d, c) \rightarrow (\mathrm{PP}, \mathrm{MK})$: On input an implicit security parameter 1^k and two system parameters (d, c), the setup algorithm **Setup** outputs a public parameter PP and a master key MK.

- **KGen**$(\gamma, \mathrm{MK}) \rightarrow (D)$: On input an attribute set γ and a master key MK, the key generation algorithm **KGen** outputs a secret key D.

- **Enc**$(\mathrm{PP}, \mathcal{T}, m) \rightarrow (E)$: On input the public parameter PP, a (d, c)-bounded access tree $\mathcal{T}$, and a message m, the encryption algorithm **Enc** outputs a ciphertext

E.

- **Dec**$(D, E) \to (m)$: On input a secret key D and a ciphertext E, if the attribute set in D satisfies the access tree in E, the decryption algorithm **Dec** decrypts the ciphertext E and returns a message m; otherwise, it outputs "$\perp$".

Selective-Tree Security Model for BCPABE. This model was first introduced by GJPS. The analogous selective-ID model can be found in [34, 37, 59, 84] A BCPABE scheme is secure in the selective-tree CPA model if no probabilistic polynomial-time adversary $\mathcal{A}$ has a non-negligible advantage in winning the following game.

Init $\mathcal{A}$ chooses an access tree $\mathcal{T}^*$ that he wishes to be challenged upon. The challenger runs **Setup** algorithm and gives $\mathcal{A}$ the resulting public parameter PP. It keeps the corresponding master key MK.

Phase 1 $\mathcal{A}$ issues queries for secret keys related with many attribute sets γ_j, where γ_j does not satisfy the access tree $\mathcal{T}^*$ for all j.

Challenge Once $\mathcal{A}$ decides that Phase 1 is over, it outputs two messages of same length, m_0 and m_1, from the message space. The challenger chooses $\mu \in \{0, 1\}$ at random and encrypts m_μ with $\mathcal{T}^*$. Then, the ciphertext C^* is given to $\mathcal{A}$.

Phase 2 The same as Phase 1.

Guess $\mathcal{A}$ outputs a guess $\mu' \in \{0, 1\}$ and wins the game if $\mu' = \mu$.

We define $\mathcal{A}$'s advantage in this game as $|\Pr[\mu' = \mu] - \frac{1}{2}|$. The selective-tree CPA model can be extended to handle chosen-ciphertext attacks by allowing for decryption queries in Phases 1 and 2, denoted as selective-tree CCA model.

3.3.3   Basic BCPABE Scheme BCP$_1$

We now proceed with the formal description of our first scheme BCP$_1$.

Setup: (d, c) This algorithm takes two parameters (d, c) as its input. Define a real attribute set $\mathcal{U} = \{1, \cdots, n\}$ and a dummy attribute set $\mathcal{U}^* = \{n+1, \cdots, n+c-1\}$. Next, we define a (d, c)-universal access tree $\mathcal{T}_u$ as explained in Section 2.4. $(d, c, \mathcal{U}, \mathcal{U}^*, \mathcal{T}_u)$ are all used in the following algorithms.

Now, the algorithm generates the public parameter for this scheme. For each real attribute $j \in \mathcal{U}$, randomly choose a set of $|\Sigma_{\mathcal{T}_u}|$ numbers $\{t_{j,x}\}_{x \in \Sigma_{\mathcal{T}_u}}$ from Z_p. Further, for each dummy attribute $j \in \mathcal{U}^*$, randomly choose a set of $|\Sigma_{\mathcal{T}_u}|$ numbers $\{t^*_{j,x}\}_{x \in \Sigma_{\mathcal{T}_u}}$ from Z_p. Finally, randomly choose $y \in Z_p$. The public parameter PP $= \langle Y = e(g, g)^y, \{T_{j,x} = g^{t_{j,x}}\}_{j \in \mathcal{U}, x \in \Sigma_{\mathcal{T}_u}}, \{T^*_{j,x} = g^{t^*_{j,x}}\}_{j \in \mathcal{U}^*, x \in \Sigma_{\mathcal{T}_u}}\rangle$. The master key MK $= \langle y, \{t_{j,x}\}_{j \in \mathcal{U}, x \in \Sigma_{\mathcal{T}_u}}, \{t^*_{j,x}\}_{j \in \mathcal{U}^*, x \in \Sigma_{\mathcal{T}_u}}\rangle$.

KGen(γ, MK): This algorithm takes an attribute set $\gamma \subseteq \mathcal{U}$ and the master key MK as its input, then it outputs a secret key D which can be used for decrypting a ciphertext encrypted under a (d, c)-bounded access tree $\mathcal{T}$ iff $\mathcal{T}(\gamma) = 1$.

Now, the algorithm generates the secret key. For each user, choose a random polynomial q_x of degree $c - 1$ for each non-leaf node x in the (d, c)-universal access tree. These polynomials are chosen in a top-down manner, satisfying $q_x(0) = q_{p(x)}(\text{index}(x))$ and $q_r(0) = y$. Once the polynomials have been fixed, it outputs the following secret key $D = \langle \gamma, \{D_{j,x} = g^{\frac{q_x(j)}{t_{j,x}}}\}_{j \in \gamma, x \in \Sigma_{\mathcal{T}_u}}, \{D^*_{j,x} = g^{\frac{q_x(j)}{t^*_{j,x}}}\}_{j \in \mathcal{U}^*, x \in \Sigma_{\mathcal{T}_u}}\rangle$.

Enc$(m, \text{PP}, \mathcal{T})$: This algorithm takes a message m, the public parameter PP, and a (d, c)-bounded access tree $\mathcal{T}$ as its input.

Now, to encrypt the message m with the access tree $\mathcal{T}$, the algorithm first sets a map from $\mathcal{T}$ to $\mathcal{T}_u$ using the method mentioned in Section 2.4. Then, for each non-leaf node $x \in \mathcal{T}$, it chooses an arbitrary $(c - k_x)$-sized set ω_x[4] of dummy child nodes of x' in $\mathcal{T}_u$, $x' = \text{map}(x)$. Following this, let $f(j, x) = 1$ if the node x in $\mathcal{T}$ has a child node associated with real attribute j; $f(j, x) = 0$ otherwise. Then, it chooses a random value $s \in Z_p$ and outputs the ciphertext $E = \langle \mathcal{T}, E' = m \cdot Y^s, \{E_{j,x} = $

---

[4]Without loss of generality, $w_x = \{n + 1, \cdots, n + c - k_x\}$.

$T^s_{j,\text{map}(x)}\}_{j\in\mathcal{U},x\in\Sigma_T,f(j,x)=1}, \{E^*_{j,x} = T^{*s}_{j,\text{map}(x)}\}_{j\in\omega_x,x\in\Sigma_T}\rangle.$

Dec(E, D): This algorithm takes a ciphertext E and a secret key D as its input. If the attribute set γ associated with D satisfies the access tree T in E, the algorithm proceeds as follows; otherwise, output $\bot$.

A recursive algorithm DecryptNode(E, D, x) takes the ciphertext E, the secret key D, and a satisfied non-leaf node x in T as its input and outputs a group element of G_T or $\bot$. For each x's child node z,

- z is a real node, let $j - \text{att}(z)$. Then, we have

  $F_{x,j} = \text{DecryptNode}(E, D, x)$

  $= \begin{cases} e(D_{j,\text{map}(x)}, E_{j,x}) = e(g,g)^{sq_{\text{map}(x)}(j)}, & \text{if } j \in \gamma; \\ \bot, & \text{otherwise.} \end{cases}$

- z is a dummy node. $j = \text{att}(z) \in w_x$. Then we have:

  $F_{x,j} = \text{DecryptNode}(E, D, x)$

  $= e(D^*_{j,\text{map}(x)}, E^*_{j,x}) = e(g,g)^{sq_{\text{map}(x)}(j)},$

From the above procedure, for each non-leaf node x in T, if $T_x(\gamma) = 1$, we have at least $k_x + c - k_x = c$ different points $F_{x,j}$ to compute $e(g,g)^{sq_{\text{map}(x)}(0)}$ using Lagrange interpolation. By recursively executing such procedure in a bottom-up manner, and finally, it obtains $E'' = e(g,g)^{sq_r(0)} = e(g,g)^{sy}$, where r is the root of T. The decryption algorithm outputs $m = E'/E''$.

3.3.4 Security Proof of BCP₁

Theorem 3.3.3. *If the DBDH assumption holds in (G, G_T), then scheme* BCP₁ *is selective-tree CPA secure in the standard model.*

Proof. Suppose a polynomial-time adversary $\mathcal{A}$ exists that can attack BCP₁ in the selective-tree CPA model with non-negligible advantage ε. We construct a simulator

$\mathcal{S}$ that can distinguish the DBDH tuple from a random tuple with non-negligible advantage $\frac{\varepsilon}{2}$.

We first let the challenger set the groups G and G_T with an efficient bilinear map e and a generator g. The challenger flips a fair binary coin ν, outside of $\mathcal{S}$'s view. If $\nu = 1$, the challenger sets $(g, A, B, C, Z) \in \mathcal{D}_{bdh}$; otherwise it sets $(g, A, B, C, Z) \in \mathcal{D}_{rand}$.

Init The simulator $\mathcal{S}$ runs $\mathcal{A}$. $\mathcal{A}$ chooses a challenge (d, c)-bounded access tree $\mathcal{T}^*$. The simulator $\mathcal{S}$ sets $Y = e(A, B) = e(g, g)^{ab}$. Then, $\mathcal{S}$ generates a (d, c)-universal access tree $\mathcal{T}_u$ and a map from $\mathcal{T}^*$ to $\mathcal{T}_u$. Randomly choose

$$\{r_{j,x}\}_{j \in \mathcal{U}, x \in \Sigma_{\mathcal{T}_u}}, \{r^*_{j,x}\}_{j \in \mathcal{U}^*, x \in \Sigma_{\mathcal{T}_u}}$$

from $\mathcal{Z}_p$.

For $j \in \mathcal{U}, x \in \Sigma_{\mathcal{T}_u}$,

$$T_{j,x} = \begin{cases} g^{r_{j,x}}, & \text{if } x = \text{map}(x'), x' \in \Sigma_{\mathcal{T}^*}, f(j, x') = 1; \\ B^{r_{j,x}}, & \text{otherwise.} \end{cases}$$

For $j \in \mathcal{U}^*, x \in \Sigma_{\mathcal{T}_u}$,

$$T^*_{j,x} = \begin{cases} g^{r^*_{j,x}}, & \text{if } j \in w_{x'}, x = \text{map}(x'), x' \in \Sigma_{\mathcal{T}^*}; \\ B^{r^*_{j,x}}, & \text{otherwise.} \end{cases}$$

Phase 1 $\mathcal{A}$ adaptively makes query for a secret key related to an attribute set γ such that $\mathcal{T}^*(\gamma) = 0$. To generate the secret key, $\mathcal{S}$ needs to assign a polynomial q_x for every non-leaf node in $\mathcal{T}$ and output a piece of secret key according to each non-leaf node.

With an attribute set γ as input, for any node $x \in \mathcal{T}_u$, we call a node x: an unsatisfied node iff there exists an unsatisfied node x' in $\mathcal{T}^*$ such that $\text{map}(x') = x$; a satisfied node iff there exists a satisfied node x' in $\mathcal{T}^*$ such that $\text{map}(x') = x$; a non-mapped node iff there exists no node x' in $\mathcal{T}^*$ so that $\mathrm{map}(x') = x$. We define the following three procedures: PolyUnsat, PolySat, and PolyNotCare.

For $j \in \gamma$,
$$D_{j,x} = \begin{cases} g^{\frac{bq_x(j)}{r_{j,x}}} = B^{\frac{q_x(j)}{r_{j,x}}}, & \text{if } f(j,x') = 1 \\ g^{\frac{bq_x(j)}{br_{j,x}}} = (g^{q_x(j)})^{\frac{1}{r_{j,x}}}, & \text{otherwise.} \end{cases}$$

For $j \in \mathcal{U}^*$,
$$D_{j,x}^* = \begin{cases} g^{\frac{bq_x(j)}{r_{j,x}^*}} = B^{\frac{q_x(j)}{r_{j,x}^*}}, & \text{if } j \in \omega_{x'}; \\ g^{\frac{bq_x(j)}{br_{j,x}^*}} = (g^{q_x(j)})^{\frac{1}{r_{j,x}^*}}, & \text{otherwise.} \end{cases}$$

Then, for each non-leaf child node z of x in $\mathcal{T}_u$,
If z is a non-mapped node,
  $\mathrm{PolyNotCare}(\mathcal{T}_z, \gamma, g^{q_x(\mathrm{index}(z))})$;
If z is a satisfied node,
  $\mathrm{PolySat}(\mathcal{T}_z, \gamma, q_x(\mathrm{index}(z)))$;
If z is an unsatisfied node,
  $\mathrm{PolyUnsat}(\mathcal{T}_z, \gamma, g^{q_x(\mathrm{index}(z))})$.

Figure 3.6: $\mathrm{PolyUnsat}(\mathcal{T}_x, \gamma, g^{\lambda_x})$

$\mathrm{PolyUnsat}(\mathcal{T}_x, \gamma, g^{\lambda_x})$ for an unsatisfied node $x \in \Sigma_{\mathcal{T}}$ is defined as follows:

This procedure generates a polynomial q_x for an unsatisfied node x. We have an unsatisfied node x' such that $\mathrm{map}(x') = x$. γ does not satisfy this access tree $\mathcal{T}_{x'}^*$, denoted as $\mathcal{T}_{x'}^*(\gamma) = 0$, where $\mathcal{T}_{x'}^*$ is a subtree of $\mathcal{T}^*$. λ_x is an integer from Z_p. The unsatisfied root node x' has at most $k_{x'} - 1$ satisfied child node. Thus, it could implicitly sets $q_x(0) = \lambda_x$, and chooses $c-1$ other points at random to fix q_x completely, including

- $c - k_{x'}$ points as $q_x(j)$ for the dummy nodes of x, where $j \in w_{x'}$;

- At most $k_{x'} - 1$ points as $q_x(\mathrm{index}(z'))$ if z' is a satisfied leaf child node of x' or $q_x(\mathrm{index}(z') + n + c - 1)$ where z' is a satisfied non-leaf child node of x'.

It executes the following steps in Figure 3.6.

$\mathrm{PolySat}(\mathcal{T}_x, \gamma, \lambda_x)$ for a satisfied node $x \in \Sigma_{\mathcal{T}}$ is defined as follows:

This procedure generates a polynomial q_x for a satisfied node x. We have a satisfied node x' such that $\text{map}(x') = x$. λ_x is an integer from Z_p. It sets $q_x(0) = \lambda_x$, and chooses $c - 1$ other points at random to completely fix q_x. Thus, for each $j \in \mathcal{U} \cup \mathcal{U}^* \cup \{n + c, \cdots, n + 2c - 1\}$, we can obtain $q_x(j)$.

It executes the following steps in Figure 3.7.

$$
\begin{array}{|l|}
\hline
\text{For } j \in \gamma, \\[2pt]
D_{j,x} = \begin{cases} g^{\frac{bq_x(j)}{r_{j,x}}} = B^{\frac{q_x(j)}{r_{j,x}}}, & \text{if } f(j, x') = 1; \\ g^{\frac{bq_x(j)}{br_{j,x}}} = g^{\frac{q_x(j)}{r_{j,x}}}, & \text{otherwise.} \end{cases} \\[2pt]
\text{For } j \in \mathcal{U}^*, \\[2pt]
D^*_{j,x} = \begin{cases} g^{\frac{bq_x(j)}{r^*_{j,x}}} = B^{\frac{q_x(j)}{r^*_{j,x}}}, & \text{if } j \in \omega_{x'}; \\ g^{\frac{bq_x(j)}{br^*_{j,x}}} = g^{\frac{q_x(j)}{r^*_{j,x}}}, & \text{otherwise.} \end{cases} \\[2pt]
\text{Then, for each non-leaf child node } z \text{ of } x \text{ in } \mathcal{T}_u, \\
\text{If } z \text{ is a non-mapped node,} \\
\quad \text{PolyNotCare}(\mathcal{T}_z, \gamma, g^{q_x(\text{index}(z))}); \\
\text{If } z \text{ is a satisfied node,} \\
\quad \text{PolySat}(\mathcal{T}_z, \gamma, q_x(\text{index}(z))); \\
\text{If } z \text{ is an unsatisfied node,} \\
\quad \text{PolyUnsat}(\mathcal{T}_z, \gamma, g^{q_x(\text{index}(z))}). \\
\hline
\end{array}
$$

Figure 3.7: PolySat$(\mathcal{T}_x, \gamma, \lambda_x)$

PolyNotCare$(\mathcal{T}_x, \gamma, g^{\lambda_x})$ for a non-mapped node $x \in \Sigma_{\mathcal{T}}$ is defined as follows:

This procedure generates a polynomial q_x for a non-mapped node x. It implicitly sets $q_x(0) = \lambda_x$, and chooses $c - 1$ other points at random to fix q_x implicitly . Thus, for $j \in \mathcal{U} \cup \mathcal{U}^* \cup \{n + c, \cdots, n + 2c - 1\}$, we could obtain $g^{q_x(j)}$. It outputs the following secret keys:

$$\{D_{j,x} = g^{\frac{bq_x(j)}{br_{j,x}}} = (g^{q_x(j)})^{\frac{1}{r_{j,x}}}\}_{j \in \gamma}, \{D^*_{j,x} = g^{\frac{bq_x(j)}{br^*_{j,x}}} = (g^{q_x(j)})^{\frac{1}{r^*_{j,x}}}\}_{j \in \mathcal{U}^*}$$

Then, it calls PolyNotCare$(\mathcal{T}_z, \gamma, g^{q_x(\text{index}(z))})$ for each non-leaf child node z of x in $\mathcal{T}_u$.

To give a secret key for an attribute set γ, $\mathcal{S}$ first runs PolyUnsat$(\mathcal{T}_r = \mathcal{T}, \gamma, A)$. Notice that we implicitly set $y = ab$ by $Y = e(A, B) = e(g, g)^y$. The secret key corresponding to each non-leaf node is recursively given by the above three procedures. Finally, it outputs

$$D = \langle \gamma, \{D_{j,x}\}_{j \in \gamma, x \in \Sigma_{T_u}}, \{D^*_{j,x}\}_{j \in U^*, x \in \Sigma_{T_u}} \rangle$$

Therefore, S can answer each secret key query with an attribute set γ, where $T^*(\gamma) = 0$. The distribution of these secret keys are identical to those in the real environment.

Challenge The adversary A will submit two challenge messages m_0 and m_1 to S. Then, S chooses $\mu \in \{0, 1\}$ at random, and returns an encryption of m_μ under the challenge access tree T^*. The challenge ciphertext E is formed as:

$$\langle T^*, E' = m_\mu \cdot Z, \{E_{j,x} = C^{r_{j,\text{map}(x)}}\}_{j \in U, x \in \Sigma_{T^*}, f(j,x)=1},$$

$$\{E^*_{j,x} = C^{r^*_{j,\text{map}(x)}}\}_{j \in \omega_x, x \in \Sigma_{T^*}} \rangle$$

If $(g, A, B, C, Z) \in \mathcal{D}_{bdh}$ and we let $s = c$, we have $Y^s = e(g,g)^{abc}$ and

$$E_{j,x} = C^{r_{j,\text{map}(x)}} = (g^{r_{j,\text{map}(x)}})^s, E^*_{j,x} = C^{r^*_{j,\text{map}(x)}} = (g^{r^*_{j,\text{map}(x)}})^s.$$

Therefore, the ciphertext is a valid random encryption of message m_μ.

Otherwise, if $(g, A, B, C, Z) \in \mathcal{D}_{rand}$, we have $E' = m_\mu \cdot Z$. Since Z is randomly chosen from G_T, E' will be a random element of G_T from the adversary's view and the ciphertext doesn't contain any information about m_μ.

Phase 2 The simulator S acts exactly as it did in **Phase 1**.

Guess S outputs $\nu' = 1$ to indicate that it was given a tuple from $\mathcal{D}_{bdh}$ if A gives a correct guess $\mu' = \mu$; otherwise output $\nu' = 0$ to indicate that it was given a tuple from $\mathcal{D}_{rand}$.

Let us compute the success probability of $\mathcal{S}$:

In the case of $\nu = 0$, the adversary gains no information about μ. Therefore, we have $\Pr[\mu \neq \mu' | \nu = 0] = \frac{1}{2}$. Since the simulator guesses $\nu' = 0$ when $\mu \neq \mu'$, we have $\Pr[\nu' = \nu | \nu = 0] = \Pr[\nu' = 0 | \nu = 0] = \frac{1}{2}$.

In the case of $\nu = 1$, the adversary gets a valid ciphertext of m_{μ}. By definition, the adversary has probability ε to guess the correct μ', and thus $\Pr[\mu = \mu' | \nu = 1] = \frac{1}{2} + \varepsilon$. Since the simulator guesses $\nu' = 1$ when $\mu = \mu'$, we have $\Pr[\nu' = \nu | \nu = 1] = \Pr[\nu' = 1 | \nu = 1] = \frac{1}{2} + \varepsilon$.

The overall advantage of the simulator to output a correct $\nu' = \nu$ is $\Pr[\nu = \nu'] - \frac{1}{2} = \Pr[\nu = \nu', \nu = 0] + \Pr[\nu = \nu', \nu = 1] - \frac{1}{2} = \frac{1}{2} \cdot \frac{1}{2} + \frac{1}{2} \cdot (\frac{1}{2} + \varepsilon) - \frac{1}{2} = \frac{\varepsilon}{2}$ □

3.3.5   Extended BCPABE Scheme BCP_2

Now, by using one-time signature technique, we present an extended scheme BCP_2 achieving chosen ciphertext security.

The selective-tree CCA model was introduced in Section 3.3.2 and the similar security model can be found in [84].

We assume that exsiting BCPABE scheme BCP_1 is secure in the selective-tree CPA model as presented in Section 3.3.2, including four algorithms

$$(\text{BCP}_1.\textbf{Setup}, \text{BCP}_1.\textbf{KGen}, \text{BCP}_1.\textbf{Enc}, \text{BCP}_1.\textbf{Dec})$$

and a secure one-time signature ots, including three algorithms

$$(\text{ots}.\textbf{KGen}, \text{ots}.\textbf{Sig}, \text{ots}.\textbf{Ver}).$$

Assume that the verification key vk from $\mathtt{ots}$ is a bit string of length l, and we write vk_i for the i-th bit in vk. Let $\mathcal{L}$ denote $\{1, 2, \cdots, l\}$. $\mathtt{BCP}_2$ is constructed based on $\mathtt{BCP}_1$ and $\mathtt{ots}$ including the following algorithms:

Setup(d, c): This algorithm takes two system parameters (d, c) as its input. Then, it calls $\mathtt{BCP}_1.$**Setup**(d, c) to generate $\mathtt{BCP}_1$'s public parameter PP_1 and master key MK_1. In addition, it randomly chooses a set $\{t'_i\}_{i \in \{1, \cdots, 2l\}}$ from Z_p and defines $T'_i - g^{t'_i}$ Now, it outputs the public parameter $\mathrm{PP} = \langle \mathrm{PP}_1, \{T'_i\}_{i \in \{1, \cdots, 2l\}} \rangle$ and keeps the master key $\mathrm{MK} = \langle \mathrm{MK}_1, \{t'_i\}_{i \in \{1, \cdots, 2l\}} \rangle$.

KGen(γ, MK): This algorithm takes an attribute set γ and the master key MK as its input. Then, it randomly chooses r' from Z_p and calls $\mathtt{BCP}_1.$**KGen**(γ, MK_1) to generate a user's secret key D_1 by using r' instead of y in MK_1 (i.e., $q_r(0) = r'$). For every $i \in \mathcal{L}$, let $D_{i,0} = g^{\frac{r_i}{t'_i}}$ and $D_{i,1} = g^{\frac{r_i}{t'_{l+i}}}$, where $\{r_i\}_{i \in \mathcal{L}}$ are randomly chosen from Z_p. Define $r = r' + \sum_{i \in \mathcal{L}} r_i$ and let $\hat{D} = g^{y-r}$. Finally, it outputs $D = \langle D_1, \{D_{i,0}, D_{i,1}\}_{i \in \mathcal{L}}, \hat{D} \rangle$.

Enc$(m, \mathrm{PP}, \mathcal{T})$: This algorithm takes a message m, the public parameter PP, and a (d, c)-bounded access tree $\mathcal{T}$ as its input.

It first calls $\mathtt{BCP}_1.$**Enc**$(m, \mathrm{PP}_1, \mathcal{T})$ and obtains a partial ciphertext E_1. Then, a key pair $\langle sk, vk \rangle$ is obtained by running $\mathtt{ots}.$**KGen**. For each $i \in \mathcal{L}$, it sets $E'_i = T'^s_i$ if $vk_i = 0$; $E'_i = T'^s_{l+i}$ otherwise. Let $\hat{E} = g^s$. [5] It runs $\mathtt{ots}.$**Sig** with input $(sk, \langle E_1, \{E'_i\}_{i \in \mathcal{L}}, \hat{E} \rangle$ and obtains σ.

The output ciphertext $E = \langle E_1, \{E'_i\}_{i \in \mathcal{L}}, \hat{E}, \sigma, vk \rangle$.

Dec(D, E): This algorithm takes a secret key D and a ciphertext E as its input.

It first checks whether σ is a valid signature on message $\langle E_1, \{E'_i\}_{i \in \mathcal{L}}, \hat{E} \rangle$ using vk.

---

[5]Here, s is consistent with the random value in E_1

If valid, it proceeds as follows; otherwise, output $\perp$.

It extracts D_1 and E_1 from tuple (D, E) and consequently decrypts $e(g, g)^{sr'}$ according to $\mathrm{BCP}_1.\textbf{Dec}$.

For each $i \in \mathcal{L}$, it computes

$$e(g, g)^{sr_i} = \begin{cases} e(D_{i,0}, T_i'^s) = e(g^{\frac{r_i}{t_i'}}, g^{t_i's}), & \text{if } vk_i = 0; \\ e(D_{i,1}, T_{l+i}'^s) = e(g^{\frac{r_i}{t_{l+i}'}}, g^{t_{l+i}'s}), & \text{if } vk_i = 1. \end{cases}$$

Finally, it computes $m = \dfrac{E'}{e(\hat{E}, \hat{D}) \cdot e(g,g)^{sr'} \cdot \prod_{i \in \mathcal{L}} e(g,g)^{sr_i}}$.

Compared with BCP_1, BCP_2's ciphertext is augmented with l elements, while public parameter and secret key are both augmented with $2l$ elements. Another method for extending BCP_1 with CCA security level has been mentioned [129]. It treats each verification key as an attribute. However, it has shorter additional size of ciphertext (1 element) but larger additional size of public parameter and secret key (2^l elements).

3.3.6   Security Proof of BCP_2

The selective-tree CCA model is introduced in Section 3.3.2. We prove the security of BCP_2 based on the strong existentially unforgeable assumption of $\mathtt{ots}$ and the DBDH assumption.

Theorem 3.3.4. *Suppose* $\mathtt{ots}$ *is a* ε_{ots}*-secure one-time signature scheme. If the DBDH assumption holds in* (G, G_T)*, then scheme* BCP_2 *is a selective-tree CCA secure in the standard model.*

Proof. Suppose there is a polynomial-time adversary $\mathcal{A}$ who can attack BCP_2 in the selective-tree CCA model with non-negligible advantage ε. We construct a simulator $\mathcal{S}$ who can distinguish the DBDH tuple from a random tuple with non-negligible advantage $\frac{\varepsilon}{2} - \varepsilon_{ots}$.

We first let the challenger set the groups G and G_T with an efficient bilinear map e

and a generator g. The challenger flips a fair binary coin ν, outside of $\mathcal{S}$'s view. If $\nu = 1$, the challenger sets $(g, A, B, C, Z) \in \mathcal{D}_{bdh}$; otherwise sets $(g, A, B, C, Z) \in \mathcal{D}_{rand}$.

Init The simulator $\mathcal{S}$ runs $\mathcal{A}$. $\mathcal{A}$ chooses a (d, c)-bounded access tree $\mathcal{T}^*$ it wishes to be challenged upon. $\mathcal{S}$ runs $\mathtt{ots.KGen}$ to obtain $\langle sk^*, vk^* \rangle$. $\mathcal{S}$ sets $Y = e(g, g)^{ab} = e(A, B)$. $\mathcal{S}$ defines a (d, c)-universal access tree $\mathcal{T}$ and a map from $\mathcal{T}^*$ to $\mathcal{T}$. Then, it generates PP_1 as **Init** step in Section 4.2.

For $i \in \mathcal{L}$, randomly choose $\eta_i, \theta_i \in Z_p$ and implicitly set

if $vk_i^* = 0$,

$$t_i' = \eta_i, T_i' = g^{\eta_i} \text{ and } t_{l+i}' = b\theta_{l+i}, T_{l+i}' = B^{\theta_{l+i}};$$

if $vk_i^* = 1$,

$$t_i' = b\eta_i, T_i' = B^{\eta_i} \text{ and } t_{l+i}' = \theta_{l+i}, T_{l+i}' = g^{\theta_{l+i}}.$$

The algorithm outputs public parameter

$$PP = \langle PP_1, \{T_i'\}_{i \in \{1, \cdots, 2l\}} \rangle$$

Phase 1 $\mathcal{A}$ is allowed to make secret key queries and decryption queries:

- Secret Key Query. $\mathcal{A}$ submits an attribute set γ such that $\mathcal{T}^*(\gamma) = 0$. $\mathcal{S}$ randomly chooses $r'', r_i' \in Z_p$ for $i \in \mathcal{L}$ and implicitly sets: $r' = ab + br'', r_i = br_i'$.

  According to the Phase 1 in Section 4, it calls PolyUnsat($\mathcal{T}_u, \gamma, A \cdot g^{r''}$) and obtains D_1. Then, it computes

$$\hat{D} = g^{y-r} = g^{ab-ab-br''-\sum_{i \in \mathcal{L}} br_i'} = \frac{1}{B^{r''+\sum_{i \in \mathcal{L}} r_i'}}$$

and for $i \in \mathcal{L}$, if $vk_i^* = 0$,

$$D_{i,0} = B^{\frac{r_i'}{\eta_i}} \text{ and } D_{i,1} = g^{\frac{r_i'}{\theta_{l+i}}};$$

if $vk_i^* = 1$

$$D_{i,0} = g^{\frac{r_i'}{\eta_i}} \text{ and } D_{i,1} = B^{\frac{r_i'}{\theta_{l+i}}}.$$

Finally, it outputs the secret key

$$D = \langle D_1, \{D_{i,0}, D_{i,1}\}_{i \in \mathcal{L}}, \hat{D} \rangle.$$

• Decryption Query. $\mathcal{A}$ submits a ciphertext $E = \langle E_1, \{E_i'\}_{i \in \mathcal{L}}, \hat{E}, \sigma, vk \rangle$ related with $\mathcal{T}$. $\mathcal{S}$ checks the signature σ using vk. If σ is invalid, $\mathcal{S}$ outputs $\bot$; otherwise $\mathcal{S}$ checks if $vk = vk^*$. If so, we call it a **forge** event and $\mathcal{S}$ outputs $\nu' = 0$ to indicate that it was given a tuple from $\mathcal{D}_{rand}$. Now, the only case is $vk \neq vk^*$. In such a case, $\mathcal{S}$ defines an attribute set γ such that $\mathcal{T}(\gamma) = 1$. If $\mathcal{T}^*(\gamma) = 0$, generate a secret key related with γ from secret key query and use it to decrypt E; otherwise $\mathcal{T}^*(\gamma) = 1$. Without loss of generality, assume $vk_j = 1, vk_j^* = 0$. $\mathcal{S}$ generates a partial secret key for decrypting E as follows:

  – $\mathcal{S}$ randomly chooses $r'', r_i' \in Z_p$ for $i \in \mathcal{L}$ and implicitly sets: $r' = br'', r_i = br_i'$ for $i \neq j$ and $r_j = ab + br_j'$.

  – According to the Phase 1 in Section 4.2, it calls PolySat($\mathcal{T}, \gamma, r''$) and obtains D_1 related with γ. Now, it can use D_1 to decrypt E_1 and obtains $e(g,g)^{sr'}$.

  – For $i \in \mathcal{L}$ and $i \neq j$, if $vk_i^* = 0$,

$$D_{i,0} = B^{\frac{r_i'}{\eta_i}} \text{ and } D_{i,1} = g^{\frac{r_i'}{\theta_{l+i}}};$$

if $vk_i^* = 1$,

$$D_{i,0} = g^{\frac{r_i'}{\eta_i}} \text{ and } D_{i,1} = B^{\frac{r_i'}{\theta_{l+i}}}.$$

- For $i = j$, it can only generate $D_{j,1} = g^{\frac{r_j}{t_{l+i}'}} = g^{\frac{ab+br_j'}{b\theta_{l+j}}} = A^{\frac{1}{\theta_{l+j}}} \cdot g^{\frac{r_j'}{\theta_{l+j}}}$.

  This is sufficient for our decryption since $vk_j = 1$ and $T_{l+j}'^s$ is involved in ciphertext E. Note that $e(D_{j,1}, T_{l+j}'^s) = e(y,y)^{sr_j}$.

- $\mathcal{S}$ then computes $\hat{D} = g^{y-r} = g^{ab-br''-ab-\sum_{i\in\mathcal{L}} br_i'} = \frac{1}{B^{r''+\sum_{i\in\mathcal{L}} r_i'}}$.

- Finally, $\mathcal{S}$ outputs $m = \frac{E'}{e(\hat{E},\hat{D}) \cdot e(g,g)^{sr'} \cdot \prod_{i\in\mathcal{L}} e(g,g)^{sr_i}}$.

Challenge The adversary $\mathcal{A}$ will submit two challenge messages m_0 and m_1 to $\mathcal{S}$. Then, $\mathcal{S}$ chooses $\mu \in \{0,1\}$ at random, and returns an encryption of m_μ under the challenge access tree $\mathcal{T}^*$ as follows:

- It first generates an encryption E_1^* according to **Challenge** in Section 4.2.

- It generates a signature

$$\sigma^* = \mathsf{ots.Sig}(sk^*, \langle E_1^*, \{C^{\eta_i}\}_{i\in\mathcal{L},vk_i^*=0}, \{C^{\theta_{l+i}}\}_{i\in\mathcal{L},vk_i^*=1}, C \rangle).$$

- It outputs the challenge ciphertext

$$E^* = \langle E_1^*, \{C^{\eta_i}\}_{i\in\mathcal{L},vk_i^*=0}, \{C^{\theta_{l+i}}\}_{i\in\mathcal{L},vk_i^*=1}, C, \sigma^*, vk^* \rangle.$$

Let $s = c$. If $(g, A, B, C, Z) \in \mathcal{D}_{bdh}$, we have

$$Y^s = e(g,g)^{abc}, E_i' = C^{\eta_i} = (g^s)^{\eta_i} = T_i'^s, E_i' = C^{\theta_{l+i}} = (g^s)^{\theta_{l+i}} = T_{l+i}'^s.$$

Therefore, the ciphertext is a valid random encryption of message m_μ.

Otherwise, if $(g, A, B, C, Z) \in \mathcal{D}_{rand}$, we have $E' = m_\mu \cdot Z$. Since Z is randomly chosen from G_T, E' will be a random element of G_T from the adversary's view. Thus the ciphertext contains no information about m_μ.

Phase 2 The simulator $\mathcal{S}$ acts exactly as it did in **Phase 1**.

Guess $\mathcal{S}$ outputs $\nu' = 1$ to indicate that it was given a tuple from $\mathcal{D}_{bdh}$ if $\mathcal{A}$ gives a correct guess $\mu' = \mu$; otherwise $\nu' = 0$ to indicate that it was given a tuple from $\mathcal{D}_{rand}$.

Let us compute the success probability of $\mathcal{S}$:

In the case of $\nu = 0$, the adversary doesn't gain any information about μ. Therefore, we have $\Pr[\mu \neq \mu' | \nu = 0] = \frac{1}{2}$. Since the simulator guesses $\nu' = 0$ when $\mu \neq \mu'$ (no **forge**), we have

$$\Pr[\nu' = \nu | \nu = 0] = \Pr[\nu' = 0 | \nu = 0] = \Pr[\mu \neq \mu', \neg\mathbf{forge} | \nu = 0] + \Pr[\mathbf{forge} | \nu = 0]$$

$$= \Pr[\mu \neq \mu' | \nu = 0] - \Pr[\mu \neq \mu', \mathbf{forge} | \nu = 0] + \Pr[\mathbf{forge} | \nu = 0]$$

$$\geq \frac{1}{2} - \Pr[\mathbf{forge} | \nu = 0] = \frac{1}{2} - \varepsilon_{ots}.$$

In the case of $\nu = 1$, the adversary gets a valid ciphertext of m_μ. By definition, the adversary has probability ε to guess the correct μ', and thus $\Pr[\mu = \mu' | \nu = 1] = \frac{1}{2} + \varepsilon$. Since the simulator guesses $\nu' = 1$ when ($\mu = \mu'$, no **forge**), we have

$$\Pr[\nu' = \nu | \nu = 1] = \Pr[\nu' = 1 | \nu = 1] = \Pr[\mu = \mu', \neg\mathbf{forge} | \nu = 1]$$

$$= \Pr[\mu = \mu' | \nu = 1] - \Pr[\mu = \mu', \mathbf{forge} | \nu = 1]$$

$$\geq \frac{1}{2} + \varepsilon - \Pr[\mathbf{forge} | \nu = 1] = \frac{1}{2} + \varepsilon - \varepsilon_{ots}.$$

The overall advantage of the simulator to output a correct $\nu' = \nu$ is

$$\Pr[\nu = \nu'] - \frac{1}{2} = \Pr[\nu = \nu', \nu = 0] + \Pr[\nu = \nu', \nu = 1] - \frac{1}{2}$$

$$\geq \frac{1}{2} \cdot \left(\frac{1}{2} - \varepsilon_{ots}\right) + \frac{1}{2} \cdot \left(\frac{1}{2} + \varepsilon - \varepsilon_{ots}\right) - \frac{1}{2} = \frac{\varepsilon}{2} - \varepsilon_{ots}.$$

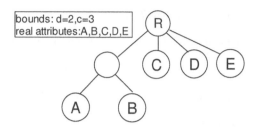

Figure 3.8: An example access tree accepted by BCP_1 but not accepted by GJPS

□

3.3.7 Comparisons

Access Policy: In this section, we compare the expressive capability of access tree of BCP_1 with that of GJPS scheme bounded by the same parameter (d, c). Actually, according to the definition of BCP_1 on the (d, c)-bounded access tree, each non-leaf node has a threshold value at the most c and no more than c non-leaf nodes share a single unique parent. Thus, one difference of the restriction on access trees between GJPS and BCP_1 is if a non-leaf node x has a non-leaf child node, in GJPS the total number of x's child node must be no more than c, while in BCP_1 the total number of x's non-leaf child node must be no more than c. Therefore, our scheme accommodates more possible access policies chosen by the sender under the same pre-set bounds. (An example is shown in Figure 3.8)

Efficiency and Parameter Size: Now, we discuss the comparisons on the computational cost of each algorithm and the sizes of parameters between GJPS and BCP_1 in Table 3.1, both of which are proved secure in the selective-tree CPA model. We assume that both schemes are initialized with same system parameters (d, c) and consider an encryption under a (d, c)-bounded access tree $\mathcal{T}$(this access tree must be chosen suitable for both schemes since there's a difference between the acceptable access trees of two schemes). $\mathcal{T}_n$ is $\mathcal{T}$'s normal form. The secret key is associated with an attribute set γ such that $\mathcal{T}(\gamma) = 1$ and $|\gamma| = x$. $\hat{\mathcal{T}}$ and $\hat{\mathcal{T}}_n$ are the γ-satisfied non-redundant trees of $\mathcal{T}$ and $\mathcal{T}_n$ with minimum non-leaf nodes, respectively. $|\mathcal{U}| = n, |\mathcal{U}^*| = c - 1, |\Sigma_{\mathcal{T}_u}| = \frac{c^d - 1}{c - 1}$. Here, TExp represents the cost of one modular exponentiation, TPair represents the cost of one bilinear pairing computing. T_{S1}, T_{K1}, T_{E1}, and T_{D1} represent the computational cost of **Setup**, **KGen**, **Enc** and **Dec** algorithms in GJPS. L_{P1}, L_{S1}, and L_{C1} represent the size of public parameter, secret key and ciphertext in GJPS. The mark with "2" indicates the counterpart of BCP$_1$.

Mark	times	Computational cost	
T_{S1}	1	TPair $+ (nc^{d-1} + c^d) \cdot$ TExp	✓
T_{S2}	1	TPair $+ (n \times \frac{c^d - 1}{c - 1} + c^d) \cdot$ TExp	
T_{K1}	1/user	$(xc^{d-1} + c^d - 1) \cdot$ TExp	✓
T_{K2}	1/user	$(x \times \frac{c^d - 1}{c - 1} + c^d - 1) \cdot$ TExp	
T_{E1}	many	$(1 + |\Theta_{\mathcal{T}_n}| + \sum_{x \in \Sigma_{\mathcal{T}_n}} (c - k_x)) \cdot$ TExp	
T_{E2}	many	$(1 + |\Theta_{\mathcal{T}}| + \sum_{x \in \Sigma_{\mathcal{T}}} (c - k_x)) \cdot$ TExp	✓
T_{D1}	many	$(|\Sigma_{\hat{\mathcal{T}}_n}| \times c - |\Sigma_{\hat{\mathcal{T}}_n}| + 1) \cdot$ TPair $+ |\Sigma_{\hat{\mathcal{T}}_n}| \times c \cdot$ TExp	
T_{D2}	many	$(|\Sigma_{\hat{\mathcal{T}}}| \times c - |\Sigma_{\hat{\mathcal{T}}}| + 1) \cdot$ TPair $+ |\Sigma_{\hat{\mathcal{T}}}| \times c \cdot$ TExp	✓
	number	Size	
L_{P1}	1	$|G_T| + (nc^{d-1} + c^d - 1) \cdot |G|$	✓
L_{P2}	1	$|G_T| + (n \times \frac{c^d - 1}{c - 1} + c^d - 1) \cdot |G|$	
L_{S1}	1/user	$(xc^{d-1} + c^d - 1) \cdot |G|$	✓
L_{S2}	1/user	$(x \times \frac{c^d - 1}{c - 1} + c^d - 1) \cdot |G|$	
L_{C1}	many	$|G_T| + (|\Theta_{\mathcal{T}_n}| + \sum_{x \in \Sigma_{\mathcal{T}_n}} (c - k_x)) \cdot |G|$	
L_{C2}	many	$|G_T| + (|\Theta_{\mathcal{T}}| + \sum_{x \in \Sigma_{\mathcal{T}}} (c - k_x)) \cdot |G|$	✓

Table 3.1: Comparisons between the scheme in GJPS and BCP$_1$

Generally, c is set to be no less than 2, then

$$x \times \frac{c^d - 1}{c - 1} + c^d - 1 < 2 \times (xc^{d-1} + c^d - 1).$$

This is because of the following deduction:

$$2 \le c \Longrightarrow 2c^{d-1} - 1 < c^d \Longrightarrow \frac{c^d - 1}{c - 1} < 2c^{d-1}.$$

Therefore, we obtain that

$$T_{S2} < 2T_{S1}, T_{K2} < 2T_{K1}, L_{P2} < 2L_{P1}, L_{S2} < 2L_{S1}.$$

3.4 Multi-Authority Encryption Schemes

When Sahai and Waters [241] introduced the notion of ABE, they also presented the following open problem: Is it possible to construct an ABE scheme in which many different authorities operate simultaneously, each handing out secret keys for a different set of attributes? This is an interesting and practical problem. In the aforementioned ABE systems, all attributes are managed by a single authority. In some applications however, this may not be desirable. For example, Alice encrypts a message with access policy ("UNIV.X.COMPUTER SCIENCE" **AND** "UNIV.X.ALUMNI" **AND** "COMP.Y.ENGINEER") so that only receivers who are the computer science alumni of University X and currently working as an engineer for Company Y, can decrypt. The authority UNIV.X Registry may only manage attributes for the students, staff, and alumni of University X, while COMP.Y Registry may be the authority handling its employees' attributes. A single-authority ABE may not be appropriate in this scenario. Another problem of single-authority ABE is the so-called *Key Escrow* problem. In a single-authority ABE system, as the single authority is responsible for issuing private keys for all attributes, it can decrypt any ciphertexts in the system, so that the authority must be fully trusted. Some multi-authority ABE systems [78, 79, 184, 198, 201, 217] have been proposed to achieve better expressiveness, efficiency, and security in the setting of multi-authority. The problems of privacy and key-escrow are also considered simultaneously in those papers. The Table 3.2 shows the properties of these systems in terms of security, expressiveness, and additional properties such as protecting privacy and preventing decryption by individual authority.

Chase [78] proposed the first multi-authority ABE system where there are one CA (Central Authority) and multiple AAs (Attribute Authorities). The CA issues identity-related keys to users and the AAs manage attributes and issue attribute-related keys. A user's keys from different AAs are linked together by the user's *global identifier*. The expressiveness of the system is limited and only "AND" policy between the AAs is sup-

Table 3.2: Existing Multi-Authority ABE Systems

	Adaptively Secure	Standard Model	Prevent Decryption by Individual Authorities	KP/CP	Expressiveness
[78]	$\times$	$\checkmark$	$\times^3$	KP	Limited[1]
[198]	$\times$	$\checkmark$	$\checkmark$	KP	Limited[1]
[79][4]	$\times$	$\checkmark$	$\checkmark$	KP	Limited[1]
[217]	$\times$	$\checkmark$	$\times^3$	CP	Expressive[2]
[184]	$\checkmark$	$\times$	Partially[3]	CP	Expressive[2]
[201]	$\checkmark$	$\checkmark$	$\checkmark$	CP	Expressive[2]

[1] [78, 79, 198] are KPABE systems. In these systems, the policy of each key is defined by some sub-policies, where each sub-policy corresponds to an authority, and only when all sub-policies of a policy are satisfied, the policy is satisfied. i.e., the policy supports *only* **AND** relation between authorities.

[2] [184, 201, 217] are CPABE systems, and the encryptor can encrypt messages with any monotone access structures defined over the *whole attribute universe*.

[3] In [78, 217], there is an authority, called Central Authority, can decrypt *all* ciphertexts. Although no individual authority in [184] can decrypt *all* ciphertexts in the system, each authority can individually decrypt the ciphertexts whose policies are satisfied by the attributes managed by the authority.

[4] The system in [79] can protect user privacy from the authorities.

ported. Also, the CA can decrypt all ciphertexts. We [198] removed the CA using a threshold technique where the set of authorities is fixed in advance and all authorities must interact during the system setup. The system cannot defend against collusion attack by m or more users where m is a system parameter chosen at setup. Chase and Chow [79] also removed the central authority using a distributed PRF (pseudo random function) technique. However, the expressiveness is as limited as the original Chase's system, and their technique does not seem applicable to CP-ABE. While [78, 79, 198] focused on KP-ABE, Müller, Katzenbeisser, and Eckert [217] proposed the first multi-authority CP-ABE system where there are one CA and multiple AAs. The AAs operate independently from each other and therefore the scheme is flexible and practical. However, the CA in the system can still decrypt all ciphertexts.

Lewko and Waters [184] proposed the first adaptively secure multi-authority CPABE system. The system is expressive, supporting any monotone access structures. There is no central authority and each authority in the system operates independently.

However, the system is proven secure in the random oracle model, and each authority can still independently decrypt ciphertexts, if the attributes, managed by the authority, satisfy the associated access policies.

We [201] proposed a new multi-authority CP-ABE system which aims to theses problems simultaneously. The system has multiple central authorities (CAs) and attribute authorities (AAs). The CAs issue identity-related keys to users but do not involve in any attribute-related operations. AAs issue attribute-related keys to users. Each AA manages a different attribute domain and operates independently from other AAs. A party may join the system to be an AA by simply registering itself to the CAs, and then publishing its attribute-related public parameters. In the proposed system, no authority can independently decrypt any ciphertext. The system is adaptively secure in the standard model which captures adaptive authority corruption. Its access policy can be any monotone access structure and the system supports large attribute universe. The efficiency of the system is also comparable to the corresponding single-authority CP-ABE system [181], which is regarded as the "state-of-the-art" single-authority CP-ABE system.

Figure 3.9 shows the architecture of the multi-authority CPABE system. The system has D Central Authorities, $CA_1, \ldots, CA_D$, and K Attribute Authorities,

$$AA_1, \ldots, AA_K.$$

Each AA manages a different domain of attributes (e.g., AA_1 manages U_1, and so on). When a user joins the system, each CA issues an identity-related key to the user. Then the user obtains an attribute-related key corresponding to the attributes that the user entitled from an AA (e.g., UNIV.X Registry). In practice, one may imagine that there could be multiple CAs run by different organizations while all of them are governed under some ordinance made by the government, then universities and companies can join the system as AAs. Each AA manages its own attribute domain and the AAs

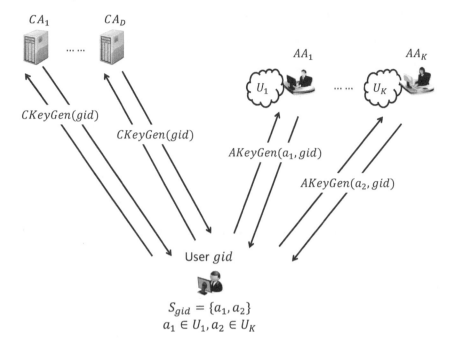

Figure 3.9: Architecture of Multi-Authority CP-ABE

operate independently from each other. The trust on the CAs by the users in the system can also be alleviated as it is unlikely to have all the CAs collude if some appropriate governmental policies and business measures are put into place to govern the practice of the CAs.

3.4.1 Security Models

Similar to [181, 184], the system is constructed over composite order groups. Let $\mathcal{G}$ be the group generator, which takes a security parameter λ and outputs

$$(p_1, p_2, p_3, G, G_T, e)$$

where p_1, p_2, and p_3 are distinct primes, G and G_T are cyclic groups of order $N = p_1 p_2 p_3$, and $e : G \times G \rightarrow G_T$ is a map such that: (1) (Bilinear) $\forall g, h \in G, a, b \in$

$\mathbb{Z}_N$, $e(g^a, h^b) = e(g, h)^{ab}$, (2) (NonDegenerate) $\exists g \in G$ such that $e(g, g)$ has order N in G_T. Assume that group operations in G and G_T as well as the bilinear map e are computable in polynomial time with respect to λ. Let G_{p_1}, G_{p_2}, and G_{p_3} be the subgroups of order p_1, p_2, and p_3 in G, respectively. Note that for any $h_i \in G_{p_i}$ and $h_j \in G_{p_j}$ where $i \neq j$, $e(h_i, h_j) = 1$.

The security of the system is based on the following three assumptions, which are also used by [183](for IBE) and [181] (for CP ABE) for obtaining full security.

For an element $T \in G$, T can (uniquely) be written as the product of elements of G_{p_1}, G_{p_2}, and G_{p_3}, and they are referred to as the "G_{p_1} part of T," "G_{p_2} part of T," and "G_{p_3} part of T," respectively. In the assumptions below, let $G_{p_1 p_2}$ and $G_{p_1 p_3}$ be the subgroups of order $p_1 p_2$ and $p_1 p_3$ in G, respectively. Similarly, an element in $G_{p_1 p_2}$ can be written as the product of elements of G_{p_1} and G_{p_2}, and an element in $G_{p_1 p_3}$ can be written as the product of elements of G_{p_1} and G_{p_3}.

Assumption 1 (Subgroup decision problem for 3 primes). [183] Given a group generator $\mathcal{G}$, define the following distribution: $\mathbb{G} = (N = p_1 p_2 p_3, G, G_T, e) \xleftarrow{R} \mathcal{G}$, $g \xleftarrow{R} G_{p_1}$, $X_3 \xleftarrow{R} G_{p_3}$, $D = (\mathbb{G}, g, X_3)$, $T_1 \xleftarrow{R} G_{p_1 p_2}$, $T_2 \xleftarrow{R} G_{p_1}$. The advantage of an algorithm $\mathcal{A}$ in breaking Assumption 1 is:

$$Adv1_{\mathcal{G},\mathcal{A}}(\lambda) := |\Pr[\mathcal{A}(D, T_1) = 1] - \Pr[\mathcal{A}(D, T_2) = 1]|.$$

Definition 3.4.1. $\mathcal{G}$ satisfies Assumption 1 if $Adv1_{\mathcal{G},\mathcal{A}}(\lambda)$ is a negligible function of λ for any polynomial time algorithm $\mathcal{A}$.

Assumption 2. [183] Given $\mathcal{G}$, define the following distribution: $\mathbb{G} = (N = p_1 p_2 p_3, G, G_T, e) \xleftarrow{R} \mathcal{G}$, $g, X_1 \xleftarrow{R} G_{p_1}$, $X_2, Y_2 \xleftarrow{R} G_{p_2}$, $X_3, Y_3 \xleftarrow{R} G_{p_3}$, $D = (\mathbb{G}, g, X_1 X_2, X_3, Y_2 Y_3)$, $T_1 \xleftarrow{R} G$, $T_2 \xleftarrow{R} G_{p_1 p_3}$. The advantage of an algorithm $\mathcal{A}$ against Assumption 2 is:

$$Adv2_{\mathcal{G},\mathcal{A}}(\lambda) := |\Pr[\mathcal{A}(D, T_1) = 1] - \Pr[\mathcal{A}(D, T_2) = 1]|.$$

Definition 3.4.2. $\mathcal{G}$ satisfies Assumption 2 if $Adv2_{\mathcal{G},\mathcal{A}}(\lambda)$ is a negligible function of λ for any polynomial time algorithm $\mathcal{A}$.

Assumption 3. [183] Given $\mathcal{G}$, define the following distribution: $\mathbb{G} = (N = p_1 p_2 p_3,$ $G, G_T, e) \xleftarrow{R} \mathcal{G},\ \alpha, s \xleftarrow{R} \mathbb{Z}_N,\ g \xleftarrow{R} G_{p_1},\ X_2, Y_2, Z_2 \xleftarrow{R} G_{p_2},\ X_3 \xleftarrow{R} G_{p_3},$ $D = (\mathbb{G}, g, g^\alpha X_2, X_3, g^s Y_2, Z_2),\ T_1 = e(g,g)^{\alpha s},\ T_2 \xleftarrow{R} G_T$. The advantage of an algorithm $\mathcal{A}$ against Assumption 3 is

$$Adv3_{\mathcal{G},\mathcal{A}}(\lambda) := |\Pr[\mathcal{A}(D, T_1) = 1] - \Pr[\mathcal{A}(D, T_2) = 1]|.$$

Definition 3.4.3. $\mathcal{G}$ satisfies Assumption 3 if $Adv3_{\mathcal{G},\mathcal{A}}(\lambda)$ is a negligible function of λ for any polynomial time algorithm $\mathcal{A}$.

Notations. There are three sets of entities in a multi-authority ciphertext-policy attribute-based encryption (MA-CP-ABE) system: (1) central authorities (CAs), (2) attribute authorities (AAs) and, (3) users. Let $CA_1, \ldots, CA_D$ be central authorities and $\mathbb{D} = \{1, \ldots, D\}$ the index set of the CAs, that is, using $d \in \mathbb{D}$ to denote the index of central authority CA_d. Let $AA_1, \ldots, AA_K$ be attribute authorities and $\mathbb{K} = \{1, \ldots, K\}$ the index set of the AAs. Each user has a global identifier denoted as gid. The CAs are responsible for issuing keys to users according to their global identifiers. The AAs are responsible for issuing keys corresponding to attributes, and each AA manages a different attribute domain (e.g., AA_i manages attributes for a university registry and AA_j manages attributes for a company registry, etc.). Let U_k be the attribute domain managed by AA_k where $U_i \cap U_j = \emptyset$ for all $i \neq j \in \mathbb{K}$, and $U = \bigcup_{k=1}^{K} U_k$ be the attribute universe.

Definition. A MA-CPABE system consists of the following seven algorithms:

GlobalSetup(λ) $\rightarrow$ (GPK): The algorithm takes as input the security parameter λ and outputs the global public parameter GPK of the system.

CASetup(GPK, d) $\rightarrow$ (CPK$_d$, CAPK$_d$, CMSK$_d$): Each CA_d runs the algorithm with GPK and its index d as input, and produces master secret key CMSK$_d$ and public parameters (CPK$_d$, CAPK$_d$). CAPK$_d$ will be used by AAs only.

AASetup(GPK, k, U_k) $\rightarrow$ (APK$_k$, ACPK$_k$, AMSK$_k$): Each AA_k runs the algorithm with GPK, its index k and its attribute domain U_k as input, and produces master secret key AMSK$_k$ and public parameters (APK$_k$, ACPK$_k$). ACPK$_k$ will be used by CAs only.

Encrypt(M, $\mathbb{A}$, GPK, {CPK$_d$|$d \in \mathbb{D}$}, {APK$_k$}) $\rightarrow$ CT: The algorithm takes as input a message M, an access policy $\mathbb{A}$ defined over the attribute universe U, the global public parameter GPK, CAs' public parameters {CPK$_d$|$d \in \mathbb{D}$}, and the related AAs' public parameters {APK$_k$}. It outputs a ciphertext CT which contains the access policy $\mathbb{A}$.

CKeyGen(gid, GPK, {ACPK$_k$|$k \in \mathbb{K}$}, CMSK$_d$) $\rightarrow$ (ucsk$_{gid,d}$, ucpk$_{gid,d}$): When a user with global identifier gid visits CA_d for obtaining a key, CA_d runs the algorithm, which takes as input gid, GPK, {ACPK$_k$|$k \in \mathbb{K}$}, and CA_d's master secret key CMSK$_d$. It outputs a *user-central-key* (ucsk$_{gid,d}$, ucpk$_{gid,d}$), where ucpk$_{gid,d}$ is called *user-central-public-key*.

AKeyGen(att, {ucpk$_{gid,d}$|$d \in \mathbb{D}$}, GPK, {CAPK$_d$|$d \in \mathbb{D}$}, AMSK$_k$) $\rightarrow$ uask$_{att,gid}$ or $\perp$: When a user requests a secret key for attribute att from AA_k, AA_k runs the algorithm, which takes as input att, {ucpk$_{gid,d}$|$d \in \mathbb{D}$}, GPK, {CAPK$_d$|$d \in \mathbb{D}$} and AMSK$_k$. If all ucpk$_{gid,d}$s are valid, the algorithm outputs a *user-attribute-key* uask$_{att,gid}$, otherwise it outputs $\perp$. For a user gid with attribute set S_{gid}, the user's *decryption-key* is defined as

$$DK_{gid} = (\{ucsk_{gid,d}, ucpk_{gid,d}|d \in \mathbb{D}\}, \{uask_{att,gid}|att \in S_{gid}\}).$$

Decrypt(CT, GPK, $\{\mathsf{APK}_k\}$, DK_{gid}) $\rightarrow M$ or $\perp$: The algorithm takes as input a
ciphertext CT associated with access policy $\mathbb{A}$, GPK, the related attribute authorities' public parameters $\{\mathsf{APK}_k\}$, and a decryption-key DK_{gid} with attribute
set S_{gid}. If S_{gid} satisfies the access policy $\mathbb{A}$, the algorithm outputs the message
M, otherwise it outputs $\perp$.

Security Model : The security of MA-CP-ABE is defined by the following game
run between a challenger $\mathcal{B}$ and an adversary $\mathcal{A}$. $\mathcal{A}$ can corrupt CAs and AAs by
specifying $\mathbb{K}_c \subset \mathbb{K}$ and $\mathbb{D}_c \subset \mathbb{D}$ after seeing the public parameters[6], where $\mathbb{D} \setminus \mathbb{D}_c \neq \emptyset$
and $\mathbb{K} \setminus \mathbb{K}_c \neq \emptyset$. Without loss of generality, we assume that $\mathcal{A}$ corrupts all CAs but
one, i.e., $|\mathbb{D} \setminus \mathbb{D}_c| = 1$.

Setup.

- GlobalSetup, CASetup(GPK, d) ($d = 1, \ldots, D$) and AASetup(GPK, k,
  U_k) ($k = 1, \ldots, K$) are run by the challenger $\mathcal{B}$. GPK, $\{\mathsf{CPK}_d, \mathsf{CAPK}_d | d \in$
  $\mathbb{D}\}$ and $\{\mathsf{APK}_k, \mathsf{ACPK}_k | k \in \mathbb{K}\}$ are given to the adversary $\mathcal{A}$.

- $\mathcal{A}$ specifies an index $d^* \in \mathbb{D}$ as the only uncorrupted CA and specifies a set
  $\mathbb{K}_c \subset \mathbb{K}$ of AAs to be corrupted where $\mathbb{K} \setminus \mathbb{K}_c \neq \emptyset$. Let $\mathbb{D}_c = \mathbb{D} \setminus \{d^*\}$.
  $\{\mathsf{CMSK}_d | d \in \mathbb{D}_c\}$ and $\{\mathsf{AMSK}_k | k \in \mathbb{K}_c\}$ are given to $\mathcal{A}$.

Key Query Phase 1. User-central-key and user-attribute-key can be obtained by querying the following oracles:

CKQ(gid, d) where $d = d^*$: $\mathcal{A}$ queries with a pair (gid, d), where gid is a
global identifier and $d = d^*$, and obtains the corresponding user-central-
key (ucsk_{gid,d^*}, ucpk_{gid,d^*}).

AKQ(att, $\{\mathsf{ucpk}_{gid,d} | d \in \mathbb{D}\}$, k) where $k \in \mathbb{K} \setminus \mathbb{K}_c$: $\mathcal{A}$ queries with (att,
$\{\mathsf{ucpk}_{gid,d} | d \in \mathbb{D}\}$, k), where $k \in \mathbb{K} \setminus \mathbb{K}_c$ is the index of an uncorrupted

---

[6]This is stronger than the static corruption model used in [78, 79, 184], where the adversary has to specify
the authorities to corrupt before seeing the public parameters. But on the other aspect, it is weaker than the
model in [184], where the corrupted authorities are set by the adversary.

AA, $\{\text{ucpk}_{gid,d}|d \in \mathbb{D}\}$ are gid's user-central-public-keys, and att is an attribute in U_k. The oracle returns a user-attribute-key $\text{uask}_{att,gid}$ or $\perp$ if $\{\text{ucpk}_{gid,d}\}$ is invalid.

Challenge Phase. $\mathcal{A}$ submits two equal-length messages M_0 and M_1, and an access policy $\mathbb{A}$. $\mathcal{B}$ flips a random coin $\beta \in \{0, 1\}$ and sends to $\mathcal{A}$ an encryption of M_β under $\mathbb{A}$.

Phase 2. $\mathcal{A}$ further queries as in **Key Query Phase 1**.

Guess. $\mathcal{A}$ submits a guess β' for β.

For a gid, the related attribute set is defined as

$$S_{gid} = \{att \mid \text{AKQ}(att, \{\text{ucpk}_{gid,d}|d \in \mathbb{D}\}, k) \text{ is made by } \mathcal{A}\}.$$

$\mathcal{A}$ wins the game if $\beta' = \beta$ under the **restriction** that there is no S_{gid} such that $S_{gid} \cup (\bigcup_{k_c \in \mathbb{K}_c} U_{k_c})$ can satisfy the challenge access policy $\mathbb{A}$. The advantage of $\mathcal{A}$ is defined as $|\Pr[\beta = \beta'] - 1/2|$.

Definition 3.4.4. An MA-CP-ABE system is secure if for all polynomial-time adversary $\mathcal{A}$ in the game above, the advantage of $\mathcal{A}$ is negligible.

Remarks : It is assumed that a user with global identifier gid requests for the central key from each CA_d only once, i.e., for each gid there is only one set of user-central-keys, $\{\text{ucpk}_{gid,d}|d \in \mathbb{D}\}$. This is not a restriction, but can help simplify the system description. Using obscure notations such as $\text{ucpk}_{gid,d,t}$ and $S_{gid,d,t}$ where t is a time stamp can remove this assumption. In the security model above, $\mathcal{A}$ has the master secret keys $\{\text{CMSK}_d|d \in \mathbb{D}_c\}$, so the user only needs to query $\text{CKQ}(gid, d^*)$ for getting $(\text{ucsk}_{gid,d^*}, \text{ucpk}_{gid,d^*})$, and he/she can generate $\{(\text{ucsk}_{gid,d}, \text{ucpk}_{gid,d})|d \in \mathbb{D}_c\}$ if they are needed for querying AKQ.

3.4.2 Construction

GlobalSetup(λ) $\rightarrow$ (GPK): Let G be a bilinear group of order $N = p_1p_2p_3$ (3 distinct primes), and G_{p_i} be the subgroup of order p_i in G. The algorithm randomly chooses $g, h \in G_{p_1}$. Let X_3 be a generator of G_{p_3}.

The global public parameter is published as GPK $= (N, g, h, X_3, \Sigma_{sign})$, where $\Sigma_{sign} = $ (KeyGen, Sign, Verify) is the description of a UF-CMA secure signature scheme.

CASetup(GPK, d) $\rightarrow$ (CPK$_d$, CAPK$_d$, CMSK$_d$): CA_d runs the KeyGen algorithm of Σ_{sign} to generate sign key pair (SignKey$_d$, VerifyKey$_d$), and chooses a random exponent $\alpha_d \in \mathbb{Z}_N$.

CA_d publishes its public parameter CPK$_d = e(g, g)^{\alpha_d}$, CAPK$_d = $ VerifyKey$_d$.

CA_d sets its master secret key CMSK$_d = (\alpha_d, $ SignKey$_d$).

AASetup(GPK, k, U_k) $\rightarrow$ (APK$_k$, ACPK$_k$, AMSK$_k$): For each $att \in U_k$, AA_k randomly chooses $s_{att} \in \mathbb{Z}_N$ and sets $T_{att} = g^{s_{att}}$. For each $d \in \mathbb{D}$, AA_k randomly chooses $v_{k,d} \in \mathbb{Z}_N$ and sets $V_{k,d} = g^{v_{k,d}}$.

AA_k publishes its public parameter APK$_k = \{T_{att}|att \in U_k\}$, ACPK$_k = \{V_{k,d}|d \in \mathbb{D}\}$.

AA_k sets its master secret key AMSK$_k = (\{s_{att}|att \in U_k\}, \{v_{k,d}|d \in \mathbb{D}\})$.

Encrypt(M, $\mathbb{A} = (A, \rho)$, GPK, $\{$CPK$_d|d \in \mathbb{D}\}$, $\{$APK$_k\}$) $\rightarrow$ CT: M is the message to be encrypted, $\mathbb{A}$ is the access policy which is expressed by an LSSS matrix (A, ρ), where A is an $l \times n$ matrix and ρ maps each row A_x of A to an attribute $\rho(x)$. Here it is required that ρ will not map two different rows to a same attribute. The algorithm chooses a random vector $\vec{v} = (s, v_2, \ldots, v_n) \in \mathbb{Z}_N^n$, and for each $x \in \{1, 2, \ldots l\}$, it randomly picks $r_x \in \mathbb{Z}_N$. Let $A_x \cdot \vec{v}$ be the inner product of the xth row of A and the vector $\vec{v}$. The ciphertext is

$$C = M \cdot \prod_{d=1}^{D} e(g,g)^{\alpha_d \cdot s}, \ C' = g^s,$$

$$\{C_x = h^{A_x \cdot \vec{v}} T_{\rho(x)}^{-r_x}, C'_x = g^{r_x} \mid x \in \{1, 2, \ldots l\}\}$$

along with the access policy $\mathbb{A} = (A, \rho)$.

CKeyGen$(gid, \mathsf{GPK}, [V_{k,d}|k \subset \mathbb{K}\}, \mathsf{CMSK}_d) \rightarrow (\mathsf{ucsk}_{gid,d}, \mathsf{ucpk}_{gid,d})$: When a user submits his gid to CA_d to request the user-central-key, CA_d randomly chooses $r_{gid,d} \in \mathbb{Z}_N$ and $R_{gid,d}, R'_{gid,d} \in G_{p_3}$, then sets

$$\mathsf{ucsk}_{gid,d} = g^{\alpha_d} h^{r_{gid,d}} R_{gid,d}, \ L_{gid,d} = g^{r_{gid,d}} R'_{gid,d}.$$

For $k = 1$ to K, CA_d randomly picks $R_{gid,d,k} \in G_{p_3}$ and computes

$$\Gamma_{gid,d,k} = V_{k,d}^{r_{gid,d}} R_{gid,d,k}.$$

CA_d computes $\sigma_{gid,d} = \mathsf{Sign}(\mathsf{SignKey}_d, gid||d||L_{gid,d}||\Gamma_{gid,d,1}||\cdots||\Gamma_{gid,d,K})$. Let $\mathsf{ucpk}_{gid,d} = (gid, d, L_{gid,d}, \{\Gamma_{gid,d,k} \mid k \in \mathbb{K}\}, \sigma_{gid,d})$.

AKeyGen$(att, \{\mathsf{ucpk}_{gid,d}|d \in \mathbb{D}\}, \mathsf{GPK}, \{\mathsf{VerifyKey}_d|d \in \mathbb{D}\}, \mathsf{AMSK}_k) \rightarrow$ $\mathsf{uask}_{att,gid}$ or $\perp$: When a user submits his $\{\mathsf{ucpk}_{gid,d}|d \in \mathbb{D}\}$ to AA_k to request the user-attribute-key for attribute $att \in U_k$,

1. For $d = 1$ to D, AA_k parses $\mathsf{ucpk}_{gid,d}$ into $(gid, d, L_{gid,d}, \{\Gamma_{gid,d,k}|k \in \mathbb{K}\}, \sigma_{gid,d})$ and checks whether

$$valid \leftarrow \mathsf{Verify}(\mathsf{VerifyKey}_d, gid||d||L_{gid,d}||\Gamma_{gid,d,1}||\cdots||\Gamma_{gid,d,K}, \sigma_{gid,d}) \tag{3.1}$$

$$e(g, \Gamma_{gid,d,k}) = e(V_{k,d}, L_{gid,d}) \neq 1. \tag{3.2}$$

If fails, AA_k outputs $\perp$ to the user to imply that the submitted $\{\mathsf{ucpk}_{gid,d}|d \in \mathbb{D}\}$ are invalid.

2. For $d = 1$ to D, AA_k randomly picks $R'_{att,gid,d} \in G_{p_3}$, and sets

$$\mathsf{uask}_{att,gid,d} = (\Gamma_{gid,d,k})^{s_{att}/v_{k,d}} R'_{att,gid,d}. \tag{3.3}$$

Note that

$$\begin{aligned}
\mathsf{uask}_{att,gid,d} &= (\Gamma_{gid,d,k})^{s_{att}/v_{k,d}} R'_{att,gid,d} \\
&= (V_{k,d}^{r_{gid,d}} R_{gid,d,k})^{s_{att}/v_{k,d}} R'_{att,gid,d} \\
&= (g^{v_{k,d} \cdot r_{gid,d}} R_{gid,d,k})^{s_{att}/v_{k,d}} R'_{att,gid,d} \\
&= T_{att}^{r_{gid,d}} (R_{gid,d,k})^{s_{att}/v_{k,d}} R'_{att,gid,d}
\end{aligned}$$

As $(R_{gid,d,k})^{s_{att}/v_{k,d}} R'_{att,gid,d}$ is in G_{p_3} and $R'_{att,gid,d}$ is randomly chosen, we can write

$$\mathsf{uask}_{att,gid,d} = T_{att}^{r_{gid,d}} R_{att,gid,d}. \tag{3.4}$$

Without knowing the value of $r_{gid,d}$, by running (3.3), AA_k can compute the value as (3.4).

3. AA_k outputs user-attribute-key $\mathsf{uask}_{att,gid}$ to user where

$$\begin{aligned}
\mathsf{uask}_{att,gid} &= \prod_{d=1}^{D} \mathsf{uask}_{att,gid,d} = \prod_{d=1}^{D} T_{att}^{r_{gid,d}} R_{att,gid,d} \\
&= T_{att}^{\sum_{d=1}^{D} r_{gid,d}} \prod_{d=1}^{D} R_{att,gid,d} \\
&= T_{att}^{\sum_{d=1}^{D} r_{gid,d}} R_{att,gid}
\end{aligned} \tag{3.5}$$

Decrypt(CT, GPK, $\{\mathsf{APK}_k\}$, DK_{gid}) $\rightarrow M$: The ciphertext CT is parsed into $\langle C, C',$ $\{C_x, C'_x | x \in \{1, 2, \ldots, l\}\}, \mathbb{A} = (A, \rho)\rangle$, and the decryption-key DK_{gid} is parsed into $(\{\mathsf{ucsk}_{gid,d}, \mathsf{ucpk}_{gid,d} | d \in \mathbb{D}\}, \{\mathsf{uask}_{att,gid} | att \in S_{gid}\})$.

The algorithm computes

- $\mathsf{ucsk}_{gid} = \prod_{d=1}^{D} \mathsf{ucsk}_{gid,d} = g^{\sum_{d=1}^{D} \alpha_d} h^{\sum_{d=1}^{D} r_{gid,d}} \prod_{d=1}^{D} R_{gid,d} =$
  $g^{\alpha} h^{r_{gid}} R_{gid}$,

  with $\alpha = \sum_{d=1}^{D} \alpha_d$, $r_{gid} = \sum_{d=1}^{D} r_{gid,d}$ and $R_{gid} = \prod_{d=1}^{D} R_{gid,d}$.

- $L_{gid} = \prod_{d=1}^{D} L_{gid,d} = g^{\sum_{d=1}^{D} r_{gid,d}} \prod_{d=1}^{D} R'_{gid,d} = g^{r_{gid}} R'_{gid}$,

  with $R'_{gid} = \prod_{d=1}^{D} R'_{gid,d}$.

Note that $\forall att \in S_{gid}$, $\mathsf{uask}_{att,gid} = T_{att}^{\sum_{d=1}^{D} r_{gid,d}} R_{att,gid} = T_{att}^{r_{gid}} R_{att,gid}$.

If S_{gid} satisfies the access policy (A, ρ), the algorithm computes constants $\omega_x \in \mathbb{Z}_N$ such that $\sum_{\rho(x) \in S_{gid}} \omega_x A_x = (1, 0, \dots, 0)$. Then it computes

$$e(C', \mathsf{ucsk}_{gid}) / \prod_{\rho(x) \in S_{gid}} \left(e(C_x, L_{gid}) \cdot e(C'_x, \mathsf{uask}_{\rho(x),gid}) \right)^{\omega_x} = e(g,g)^{\alpha s}.$$

As $C = M \cdot \prod_{d=1}^{D} e(g,g)^{\alpha_d \cdot s} = M \cdot e(g,g)^{s \sum_{d=1}^{D} \alpha_d} = M \cdot e(g,g)^{s\alpha}$, M can be recovered from $C / e(g,g)^{\alpha s}$.

In the above system, an attribute is required to appear at most once in an LSSS matrix (A, ρ). This restriction is crucial to the security proof. Such a system is called a one-use system, and it can be extended to a multi-use system by using the encoding technique [181].

3.4.3 Security Analysis

Let Π denote the main construction, we modify Π to Π' as follows:

- In the AKeyGen algorithm, it outputs $\mathsf{uask}_{att,gid} = \{\mathsf{uask}_{att,gid,d} | d \in \mathbb{D}\}$

rather than $\mathsf{uask}_{att,gid} = \prod_{d=1}^{D} \mathsf{uask}_{att,gid,d}$. i.e., gid's decryption-key is

$$
\begin{aligned}
\mathsf{DK}_{gid} &= (\{\mathsf{ucsk}_{gid,d}, \mathsf{ucpk}_{gid,d}|d \in \mathbb{D}\}, \{\mathsf{uask}_{att,gid} \mid att \in S_{gid}\}) \\
&= (\{\mathsf{ucsk}_{gid,d}, \mathsf{ucpk}_{gid,d}|d \in \mathbb{D}\}, \{\{\mathsf{uask}_{att,gid,d}|d \in \mathbb{D}\}|att \in S_{gid}\}) \\
&= (\{\mathsf{ucsk}_{gid,d}, \mathsf{ucpk}_{gid,d}|d \in \mathbb{D}\}, \{\{\mathsf{uask}_{att,gid,d}|att \in S_{gid}\}|d \in \mathbb{D}\}) \\
&= \{(\mathsf{ucsk}_{gid,d}, \mathsf{ucpk}_{gid,d}, \{\mathsf{uask}_{att,gid,d}|att \in S_{gid}\})|d \in \mathbb{D}\} \\
&= \{\mathsf{usk}_{gid,d}|d \in \mathbb{D}\}
\end{aligned}
$$

where $\mathsf{usk}_{gid,d} = (\mathsf{ucsk}_{gid,d}, \mathsf{ucpk}_{gid,d}, \{\mathsf{uask}_{att,gid,d}|att \in S_{gid}\})$ is called gid's **user-key** related to d.

- In the Decrypt algorithm,

  1. For $d = 1$ to D, the algorithm uses $\mathsf{usk}_{gid,d}$ to reconstruct $e(g, g)^{\alpha_d s}$:

$$
e(C', \mathsf{ucsk}_{gid,d}) / \prod_{\rho(x) \in S_{gid}} \left(e(C_x, L_{gid,d}) \cdot e(C'_x, \mathsf{uask}_{\rho(x),gid,d}) \right)^{\omega_x} = e(g, g)^{\alpha_d s}.
$$

$$
\tag{3.6}
$$

  2. The algorithm recovers M by

$$
M = C / \prod_{d=1}^{D} e(g, g)^{\alpha_d s}. \tag{3.7}
$$

Note that the user and the attacker will get more information in Π', the security of Π' will imply the security of Π. We show the security of Π' in the following: In the security model, CA_{d^*} is the only uncorrupted central authority and no $S_{gid} \cup (\bigcup_{k_c \in \mathbb{K}_c} U_{k_c})$ can satisfy the challenge access policy. It means that the adversary could not request keys to form a usk_{gid,d^*} to reconstruct $e(g, g)^{\alpha_{d^*} s}$. In our proof, the challenger will respond the adversary as in real attack for all key queries related to $d \neq d^*$. On the key queries related to d^*, we use the proof technique [181] to provide the answers. Before we give our proof, we need to define two additional structures: semi-functional ciphertexts and keys. We choose random values $z_{att} \in \mathbb{Z}_N$ associated to the attributes.

Semi-functional Ciphertext: A semi-functional ciphertext is formed as follows. Let g_2 denote a generator of G_{p_2} and c a random exponent modulo N. Besides the random vector $\vec{v} = (s, v_2, \ldots, v_n)$ and the random values $\{r_x | x \in \{1, 2, \ldots, l\}\}$, we also choose a random vector $\vec{u} = (u_1, u_2, \ldots, u_n) \in \mathbb{Z}_N^n$ and random values $\{\gamma_x \in \mathbb{Z}_N | x \in \{1, 2, \ldots, l\}\}$. Then:

$$C' = g^s g_2^c, \{C_x = h^{A_x \cdot \vec{v}} T_{\rho(x)}^{-r_x} g_2^{A_x \vec{u} + \gamma_x z_{\rho(x)}}, C'_x = g^{r_x} g_2^{-\gamma_x} \mid x \in \{1, 2, \ldots, l\}\},$$

Semi-functional Key: For a gid, a semi-functional user-key usk_{gid,d^*} has two possible forms. Exponents $r_{gid,d^*}, \delta, b \in \mathbb{Z}_N$, $\{w_{k,d^*} \in \mathbb{Z}_N | k \in \mathbb{K}\}$, and elements $R_{gid,d^*}, R'_{gid,d^*} \in G_{p_3}, \{R_{att,gid,d^*} \in G_{p_3} | att \in S_{gid}\}, \{R_{gid,d^*,k} \in G_{p_3} | k \in \mathbb{K}\}$ are chosen randomly.

- Type 1:

  The user-central-key $(\mathsf{ucsk}_{gid,d^*}, \mathsf{ucpk}_{gid,d^*})$ is formed as

  $$\mathsf{ucsk}_{gid,d^*} = g^{\alpha_{d^*}} h^{r_{gid,d^*}} R_{gid,d^*} g_2^{\delta}, \quad L_{gid,d^*} = g^{r_{gid,d^*}} R'_{gid,d^*} g_2^b,$$
  $$\Gamma_{gid,d^*,k} = V_{k,d^*}^{r_{gid,d^*}} R_{gid,d^*,k} g_2^{bw_{k,d^*}} \quad (k = 1, 2, \ldots K),$$

  $$\sigma_{gid,d^*} = \mathsf{Sign}(\mathsf{SignKey}_{d^*}, gid||d^*||L_{gid,d^*}||\Gamma_{gid,d^*,1}||\cdots||\Gamma_{gid,d^*,K}),$$

  $$\mathsf{ucpk}_{gid,d^*} = (gid, d^*, L_{gid,d^*}, \{\Gamma_{gid,d^*,k} | k \in \mathbb{K}\}, \sigma_{gid,d^*}).$$

  $\forall att \in S_{gid}$, the derived $\mathsf{uask}_{att,gid,d^*}$ is formed as

  $$\mathsf{uask}_{att,gid,d^*} = T_{att}^{r_{gid,d^*}} R_{att,gid,d^*} g_2^{bz_{att}}.$$

- Type 2:

The user-central-key (ucsk_{gid,d^*}, ucpk_{gid,d^*}) is formed as

$$\mathsf{ucsk}_{gid,d^*} = g^{\alpha_{d^*}} h^{r_{gid,d^*}} R_{gid,d^*} g_2^\delta, \quad L_{gid,d^*} = g^{r_{gid,d^*}} R'_{gid,d^*},$$

$$\Gamma_{gid,d^*,k} = V_{k,d^*}^{r_{gid,d^*}} R_{gid,d^*,k}(k = 1, 2, \ldots K),$$

$$\sigma_{gid,d^*} = \mathsf{Sign}(\mathsf{SignKey}_{d^*}, gid||d^*||L_{gid,d^*}||\Gamma_{gid,d^*,1}||\cdots||\Gamma_{gid,d^*,K}),$$

$$\mathsf{ucpk}_{gid,d^*} = (gid, d^*, L_{gid,d^*}, \{\Gamma_{gid,d^*,k}|k \in \mathbb{K}\}, \sigma_{gid,d^*}).$$

$\forall att \in S_{gid}$, the derived $\mathsf{uask}_{att,gid,d^*}$ is formed as

$$\mathsf{uask}_{att,gid,d^*} = T_{att}^{r_{gid,d^*}} R_{att,gid,d^*}.$$

Note that both the semi-functional user-keys of types 1 and 2 satisfy (3.1) and (3.2), and that type 2 is a special case of type 1 with $b = 0$.

When a normal usk_{gid,d^*} and a semi-functional ciphertext, or a semi-functional usk_{gid,d^*} and a normal ciphertext, are used in computation (3.6), $e(g,g)^{\alpha_{d^*} s}$ is available, and this value could be used in the computation (3.7). When a semi-functional usk_{gid,d^*} and a semi-functional ciphertext are used in computation (3.6), the result is $e(g,g)^{\alpha_{d^*} s} \cdot e(g_2, g_2)^{c\delta - bu_1}$. The additional term $e(g_2, g_2)^{c\delta - bu_1}$ will hinder the computation (3.7). We call a semi-functional user-key of type 1 *nominally semi-functional* if $c\delta - bu_1 = 0$.

The security of Π' relies on Assumptions 1, 2, and 3. We use a hybrid argument over a sequence of games. The first game **Game**$_{\mathsf{Real}}$ is the real security game. In the final game **Game**$_{\mathsf{Final}}$, all user-keys related to d^*, $\{\mathsf{usk}_{gid,d^*}\}$, are semi-functional of type 2 and the ciphertext is a semi-functional encryption of a random message, independent of the two messages provided by $\mathcal{A}$.

Game$_{\mathsf{Real}}$: The challenge ciphertext is normal. All CKQs are answered with normal user-central-key. All AKQs are answered with user-attribute-key generated by running the normal AKeyGen algorithm.

Game$_0$: The challenge ciphertext is semi-functional. All CKQs are answered with normal user-central-key. All AKQs are answered with user-attribute-key generated by running the normal **AKeyGen** algorithm.

Let q denote the number of CKQ made by $\mathcal{A}$. For j from 1 to q, we consider the following games:

Game$_{j,1}$: In this game, the challenge ciphertext is semi-functional. The first j 1 CKQs are answered with semi-functional user-central-key of type 2; the jth CKQ is answered with semi-functional user-central-key of type 1; and the remaining CKQs are answered with normal user-central-key. All AKQs are answered with user-attribute-key generated by running the normal **AKeyGen** algorithm.

Game$_{j,2}$: In this game, the challenge ciphertext is semi-functional. The first $j-1$ CKQs are answered with semi-functional user-central-key of type 2; the jth CKQ is answered with semi-functional user-central-key of type 2; and the remaining CKQs are answered with normal user-central-key. All AKQs are answered with user-attribute-key generated by running the normal **AKeyGen** algorithm.

Game$_{Final}$: In this game, the challenge ciphertext is a semi-functional encryption of a random message, independent of the two messages provided by the adversary. All CKQs are answered with semi-functional user-central-key of type 2. All AKQs are answered with user-attribute-key generated by running the normal **AKeyGen** algorithm, and this step has been followed in all the games.

In the proofs, we will show that the derived $\mathsf{uask}_{att,gid,d^*}$ is decided by the corresponding user-central-key $(\mathsf{ucsk}_{gid,d^*}, \mathsf{ucpk}_{gid,d^*})$, i.e., if $(\mathsf{ucsk}_{gid,d^*}, \mathsf{ucpk}_{gid,d^*})$ is semi-functional of type 1 (respectively, type 2), then the derived $\mathsf{uask}_{att,gid,d^*}$ is also semi-functional of type 1 (respectively, type 2). Consequently, usk_{gid,d^*} is decided by the corresponding $(\mathsf{ucsk}_{gid,d^*}, \mathsf{ucpk}_{gid,d^*})$ as well. Note that in **Game$_0$** all user-central-keys related to d^* are normal and in **Game$_{q,2}$** all user-central-keys related to

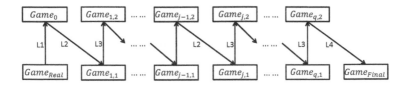

Figure 3.10: Indistinguishable games. L1 denotes Lemma 1, and so on.

d^* are semi-functional of type 2. It means that in **Game**$_0$ all user-keys usk_{gid,d^*} are normal and in **Game**$_{q,2}$ all user-keys usk_{gid,d^*} are semi-functional of type 2. We show these games are indistinguishable in the following four lemmas (see Figure 3.10).

Lemma 3.4.5. *Given a UF-CMA signature scheme Σ_{sign}, suppose a poly-time algorithm $\mathcal{A}$ exists so that $\textbf{Game}_{Real}Adv_{\mathcal{A}} - \textbf{Game}_0 Adv_{\mathcal{A}} = \epsilon$. We can construct a poly-time algorithm $\mathcal{B}$ with advantage ϵ against Assumption 1.*

Lemma 3.4.6. *Use $\textbf{Game}_{0,2}$ to denote $\textbf{Game}_0$. Given a UF-CMA signature scheme Σ_{sign}, suppose a poly-time algorithm $\mathcal{A}$ exists so that*

$$\textbf{Game}_{j-1,2}Adv_{\mathcal{A}} - \textbf{Game}_{j,1}Adv_{\mathcal{A}} = \epsilon.$$

We can construct a poly-time algorithm $\mathcal{B}$ with advantage negligibly close to ϵ against Assumption 2.

Lemma 3.4.7. *Given a UF-CMA signature scheme Σ_{sign}, suppose a poly-time algorithm $\mathcal{A}$ exists so that $\textbf{Game}_{j,1}Adv_{\mathcal{A}} - \textbf{Game}_{j,2}Adv_{\mathcal{A}} = \epsilon$. We can construct a poly-time algorithm $\mathcal{B}$ with advantage ϵ against Assumption 2.*

Lemma 3.4.8. *Given a UF-CMA signature scheme Σ_{sign}, suppose a poly-time algorithm $\mathcal{A}$ exists so that $\textbf{Game}_{q,2}Adv_{\mathcal{A}} - \textbf{Game}_{Final}Adv_{\mathcal{A}} = \epsilon$. We can construct a poly-time algorithm $\mathcal{B}$ with advantage $\frac{\epsilon}{D}$ against Assumption 3.*

Theorem 3.4.9. *If the signature scheme Σ_{sign} is UF-CMA secure and Assumptions 1, 2, and 3 hold, then our MA-CP-ABE scheme is secure.*

Proof. If Assumptions 1, 2 and 3 hold, and the signature scheme Σ_{sign} is UF-CMA secure, previous lemmas have shown that the real security game is indistinguishable from **Game**$_{\mathsf{Final}}$, in which the value of β is information-theoretically hidden from the adversary. Hence the adversary can not obtain a non-negligible advantage in breaking Π', which implies the adversary can not obtain a non-negligible advantage in breaking our MACPABE scheme Π. $\square$

3.4.4 Security Proofs

In the following proof, it should be noted that all AKQs are answered with user-attribute-key generated by running the **normal** AKeyGen algorithm. In the AKeyGen algorithm, the signature verification (3.1) used to verify gid's ucpk_{gid,d^*}, the check (3.2), and the computation (3.3) work together to ensure that the adversary could not make use of ucpk_{gid,d^*} to construct $\mathsf{ucpk}_{gid,d}(d \in \mathbb{D}_c)$.

Proof of Lemma 3.4.5.

Proof. $\mathcal{B}$ is given $\mathbb{G} = (N, G, G_T, e), g, X_3, T$. It will simulate **Game**$_{\mathsf{Real}}$ or **Game**$_0$ to $\mathcal{A}$.

$\mathcal{B}$ randomly chooses $a \in \mathbb{Z}_N$, then gives $\mathcal{A}$ GPK $= \{N, g, h = g^a, X_3\}$, $\mathbb{D} = \{1, 2, \ldots, D\}, \mathbb{K} = \{1, 2, \ldots, K\}$, the descriptions of $U_1, U_2, \ldots, U_K$, and the description of secure signature scheme Σ_{sign}.

For $d = 1$ to D, $\mathcal{B}$ generates sign key pair (SignKey$_d$, VerifyKey$_d$) and randomly chooses $\alpha_d \in \mathbb{Z}_N$, then gives $\mathcal{A}$ CPK$_d = e(g, g)^{\alpha_d}$ and CAPK$_d =$ VerifyKey$_d$.

For $k = 1$ to K, $\mathcal{B}$ randomly chooses $\{s_{att} \in \mathbb{Z}_N | att \in U_k\}$ and $\{v_{k,d} \in \mathbb{Z}_N | d \in \mathbb{D}\}$, then gives $\mathcal{A}$ APK$_k = \{T_{att} = g^{s_{att}} | att \in U_k\}$ and ACPK$_k = \{V_{k,d} = g^{v_{k,d}} | d \in \mathbb{D}\}$.

$\mathcal{A}$ gives $\mathcal{B}$ the $d^* \in \mathbb{D}$ and $\mathbb{K}_c \subset \mathbb{K}$ where $\mathbb{K} \setminus \mathbb{K}_c \neq \emptyset$. Let $\mathbb{D}_c = \mathbb{D} \setminus \{d^*\}$.

$\mathcal{B}$ gives $\mathcal{A}$ CMSK$_d = (\alpha_d, \mathsf{SignKey}_d)$ for each $d \in \mathbb{D}_c$, and AMSK$_k = (\{s_{att} | att \in U_k\}, \{v_{k,d} | d \in \mathbb{D}\})$ for each $k \in \mathbb{K}_c$.

$\mathcal{B}$ responds $\mathcal{A}$'s key queries as follows:

- $\mathcal{A}$ makes $\mathsf{CKQ}(gid, d^*)$. $\mathcal{B}$ runs the normal $\mathsf{CKeyGen}$ algorithm, since it knows CMSK_{d^*}.

- $\mathcal{A}$ makes $\mathsf{AKQ}(att, k, \{\mathsf{ucpk}_{gid,d} | d \in \mathbb{D}\})$ where $(k \in \mathbb{K} \setminus \mathbb{K}_c)$. $\mathcal{B}$ runs the normal $\mathsf{AKeyGen}$ algorithm, since it knows AMSK_k for each $k \in \mathbb{K} \setminus \mathbb{K}_c$.

$\mathcal{A}$ sends two messages M_0 and M_1 and an LSSS matrix (A^*, ρ) to $\mathcal{B}$. To make the challenge ciphertext, $\mathcal{B}$ will implicitly set g^s to be the G_{p_1} part of T (we mean that T is the product of $g^s \in G_{p_1}$ and possibly an element of G_{p_2}). $\mathcal{B}$ randomly chooses $v'_2, \ldots, v'_n \in \mathbb{Z}_N$ to define a vector $\vec{v'} = (1, v'_2, \ldots, v'_n) \in \mathbb{Z}_N^n$, and for each row x of A^* it randomly picks $r'_x \in \mathbb{Z}_N$. Then it chooses a random $\beta \in \{0, 1\}$ and sets

$$C = M_\beta \cdot \prod_{d=1}^D e(g^{\alpha_d}, T) , \; C' = T,$$

$$\{C_x = T^{a A_x^* \cdot \vec{v'}} T^{-r'_x \cdot s_{\rho(x)}} , \; C'_x = T^{r'_x} \mid x \in \{1, 2, \ldots l\}\}$$

Note that this implicitly sets $\vec{v} = s\vec{v'}$ and $r_x = sr'_x$. Modulo p_1, this $\vec{v}$ is a random vector with first coordinate s and r_x is a random value.

If $T \in G_{p_1}$ (i.e., $T = g^s$), this is a properly distributed normal ciphertext.

If $T \in G_{p_1 p_2}$, we let g_2^c denote the G_{p_2} part of T (i.e., $T = g^s g_2^c$). We then have a semi-functional ciphertext with $\vec{u} = ca\vec{v'}$, $\gamma_x = -cr'_x$, and $z_{\rho(x)} = s_{\rho(x)}$. While the values of $a, v'_2, \ldots, v'_n, r'_x, s_{\rho(x)}$ modulo p_2 are uncorrelated from their values modulo p_1 by the Chinese Remainder Theorem, this is a properly distributed semi-functional ciphertext.

Hence, neglecting the probability that Σ_{sign} is broken, $\mathcal{B}$ can use the output of $\mathcal{A}$ to break Assumption 1 with advantage ϵ. $\qquad\square$

Proof of Lemma 3.4.6.

Proof. $\mathcal{B}$ is given $\mathbb{G} = (N, G, G_T, e), g, X_1X_2, X_3, Y_2Y_3, T$. It will simulate **Game**$_{j-1,2}$ or **Game**$_{j,1}$ to $\mathcal{A}$.

$\mathcal{B}$ randomly chooses $a \in \mathbb{Z}_N$, then gives $\mathcal{A}$ GPK $= \{N, g, h = g^a, X_3\}$, $\mathbb{D} = \{1, 2, \ldots, D\}$, $\mathbb{K} = \{1, 2, \ldots, K\}$, the descriptions of $U_1, U_2, \ldots, U_K$, and the description of secure signature scheme Σ_{sign}.

For $d - 1$ to D, $\mathcal{B}$ generates sign key pair $(\mathsf{SignKey}_d, \mathsf{VerifyKey}_d)$ and randomly chooses $\alpha_d \in \mathbb{Z}_N$, then gives $\mathcal{A}$ CPK$_d = e(g, g)^{\alpha_d}$ and CAPK$_d = \mathsf{VerifyKey}_d$.

For $k = 1$ to K, $\mathcal{B}$ randomly chooses $\{s_{att} \in \mathbb{Z}_N | att \in U_k\}$ and $\{v_{k,d} \in \mathbb{Z}_N | d \in \mathbb{D}\}$, then gives $\mathcal{A}$ APK$_k = \{T_{att} = g^{s_{att}} | att \in U_k\}$ and ACPK$_k = \{V_{k,d} = g^{v_{k,d}} | d \in \mathbb{D}\}$.

$\mathcal{A}$ gives $\mathcal{B}$ the $d^* \in \mathbb{D}$ and $\mathbb{K}_c \subset \mathbb{K}$ where $\mathbb{K} \setminus \mathbb{K}_c \neq \emptyset$. Let $\mathbb{D}_c = \mathbb{D} \setminus \{d^*\}$.

$\mathcal{B}$ gives $\mathcal{A}$ CMSK$_d = (\alpha_d, \mathsf{SignKey}_d)$ for each $d \in \mathbb{D}_c$, and AMSK$_k = (\{s_{att} | att \in U_k\}, \{v_{k,d} | d \in \mathbb{D}\})$ for each $k \in \mathbb{K}_c$.

$\mathcal{B}$ responds $\mathcal{A}$'s key queries as follows:

- $\mathcal{A}$ makes CKQ(gid, d^*). Assume it is the ith CKQ and the corresponding user-key will be usk$_{gid,d^*}$,

  - If $i < j$, $\mathcal{B}$ randomly chooses $r_{gid,d^*} \in \mathbb{Z}_N, R'_{gid,d^*} \in G_{p_3}, \{R_{gid,d^*,k} \in G_{p_3} | k \in \mathbb{K}\}$, and sets

    $$\mathsf{ucsk}_{gid,d^*} = g^{\alpha_{d^*}} h^{r_{gid,d^*}} (Y_2Y_3)^{r_{gid,d^*}}, \quad L_{gid,d^*} = g^{r_{gid,d^*}} R'_{gid,d^*},$$

    $$\Gamma_{gid,d^*,k} = V_{k,d^*}^{r_{gid,d^*}} R_{gid,d^*,k} (k = 1, 2, \ldots K),$$

    $$\sigma_{gid,d^*} = \mathsf{Sign}(\mathsf{SignKey}_{d^*}, gid||d^*||L_{gid,d^*}||\Gamma_{gid,d^*,1}|| \ldots ||\Gamma_{gid,d^*,K}).$$

    $$(3.8)$$

    Let $\mathsf{ucpk}_{gid,d^*} = (gid, d^*, L_{gid,d^*}, \{\Gamma_{gid,d^*,k} \mid k \in \mathbb{K}\}, \sigma_{gid,d^*})$, $\mathcal{B}$ answers $\mathcal{A}$ with $(\mathsf{ucsk}_{gid,d^*}, \mathsf{ucpk}_{gid,d^*})$. Note that the values of r_{gid,d^*} modulo p_2 and p_3 are uncorrelated to its value modulo

p_1, $(\mathsf{ucsk}_{gid,d^*}, \mathsf{ucpk}_{gid,d^*})$ is properly distributed semi-functional user-central-key of type 2.

- If $i > j$, $\mathcal{B}$ runs the normal $\mathsf{CKeyGen}$ algorithm, since it knows CMSK_{d^*}.

- If $i = j$, $\mathcal{B}$ implicitly sets $g^{r_{gid,d^*}}$ equal to the G_{p_1} part of T. $\mathcal{B}$ randomly chooses $R_{gid,d^*}, R'_{gid,d^*} \in G_{p_3}, \{R_{gid,d^*,k} \in G_{p_3} | k \in \mathbb{K}\}$, and sets

$$\mathsf{ucsk}_{gid,d^*} = g^{\alpha_{d^*}} T^a R_{gid,d^*}, \; L_{gid,d^*} = T R'_{gid,d^*},$$

$$\Gamma_{gid,d^*,k} = T^{v_{k,d^*}} R_{gid,d^*,k}(k = 1, 2, \ldots, K),$$

$$\sigma_{gid,d^*} = \mathsf{Sign}(\mathsf{SignKey}_{d^*}, gid\|d^*\|L_{gid,d^*}\|\Gamma_{gid,d^*,1}\|\cdots\|\Gamma_{gid,d^*,K}).$$

$$(3.9)$$

Let $\mathsf{ucpk}_{gid,d^*} = (gid, d^*, L_{gid,d^*}, \{\Gamma_{gid,d^*,k} \mid k \in \mathbb{K}\}, \sigma_{gid,d^*})$, $\mathcal{B}$ answers $\mathcal{A}$ with $(\mathsf{ucsk}_{gid,d^*}, \mathsf{ucpk}_{gid,d^*})$.

If $T \in G_{p_1 p_3}$, $(\mathsf{ucsk}_{gid,d^*}, \mathsf{ucpk}_{gid,d^*})$ is properly distributed normal user-central-key .

If $T \in G$, $(\mathsf{ucsk}_{gid,d^*}, \mathsf{ucpk}_{gid,d^*})$ is properly distributed semi-functional user-central-key of type 1. In this case, we have implicitly set $w_{k,d^*} = v_{k,d^*}$. If we let g_2^b denote the G_{p_2} part of T, we have that $\delta = ba$ modulo p_2, i.e., the G_{p_2} part of ucsk_{gid,d^*} is g_2^{ba}, the G_{p_2} part of L_{gid,d^*} is g_2^b, and the G_{p_2} part of $\Gamma_{gid,d^*,,k}$ is $g_2^{bv_{k,d^*}}$. Note that the value of v_{k,d^*} modulo p_2 is uncorrelated from the value of v_{k,d^*} modulo p_1.

- $\mathcal{A}$ makes $\mathsf{AKQ}(att, k, \{\mathsf{ucpk}_{gid,d} | d \in \mathbb{D}\})$ where $(k \in \mathbb{K} \setminus \mathbb{K}_c)$. $\mathcal{B}$ responds $\mathcal{A}$ as in the real security game by running the normal $\mathsf{AKeyGen}$ algorithm, since it knows AMSK_k for each $k \in \mathbb{K} \setminus \mathbb{K}_c$.

  On the $\mathsf{uask}_{att,gid,d^*}$, the corresponding ucpk_{gid,d^*} must be the answer that $\mathcal{A}$ gets from $\mathcal{B}$ by making CKQ because of the signature verification (3.1). Assume the corresponding CKQ is i^{th} CKQ, we have

– If $i < j$,

$$\mathsf{uask}_{att,gid,d^*} = (\Gamma_{gid,d^*,k})^{s_{att}/v_{k,d^*}} R'_{att,gid,d^*}$$

$$= (V_{k,d^*}^{r_{gid,d^*}} R_{gid,d^*,k})^{s_{att}/v_{k,d^*}} R'_{att,gid,d^*}$$

$$\text{By } \Gamma_{gid,d^*,k} \text{ in } (3.8)$$

$$= T_{att}^{r_{gid,d^*}} R_{att,gid,d^*}.$$

It is properly distributed semi-functional of type 2. Note the corresponding $(\mathsf{ucsk}_{gid,d^*}, \mathsf{ucpk}_{gid,d^*})$ is properly distributed semi-functional of type 2, we have usk_{gid,d^*} as a semi-functional user-key of type 2.

– If $i > j$, note that the corresponding $(\mathsf{ucsk}_{gid,d^*}, \mathsf{ucpk}_{gid,d^*})$ is generated by running the normal **CKeyGen** algorithm, we have usk_{gid,d^*} as a normal user-key.

– If $i = j$,

$$\mathsf{uask}_{att,gid,d^*} = (\Gamma_{gid,d^*,k})^{s_{att}/v_{k,d^*}} R'_{att,gid,d^*}$$

$$= (T^{v_{k,d^*}} R_{gid,d^*,k})^{s_{att}/v_{k,d^*}} R'_{att,gid,d^*}$$

$$\text{By } \Gamma_{gid,d^*,k} \text{ in } (3.9)$$

$$= T^{s_{att}} R_{att,gid,d^*}.$$

If $T \in G_{p_1 p_3}$, $\mathsf{uask}_{att,gid,d^*}$ is a properly distributed normal, where $g^{r_{gid,d^*}}$ equals to the G_{p_1} part of T. Note that the corresponding $(\mathsf{ucsk}_{gid,d^*}, \mathsf{ucpk}_{gid,d^*})$ is properly distributed normal, we have usk_{gid,d^*} as a normal user-key.

If $T \in G$, $\mathsf{uask}_{att,gid,d^*}$ is a properly distributed semi-functional of type 1. In this case, we have implicitly set $z_{att} = s_{att}$. The G_{p_2} part of $\mathsf{uask}_{att,gid,d^*}$ is $g_2^{bs_{att}}$, and the value of s_{att} modulo p_2 is uncorrelated from the value of s_{att} modulo p_1. Note that the corresponding $(\mathsf{ucsk}_{gid,d^*},$

ucpk_{gid,d^*}) is properly distributed semi-functional of type 1, we have usk_{gid,d^*} as a semi-functional user-key of type 1.

$\mathcal{A}$ sends two messages M_0 and M_1 and an LSSS matrix (A^*, ρ) to $\mathcal{B}$. To make the semi-functional challenge ciphertext, $\mathcal{B}$ implicitly sets $g^s = X_1$ and $g_2^c = X_2$. It randomly chooses $u_2, \ldots, u_n \in \mathbb{Z}_N$ to define a vector $\vec{u'} = (a, u_2, \ldots, u_n)$, and for each row x of A^* it randomly picks $r'_x \in \mathbb{Z}_N$. Then it chooses a random $\beta \in \{0, 1\}$ and sets

$$C = M_\beta \cdot \prod_{d=1}^{D} e(g^{\alpha_d}, X_1 X_2) \,, \; C' = X_1 X_2,$$

$$\{C_x = (X_1 X_2)^{A_x^* \cdot \vec{u'}} (X_1 X_2)^{-r'_x \cdot s_{\rho(x)}} \,, \; C'_x = (X_1 X_2)^{r'_x} \mid x \in \{1, 2, \ldots, l\}\}.$$

Note that this implicitly sets $\vec{v} = sa^{-1}\vec{u'}$, $\vec{u} = c\vec{u'}$, $r_x = sr'_x$, and $\gamma_x = -cr'_x$. The values $z_{\rho(x)} = s_{\rho(x)}$ match those in the jth usk_{gid,d^*} if it is a semi-functional user-key of type 1, as required.

The challenge ciphertext and the usk_{gid,d^*} are *almost* properly distributed, except for the fact that the first coordinate of $\vec{u}$ (which equals ca) is correlated with the value of a modulo p_2 that also appears in ucsk_{gid,d^*} if it is semi-functional. In fact, if the usk_{gid,d^*} can be used to reconstruct $e(g, g)^{\alpha_{d^*} s}$, we would have $c\delta - bu_1 = cba - bca = 0$ modulo p_2, so usk_{gid,d^*} is either normal or nominally semi-functional. We must argue that this is hidden to $\mathcal{A}$, who *cannot* request any user-keys to reconstruct $e(g, g)^{\alpha_{d^*} s}$ and then decrypt the challenge ciphertext.

To argue that the value being shared in G_{p_2} in the challenge ciphertext is information theoretically hidden, we appeal to our restriction that attributes are only used once in labeling the rows of the matrix. Since $S_{gid} \cup (\bigcup_{k \in \mathbb{K}_c} U_k)$ could not satisfy (A^*, ρ), $R = span(\{A_x^* | \rho(x) \in S_{gid} \cup (\bigcup_{k \in \mathbb{K}_c} U_k)\})$ does not include the vector $(1, 0, \ldots, 0)$. Then there exists a vector $\vec{w}$ such that $w_1 = \vec{w} \cdot (1, 0, \ldots, 0) \neq 0$ and $\forall \vec{v_r} \in R, \vec{w} \cdot \vec{v_r} = 0$. We can write $\vec{u} = \theta\vec{w} + \vec{u''}$ where $\theta \in \mathbb{Z}_N$ and $\vec{u''} \neq \vec{w}$ is distributed uniformly random in R. We note that $\vec{u}''$ reveals no information about θ and that $u_1 = \vec{u} \cdot (1, 0, \ldots, 0) = \theta w_1 + \vec{u}'' \cdot (1, 0, \ldots, 0)$ can not be determined from $\vec{u}''$ alone. However, the shares corresponding to rows whose attributes are in $S_{gid} \cup (\bigcup_{k \in \mathbb{K}_c} U_k)$ only reveal information about $\vec{u}''$ since $\forall \vec{v_r} \in R, \vec{w} \cdot \vec{v_r} = 0$.

$\theta \vec{w}$ appears only in the equations of the form $A_x^* \cdot \vec{u} + \gamma_x z_{\rho(x)}$ where the attribute $\rho(x) \notin S_{gid} \cup (\bigcup_{k \in \mathbb{K}_c} U_k)$. While attributes are only used once in labeling the rows of the matrix, as long as each γ_x is not congruent to 0 modulo p_2, each of these equations introduces a new unknown $z_{\rho(x)}$ that appears nowhere else, and thus no information about θ can be learned by the adversary. The probability that any γ_x is congruent to 0 modulo p_2 is negligible. Thus, the ciphertext and the key usk_{gid,d^*} are properly distributed in the adversary's view with probability negligibly close to 1.

Thus, if $T \in G_{p_1 p_3}$, $\mathcal{B}$ has properly simulated $\mathbf{Game}_{j-1,2}$, and if $T \in G$ and all the $\gamma_x = -c r_x'$ values are non-zero modulo p_2, then $\mathcal{B}$ has properly simulated $\mathbf{Game}_{j,1}$. Neglecting the probability that Σ_{sign} is broken, $\mathcal{B}$ can, therefore, use the output of $\mathcal{A}$ to gain advantage negligibly close to ϵ against Assumption 2. $\qquad \square$

Proof of Lemma 3.4.7.

Proof. $\mathcal{B}$ is given $\mathbb{G} = (N, G, G_T, e), g, X_1 X_2, X_3, Y_2 Y_3, T$. It will simulate $\mathbf{Game}_{j,1}$ or $\mathbf{Game}_{j,2}$ to $\mathcal{A}$.

$\mathcal{B}$ randomly chooses $a \in \mathbb{Z}_N$, then gives $\mathcal{A}$ $\mathsf{GPK} = \{N, g, h = g^a, X_3\}$, as well as $\mathbb{D} = \{1, 2, \ldots, D\}$, $\mathbb{K} = \{1, 2, \ldots, K\}$, the descriptions of $U_1, U_2, \ldots, U_K$, and the description of secure signature scheme Σ_{sign}.

For $d = 1$ to D, $\mathcal{B}$ generates sign key pair $(\mathsf{SignKey}_d, \mathsf{VerifyKey}_d)$ and randomly chooses $\alpha_d \in \mathbb{Z}_N$, then gives $\mathcal{A}$ $\mathsf{CPK}_d = e(g, g)^{\alpha_d}$ and $\mathsf{CAPK}_d = \mathsf{VerifyKey}_d$.

For $k = 1$ to K, $\mathcal{B}$ randomly chooses $\{s_{att} \in \mathbb{Z}_N | att \in U_k\}$ and $\{v_{k,d} \in \mathbb{Z}_N | d \in \mathbb{D}\}$, then gives $\mathcal{A}$ $\mathsf{APK}_k = \{T_{att} = g^{s_{att}} | att \in U_k\}$ and $\mathsf{ACPK}_k = \{V_{k,d} = g^{v_{k,d}} | d \in \mathbb{D}\}$. $\mathcal{A}$ gives $\mathcal{B}$ the $d^* \in \mathbb{D}$ and $\mathbb{K}_c \subset \mathbb{K}$ where $\mathbb{K} \setminus \mathbb{K}_c \neq \emptyset$. Let $\mathbb{D}_c = \mathbb{D} \setminus \{d^*\}$.

$\mathcal{B}$ gives $\mathcal{A}$ $\mathsf{CMSK}_d = (\alpha_d, \mathsf{SignKey}_d)$ for each $d \in \mathbb{D}_c$, and $\mathsf{AMSK}_k = (\{s_{att} | att \in U_k\}, \{v_{k,d} | d \in \mathbb{D}\})$ for each $k \in \mathbb{K}_c$.

$\mathcal{B}$ responds $\mathcal{A}$'s key queries as follows:

- $\mathcal{A}$ makes $\mathsf{CKQ}(gid, d^*)$. Assume it is the ith CKQ and the corresponding user-key will be usk_{gid,d^*},

    - If $i < j$, $\mathcal{B}$ randomly chooses $r_{gid,d^*} \in \mathbb{Z}_N, R'_{gid,d^*} \in G_{p_3}, \{R_{gid,d^*,k} \in G_{p_3} | k \in \mathbb{K}\}$, and sets

    $$\mathsf{ucsk}_{gid,d^*} = g^{\alpha_{d^*}} h^{r_{gid,d^*}} (Y_2 Y_3)^{r_{gid,d^*}} \,, \; L_{gid,d^*} = g^{r_{gid,d^*}} R'_{gid,d^*},$$

    $$\Gamma_{gid,d^*,k} = V_{k,d^*}^{r_{gid,d^*}} R_{gid,d^*,k}(k = 1, 2, \ldots K),$$

    $$\sigma_{gid,d^*} = \mathsf{Sign}(\mathsf{SignKey}_{d^*}, gid||d^*||L_{gid,d^*}||\Gamma_{gid,d^*,1}|| \cdots ||\Gamma_{gid,d^*,K}).$$

    $$(3.10)$$

    Let $\mathsf{ucpk}_{gid,d^*} = (gid, d^*, L_{gid,d^*}, \{\Gamma_{gid,d^*,k} \mid k \in \mathbb{K}\}, \sigma_{gid,d^*})$, $\mathcal{B}$ answers $\mathcal{A}$ with $(\mathsf{ucsk}_{gid,d^*}, \mathsf{ucpk}_{gid,d^*})$. Note that the values of r_{gid,d^*} modulo p_2 and p_3 are uncorrelated to its value modulo p_1, $(\mathsf{ucsk}_{gid,d^*}, \mathsf{ucpk}_{gid,d^*})$ is properly distributed semi-functional user-central-key of type 2.

    - If $i > j$, $\mathcal{B}$ runs the normal $\mathsf{CKeyGen}$ algorithm, since it knows CMSK_{d^*}.

    - If $i = j$, $\mathcal{B}$ implicitly sets $g^{r_{gid,d^*}}$ equal to the G_{p_1} part of T. $\mathcal{B}$ randomly chooses $R_{gid,d^*}, R'_{gid,d^*} \in G_{p_3}, \{R_{gid,d^*,k} \in G_{p_3} | k \in \mathbb{K}\}, \eta \in \mathbb{Z}_N$, and sets

    $$\mathsf{ucsk}_{gid,d^*} = g^{\alpha_{d^*}} T^a R_{gid,d^*} (Y_2 Y_3)^\eta \,, \; L_{gid,d^*} = T R'_{gid,d^*},$$

    $$\Gamma_{gid,d^*,k} = T^{v_{k,d^*}} R_{gid,d^*,k}(k = 1, 2, \ldots, K),$$

    $$\sigma_{gid,d^*} = \mathsf{Sign}(\mathsf{SignKey}_{d^*}, gid||d^*||L_{gid,d^*}||\Gamma_{gid,d^*,1}|| \cdots ||\Gamma_{gid,d^*,K}).$$

    $$(3.11)$$

Let $\mathsf{ucpk}_{gid,d^*} = (gid, d^*, L_{gid,d^*}, \{\Gamma_{gid,d^*,k} \mid k \in \mathbb{K}\}, \sigma_{gid,d^*})$, $\mathcal{B}$ answers $\mathcal{A}$ with $(\mathsf{ucsk}_{gid,d^*}, \mathsf{ucpk}_{gid,d^*})$.

If $T \in G_{p_1p_3}$, $(\mathsf{ucsk}_{gid,d^*}, \mathsf{ucpk}_{gid,d^*})$ is properly distributed semifunctional user-central-key of type 2.

If $T \in G$, $(\mathsf{ucsk}_{gid,d^*}, \mathsf{ucpk}_{gid,d^*})$ is properly distributed semi-functional user-central-key of type 1. In this case, we have implicitly set $w_{k,d^*} = v_{k,d^*}$. If we let g_2^b denote the G_{p_2} part of T and $g_2^{y_2} = Y_2$, we have that $\delta = ba + \eta y_2$ modulo p_2. i.e., the G_{p_2} part of ucsk_{gid,d^*} is $g_2^{ba+\eta y_2}$, the G_{p_2} part of L_{gid,d^*} is g_2^b, and the G_{p_2} part of $\Gamma_{gid,d^*,k}$ is $g_2^{bv_{k,d^*}}$. Note that the value of v_{k,d^*} modulo p_2 is uncorrelated from the value of v_{k,d^*} modulo p_1.

- $\mathcal{A}$ makes $\mathsf{AKQ}(att, k, \{\mathsf{ucpk}_{gid,d} \mid d \in \mathbb{D}\})$ where $(k \in \mathbb{K} \setminus \mathbb{K}_c)$. $\mathcal{B}$ responds $\mathcal{A}$ as in the real security game by running the normal $\mathsf{AKeyGen}$ algorithm, since it knows AMSK_k for each $k \in \mathbb{K} \setminus \mathbb{K}_c$.

  On the $\mathsf{uask}_{att,gid,d^*}$, the corresponding $\mathsf{ucpk}_{gid,d}$ must be the answer that $\mathcal{A}$ gets from $\mathcal{B}$ by making CKQ because of the signature verification (3.1). Assume the corresponding CKQ is i^{th} CKQ, we have

  - If $i < j$,

  $$\begin{aligned}
  \mathsf{uask}_{att,gid,d^*} &= (\Gamma_{gid,d^*,k})^{s_{att}/v_{k,d^*}} R'_{att,gid,d^*} \\
  &= (V_{k,d^*}^{r_{gid,d^*}} R_{gid,d^*,k})^{s_{att}/v_{k,d^*}} R'_{att,gid,d^*} \\
  &\qquad\qquad\qquad \text{By } \Gamma_{gid,d^*,k} \text{ in (3.10)} \\
  &= T_{att}^{r_{gid,d^*}} R_{att,gid,d^*}.
  \end{aligned}$$

  It is properly distributed semi-functional of type 2.

  Note the corresponding $(\mathsf{ucsk}_{gid,d^*}, \mathsf{ucpk}_{gid,d^*})$ is properly distributed semi-functional of type 2, we have usk_{gid,d^*} as a semi-functional user-key of type 2.

- If $i > j$, note that the corresponding $(\mathsf{ucsk}_{gid,d^*}, \mathsf{ucpk}_{gid,d^*})$ is generated by running the normal $\mathsf{CKeyGen}$ algorithm, we have usk_{gid,d^*} as a normal user-key.

- If $i = j$,

$$
\begin{aligned}
\mathsf{uask}_{att,gid,d^*} &= (\Gamma_{gid,d^*,k})^{s_{att}/v_{k,d^*}} R'_{att,gid,d^*} \\
&= (T^{v_{k,d^*}} R_{gid,d^*,k})^{s_{att}/v_{k,d^*}} R'_{att,gid,d^*} \\
&\qquad\qquad\qquad \text{By } \Gamma_{gid,d^*,k} \text{ in (3.11)} \\
&= T^{s_{att}} R_{att,gid,d^*}.
\end{aligned}
$$

If $T \in G_{p_1 p_3}$, $\mathsf{uask}_{att,gid,d^*}$ is properly distributed semi-functional of type 2, where $g^{r_{gid,d^*}}$ equals to the G_{p_1} part of T. Note that the corresponding $(\mathsf{ucsk}_{gid,d^*}, \mathsf{ucpk}_{gid,d^*})$ is properly distributed semi-functional of type 2, we have usk_{gid,d^*} as a semi-functional user-key of type 2.

If $T \in G$, $\mathsf{uask}_{att,gid,d^*}$ is properly distributed semi-functional of type 1. In this case, we have implicitly set $z_{att} = s_{att}$. The G_{p_2} part of $\mathsf{uask}_{att,gid,d^*}$ is $g_2^{bs_{att}}$, and the value of s_{att} modulo p_2 is uncorrelated from the value of s_{att} modulo p_1. Note that the corresponding $(\mathsf{ucsk}_{gid,d^*}, \mathsf{ucpk}_{gid,d^*})$ is properly distributed semi-functional of type 1, we have usk_{gid,d^*} as a semi-functional user-key of type 1.

$\mathcal{A}$ sends two messages M_0 and M_1 and an LSSS matrix (A^*, ρ) to $\mathcal{B}$. To make the semi-functional challenge ciphertext, $\mathcal{B}$ implicitly sets $g^s = X_1$ and $g_2^c = X_2$. It randomly chooses $u_2, \ldots, u_n \in \mathbb{Z}_N$ to define a vector $\vec{u'} = (a, u_2, \ldots, u_n)$, and for each row x of A^* it randomly picks $r'_x \in \mathbb{Z}_N$. Then it chooses a random $\beta \in \{0,1\}$ and sets

$$C = M_\beta \cdot \prod_{d=1}^{D} e(g^{\alpha_d}, X_1 X_2) , \; C' = X_1 X_2,$$

$$\{C_x = (X_1 X_2)^{A_x^* \cdot \vec{u}'} (X_1 X_2)^{-r'_x \cdot s_{\rho(x)}} , \; C'_x = (X_1 X_2)^{r'_x} \mid x \in \{1, 2, \ldots, l\}\}$$

We note that this implicitly sets $\vec{v} = sa^{-1}\vec{u}'$, $\vec{u} = c\vec{u}'$, $r_x = sr'_x$ and $\gamma_x = -cr'_x$. The values $z_{\rho(x)} = s_{\rho(x)}$ match those in the jth usk_{gid, a^*} if it is a semi-functional user-key of type 1, as required.

The challenge ciphertext and the usk_{gid, d^*} are properly distributed because the G_{p_2} part of ucsk_{gid, d^*} is randomized by η.

Thus, if $T \in G_{p_1 p_3}$, $\mathcal{B}$ has properly simulated $\mathbf{Game}_{j,2}$, and if $T \in G$, $\mathcal{B}$ has properly simulated $\mathbf{Game}_{j,1}$. Neglecting the probability that Σ_{sign} is broken, $\mathcal{B}$ can, therefore, use the output of $\mathcal{A}$ to gain advantage ϵ against Assumption 2. $\qquad\square$

Proof of Lemma 3.4.8.

Proof. $\mathcal{B}$ is given $\mathbb{G} = (N, G, G_T, e), g, g^\alpha X_2, X_3, g^s Y_2, Z_2, T$. It will simulate $\mathbf{Game}_{q,2}$ or $\mathbf{Game}_{\mathsf{Final}}$ to $\mathcal{A}$.

$\mathcal{B}$ randomly chooses $a \in \mathbb{Z}_N$, then gives $\mathcal{A}$ $\mathsf{GPK} = \{N, g, h = g^a, X_3\}$, as well as $\mathbb{D} = \{1, 2, \ldots, D\}$, $\mathbb{K} = \{1, 2, \ldots, K\}$, the descriptions of $U_1, U_2, \ldots, U_K$, and the description of secure signature scheme Σ_{sign}. $\mathcal{B}$ randomly chooses $d' \in \mathbb{D}$.

For $d = 1$ to D, $\mathcal{B}$ generates sign key pair ($\mathsf{SignKey}_d$, $\mathsf{VerifyKey}_d$). If $d \neq d'$, $\mathcal{B}$ randomly chooses $\alpha_d \in \mathbb{Z}_N$ and sets $\mathsf{CPK}_d = e(g, g)^{\alpha_d}$, otherwise $\mathcal{B}$ sets $\mathsf{CPK}_d = e(g, g^\alpha X_2)$. $\mathcal{B}$ gives $\mathcal{A}$ CPK_d and $\mathsf{CAPK}_d = \mathsf{VerifyKey}_d$. This implicitly sets $\alpha_{d'} = \alpha$. For $k = 1$ to K, $\mathcal{B}$ randomly chooses $\{s_{att} \in \mathbb{Z}_N | att \in U_k\}$ and $\{v_{k,d} \in \mathbb{Z}_N | d \in \mathbb{D}\}$, then gives $\mathcal{A}$ $\mathsf{APK}_k = \{T_{att} = g^{s_{att}} | att \in U_k\}$ and $\mathsf{ACPK}_k = \{V_{k,d} = g^{v_{k,d}} | d \in \mathbb{D}\}$. $\mathcal{A}$ gives $\mathcal{B}$ the $d^* \in \mathbb{D}$ and $\mathbb{K}_c \subset \mathbb{K}$ where $\mathbb{K} \setminus \mathbb{K}_c \neq \emptyset$. If $d^* \neq d'$, it means $\mathcal{B}$ guesses a wrong d^*, $\mathcal{B}$ aborts. Note that the public parameters are properly distributed, the probability $\mathcal{B}$ guesses the right b^* is $\frac{1}{D}$.

$\mathcal{B}$ gives $\mathcal{A}$ $\mathsf{CMSK}_d = (\alpha_d, \mathsf{SignKey}_d)$ for each $d \in \mathbb{D}_c = \mathbb{D} \setminus \{d^*\}$, and $\mathsf{AMSK}_k = (\{s_{att} | att \in U_k\}, \{v_{k,d} | d \in \mathbb{D}\})$ for each $k \in \mathbb{K}_c$.

$\mathcal{B}$ responds $\mathcal{A}$'s key queries as follows:

- $\mathcal{A}$ makes $\mathsf{CKQ}(gid, d^*)$. Assume the corresponding user-key will be usk_{gid,d^*}, $\mathcal{B}$ randomly chooses $r_{gid,d^*} \in \mathbb{Z}_N, R_{gid,d^*}, R'_{gid,d^*} \in G_{p_3}, \{R_{gid,d^*,k} \in G_{p_3} | k \in \mathbb{K}\}$, and sets

$$\mathsf{ucsk}_{gid,d^*} = (g^\alpha X_2) h^{r_{gid,d^*}} (Z_2)^{r_{gid,d^*}} R_{gid,d^*}, \quad L_{gid,d^*} = g^{r_{gid,d^*}} R'_{gid,d^*},$$

$$\Gamma_{gid,d^*,k} = V_{k,d^*}^{r_{gid,d^*}} R_{gid,d^*,k} (k = 1, 2, \ldots K),$$

$$\sigma_{gid,d^*} = \mathsf{Sign}(\mathsf{SignKey}_{d^*}, gid || d^* || L_{gid,d^*} || \Gamma_{gid,d^*,1} || \cdots || \Gamma_{gid,d^*,K}).$$

(3.12)

Let $\mathsf{ucpk}_{gid,d^*} = (gid, d^*, L_{gid,d^*}, \{\Gamma_{gid,d^*,k} | k \in \mathbb{K}\}, \sigma_{gid,d^*})$, $\mathcal{B}$ answers $\mathcal{A}$ with $(\mathsf{ucsk}_{gid,d^*}, \mathsf{ucpk}_{gid,d^*})$. Note that the value of r_{gid,d^*} modulo p_2 is uncorrelated to its value modulo p_1, $(\mathsf{ucsk}_{gid,d^*}, \mathsf{ucpk}_{gid,d^*})$ is properly distributed semi-functional user-central-key of type 2.

- $\mathcal{A}$ makes $\mathsf{AKQ}(att, k, \{\mathsf{ucpk}_{gid,d} | d \in \mathbb{D}\})$ where $(k \in \mathbb{K} \setminus \mathbb{K}_c)$. $\mathcal{B}$ responds $\mathcal{A}$ as in the real security game by running the normal $\mathsf{AKeyGen}$ algorithm, since it knows AMSK_k for each $k \in \mathbb{K} \setminus \mathbb{K}_c$.

On the $\mathsf{uask}_{att,gid,d^*}$, the corresponding ucpk_{gid,d^*} must be the answer that $\mathcal{A}$ gets from $\mathcal{B}$ by making CKQ because of the signature verification (3.1). Then we have

$$\mathsf{uask}_{att,gid,d^*} = (\Gamma_{gid,d^*,k})^{s_{att}/v_{k,d^*}} R'_{att,gid,d^*}$$

$$= (V_{k,d^*}^{r_{gid,d^*}} R_{gid,d^*,k})^{s_{att}/v_{k,d^*}} R'_{att,gid,d^*} \quad \text{By } \Gamma_{gid,d^*} \text{ in (3.12)}$$

$$= T_{att}^{r_{gid,d^*}} R_{att,gid,d^*}.$$

It is properly distributed semi-functional of type 2. Note that the corresponding $(\mathsf{ucsk}_{gid,d^*}, \mathsf{ucpk}_{gid,d^*})$ is properly distributed semi-functional of type 2, we have usk_{gid,d^*} as a semi-functional user-key of type 2.

$\mathcal{A}$ sends two messages M_0, M_1 and an LSSS matrix (A^*, ρ) to $\mathcal{B}$. To make the semi-functional challenge ciphertext, $\mathcal{B}$ implicitly takes s from the assumption term $g^s Y_2$ and sets $g_2^c = Y_2$. It randomly chooses $u_2, \ldots, u_n \in \mathbb{Z}_N$ to define a vector $\vec{u'} = (u, u_2, \ldots, u_n)$, and for each row x of A^* it randomly picks $r'_x \in \mathbb{Z}_N$. Then it chooses a random $\beta \in \{0,1\}$ and sets

$$C = M_\beta \cdot T \cdot \prod_{d \in \mathbb{D}, d \neq d^*} e(g,g)^{\alpha_d \cdot s} , \; C' = g^s Y_2,$$

$$\{C_x = (g^s Y_2)^{A_x^* \cdot \vec{u'}} (g^s Y_2)^{-r'_x \cdot s_{\rho(x)}} , \; C'_x = (g^s Y_2)^{r'_x} \mid x \in \{1, 2, \ldots l\}\}.$$

We note that this implicitly sets $\vec{v} = sa^{-1}\vec{u'}$, $\vec{u} = c\vec{u'}$, $r_x = sr'_x$ and $\gamma_x = -cr'_x$.

If $T = e(g,g)^{\alpha s}$, this is a properly distributed semi-functional ciphertext of the encryption of M_β. Otherwise, it is a properly distributed semi-functional ciphertext of the encryption of a random message in G_T. Thus, neglecting the probability that Σ_{sign} is broken, $\mathcal{B}$ can use the output of $\mathcal{A}$ to gain advantage $\frac{\epsilon}{D}$ against Assumption 3. $\qquad \square$

3.5 Interval Encryption Schemes

A broadcast encryption (BE) scheme enables a broadcaster to choose a subset S of n users, who are listening to the broadcast channel, and encrypt a message for this subset. Any user in S is allowed to successfully decrypt the message. Even if all the users outside of S collude together, they can not obtain any useful information of the broadcast message. In the following, we also use r to represent the number of revoked users, i.e., $r = n - |S|$ where $|S|$ is the size of S. Compared with a private key BE scheme [13, 288], a public key broadcast encryption has the benefit that users are not required to pre-share any private information. Therefore, in this section, we mainly focus on pubic key broadcast encryption. Three efficiency parameters of a BE scheme are of our major concern: the transmission cost, user storage, and the decryption time.

The transmission cost of most current public key BE constructions will grow with the increase of the revocation number r. Naor et al. [221] presented a BE construction (NNL method) with an average ciphertext size of $1.38r$ and private key size $\mathcal{O}(\log^2 n)$. The private key size is further improved to $\mathcal{O}(\log^{1+\epsilon} n), 0 < \epsilon < 1$ in the HS construction [134], where the ciphertext size blows up with a $\frac{1}{\epsilon}$ factor. The private key size is further improved to $\mathcal{O}(\log n)$ by Goodrich et al. [126]. Dodis and Fazio [101] presented a generic method (DF transformation) to transform the NNL method and HS construction into a public key broadcast system using hierarchical identity-based encryption (HIBE). The transmission overload remains unchanged and the private key consists of $\mathcal{O}(\log^2 n)$ and $\mathcal{O}(\log^{1+\epsilon} n)$ HIBE node secret keys if DF transformation is instantiated with BBG HIBE [35]. The security is reduced to standard Decisional BDHE assumption and the decryption time cost is $\mathcal{O}(\log n)$. The decryption time is then improved to constant by Liu and Teng [200]. However, their security is reduced to decisional BDH assumption in the random oracle model. Recently, Sahai and Waters proposed a BE system with a transmission cost linearly dependent on r and constant storage cost. However, the decryption cost is linearly dependent on r and the security is reduced to a complex assumption called q-MEBDH assumption. Actually, it has been pointed out [159] that at least a single key per revoked user should be included in the transmission cost and hence, r might be the lower bound of the transmission overload in any BE scheme with reasonable decryption computational and storage cost. Therefore, constructing a BE system with a transmission overload lower than r as well as reasonable user storage and computational cost is still an open problem, which is one of the major issues of this section.

On the other hand, there are two major application scenarios [39] for BE: applications where we broadcast large sets, namely sets of size $n-r$ for $r \ll n$ and applications where we broadcast small sets, namely sets of size $|S| \ll n$. Apparently, a BE system with a transmission cost dependent on r is not efficient when r grows, and especially it fails to be an optimal choice for the second kind of application where r is very close to n. Before BGW proposed their construction [39], the only suitable solution for the latter scenario is the trivial solution, i.e., encrypting the message under each recipient's key.

In order to construct a BE scheme suitable for *arbitrary* receiver sets, we need to break the barrier of r. BGW [39] proposed an elegant BE scheme with constant size ciphertext as the first attempt to solve this problem. Although the ciphertext and private key size of their construction is constant, the size of public keys is linearly dependent on n. The public key must be accessible to any decryptor, which implies a high storage cost of size $\mathcal{O}(n)$. This makes their system unsuitable for the application scenario where users have only limited storage capability [240]. Their underlying assumption is the standard Decisional BDHE assumption. Later, Delerablee [95] proposed a BE construction where the public key size depends on the maximum size of S while both ciphertext and private key remain constant size. However, this still does not serve as an efficient solution for applications where the receiver set is large, namely $r \ll n$. The security of this construction is reduced to a complex assumption called GDHE assumption in the random oracle model. Besides, the decryption of both constructions is not efficient. The decryption cost of the BGW construction depends on n, and the decryption of Delerablee's construction requires $\mathcal{O}(|S|)$ operations.

In this section, we consider this problem from a brand-new angle and a more practical point of view. The basic motivation comes from the following observation: in a BE system with n users, where each user is assigned with an index $i \in [1, n]$. The receiver set S can be regarded as a collection of k intervals. Considering the fact that the number of intervals containing in S is always less than $r + 1$ and in the best cases k could even be much less than r, the system performance can be dramatically increased if the transmission overhead of the BE system is only determined by the interval number k while irrelevant of r. In this study, we will use more detailed performance analysis and simulation to show that a BE construction based on k is always more efficient than the previous scheme dependent on r, and suitable for more cases in practice.

In order to realize a BE system with a transmission overload dependent on k, we propose a new type of encryption called interval encryption. In interval encryption, a message is encrypted under a collection of natural intervals $S = \bigcup_{j=1}^{k} NI_j$, where NI_j is a natural interval in $[1, n]$. Each receiver is identified by a unique natural number $i \in [1, n]$ and assigned with the respective private key. The decryption is successful if and only if the natural number i belongs to S.

We present a generic methodology which can transform a series of binary tree encryptions into interval encryptions. We illustrate the basic methodology using the BBG HIBE scheme [35]. The construction achieves a ciphertext size of $\mathcal{O}(k)$, and $\mathcal{O}(\log n)$ private storage. The decryption is dominated by at most $\mathcal{O}(\log n)$ group operations. The security is reduced to the Decisional BDHE assumption. We note that one of the best public key BE schemes under this assumption is the DF transformation of the HS construction which requires a transmission cost of $\mathcal{O}(r/\epsilon)$ size and the private key consists of $\mathcal{O}(\log^{1+\epsilon} n)$ HIBE node secret keys, where $0 < \epsilon < 1$.

We also apply our basic methodology to the fully secure HIBE [182] scheme proposed by Lewko and Waters to present an adaptively secure interval encryption scheme. Gentry and Waters [123] proposed the first adaptively secure BE scheme under a complex bilinear assumption. The public parameter size of their construction is of $\mathcal{O}(|S|)$. The private key size is constant, and the ciphertext size of their construction is of $\mathcal{O}(max|S|)$. Later, Waters [290] gave the first short ciphertext adaptively secure BE system under static (i.e. non q-based) assumptions. However, both of the public parameter and private key size are linearly dependent on n. The public parameter of our construction is of size $\mathcal{O}(\log n)$ and the ciphertext size is of $\mathcal{O}(k)$. It only requires

$\mathcal{O}(\log n)$ private storage. In other words, our construction serves as one of the most efficient adaptively secure BE systems. Besides, our construction also reduces its security to static assumptions.

Since we consider the proposal of this new concept and the corresponding methodology of one of our major contributions, an inclusive extended interval encryption is proposed as another illustration of the power of our basic methodology. A message is encrypted under a collection of intervals $S - \bigcup_{j=1}^{k} NI_j$ in this extended construction. A user's private key corresponds to a certain interval NI_ω. The decryption is successful if and only if there's at least one interval $NI_j, j \in [1, k]$ such that $NI_\omega \subseteq NI_j$. The construction also provides user with delegation capability. We also discuss several interesting applications of interval encryption. In particular, we propose a useful concept of range attribute-based encryption and present an efficient construction from interval encryption.

3.5.1 Definitions

Assumptions. Bilinear maps [223] are crucial for our construction. A pairing is an efficiently computable, nondegenerate function, $\hat{e} : \mathbb{G}_1 \times \mathbb{G}_1 \to \mathbb{G}_2$, with the bilinearity property that $\hat{e}(g^r, g^s) = \hat{e}(g, g)^{rs}$. Here, $\mathbb{G}_1$, and $\mathbb{G}_2$ are all multiplicative groups of prime order p, respectively, generated by g and $\hat{e}(g, g)$.

The security proof of our constructions relies on the Decisional $d + 1$ BDHE assumption, which can be stated as [47]: Given a tuple $[h, g, g^\alpha, g^{(\alpha^2)}, \cdots, g^{\alpha^d}, g^{(\alpha^{d+2})}, \cdots, g^{(\alpha^{2d})}, Z] \in \mathbb{G}_1^{2d+1} \times \mathbb{G}_2$ for a random exponent $\alpha \in \mathbb{Z}_p$, decide whether $Z = \hat{e}(g, h)^{\alpha^{d+1}}$.

Notations. We inherit most notations from the underlying BTE and FSE [62] construction. Recall that d denotes the depth of the tree, and $n = 2^d$ is the number of leaf nodes. We set the root node to be ε by convention. The other nodes on the tree have an associated name chosen from $\{0, 1\}^{\leq d}$. The left child of a node is concatenated

with 0, and the right child is concatenated with 1. Therefore, each leaf node will also
have an associated binary name $[\omega_1\omega_2 \cdots \omega_d]$. We also let a natural number $\omega \in [1, n]$
associated with the ω-th leaf node of the binary tree (starting from left to right). We
implicitly let $\omega = [\omega_1\omega_2 \cdots \omega_d]$ in the remainder of this section. The j-bit prefix of a
string $\omega = [\omega_1\omega_2 \cdots \omega_d]$ is denoted by $\omega|_j$, namely $\omega|_j = [\omega_1\omega_2 \cdots \omega_j]$. We implicitly set
$\omega|_0 = \varepsilon$ and $\omega|_d = \omega$. It is easy to observe that a set of nodes $\omega|_j, j \in [1, d]$ corresponds
to the nodes on the path from the root to the leaf node ω (see Figure 3.11. (b)). Besides,
we use $\omega|_{j,\text{(RS)}}$ or $\omega|_{j,\text{(LS)}}$ to denote the right or left sibling of $\omega|_j$, respectively, if $\omega|_j$ has
such a sibling. Namely, $\omega|_{j,\text{(RS)}} = [\omega_1\omega_2 \cdots \omega_{j-1}1]$ or $\omega|_{j,\text{(LS)}} = [\omega_1\omega_2 \cdots \omega_{j-1}0]$.

Generally, our BE system consists of two parallel BTE systems: the right BTE
system and the left BTE system. The right BTE system covers all the leaf nodes in the
interval $[\omega, n]$ and the left BTE system covers all the leaf nodes in the interval $[1, \omega]$.
User ω is assigned with a unique right master key and a left master key. All the node
secret keys or private keys for ω in the right BTE system are derived from the right
master key and in the left BTE system are derived from the left master key. We use two
different subindexes (L) or (R) in the notations of all these keys to distinguish the left
or right BTE system they correspond to respectively.

3.5.2   Security Models

Our construction is a key encapsulation mechanism (KEM)[7], thus long messages
can be encrypted under a short symmetric key. An interval encryption scheme is made
up of four randomized algorithms:

Setup(n): Take as input a natural interval $[1, n]$. It outputs a public key PK and the
system master key SK_ε.

PvkGen(ω, SK_ε): Take as input a natural number $\omega \in [1, n]$ and the system master

---

[7]We adopt KEM for the ease of comparison since all the BE constructions in the literature employ the
same mechanism.

key SK_ε. It outputs a private key D_ω.

Encrypt(S, PK): Take as input a public key PK, and a k-wise natural interval set $S = \bigcup_{j=1}^k NI_j$ where $NI_j = [l_j, r_j]$ satisfying $1 \le l_1 \le r_1 < l_2 \le r_2 \cdots < l_k \le r_k \le n$. For $j \in [1, k]$, it outputs k pairs $\{Hdr_j, K_j\}$. We call Hdr $= \{Hdr_j\}_{j=1}^k$ the header and $K = \{K_j\}_{j=1}^k$ the message encryption keys.

Let M be a message that should be decipherable precisely by the receivers holding the private key corresponding to $\omega \in S$. For $j \subset [1, k]$, let C_j be the encryption of M under the message encryption key K_j. Let C_M be the collection of these encryption, namely $C_M = \{C_j\}_{j=1}^k$. The whole ciphertext consists of (S, Hdr, C_M).

Decrypt $(S, \omega, D_\omega, Hdr, PK)$: Take as input a k-wise natural interval set $S = \bigcup_{j=1}^k NI_j$ and the private key D_ω for a natural number $\omega \in [1, n]$, a header Hdr, a public key PK. If $\omega \in NI_j, 1 \le j \le k$, the algorithm outputs the corresponding message encryption key $K_j \in \mathcal{K}$.

The system is considered correct, if for all k-wise natural interval sets $S = \bigcup_{j=1}^k NI_j$ and natural numbers $\omega \in NI_j$ (where $j \in [1, k]$), if

$$PK \xleftarrow{R} \textbf{Setup}(n), D_\omega \xleftarrow{R} \textbf{PvkGen}(\omega, SK_\varepsilon), (Hdr, K) \xleftarrow{R} \textbf{Encrypt}(S, PK),$$

then **Decrypt**$(S, \omega, D_\omega, Hdr, PK){=}K_j$. The concept of interval encryption is close to private linear BE (PLBE) mentioned in [42], and can be viewed as an extension of PLBE.

Semantic Security(IND-sI-CPA): The selective interval game is very similar to that of BE [95], and it forms as follow:

Init The adversary outputs a k-wise natural interval set $S^* = \bigcup_{j=1}^k NI_j^*$, where $NI_j^* = [l_j^*, r_j^*]$ satisfying $1 \le l_1^* \le r_1^* < l_2^* \le r_2^* \cdots < l_k^* \le r_k^* \le n$, which it wishes to attack.

Setup The challenger runs **Setup**(n) to obtain a public key PK for the adversary.

Phase 1 The adversary issues query for private key of $\omega \notin S^*$.

Challenge The challenger runs algorithm **Encrypt** to obtain $(\text{Hdr}^*, K) \xleftarrow{R}$ **Encrypt** (S^*, PK) where $K \in \mathcal{K}^k$. Next, the challenger picks a random $\beta \in \{0, 1\}$. It sets $K^* = K$ if $\beta = 1$ and sets K^* to a random string of length equal to $|K|$ otherwise. It then sends Hdr^*, K^* to the adversary.

Phase 2 Same as phase 1.

Guess The adversary outputs its guess $\beta' \in \{0, 1\}$ for β and wins the game if $\beta' = \beta$.

The adversary's advantage is the absolute value of the difference between its success probability and $\frac{1}{2}$.

Definition 3.5.1. An interval encryption scheme is selective-interval chosen plaintext secure (IND-sI-CPA) if all polynomial time adversaries have at most a negligible advantage in winning the above security game.

The adaptive CPA security can be defined in a similar way except that there is no **Init** stage in the adaptive game and the challenge interval S^* in the **Challenge** stage should be provided under the restriction that none of the identities ω for the key queries of **Phase 1** and **Phase 2** belongs to S^*, i.e., $\omega \notin S^*$.

The ultimate security goal is to realize IND-CCA security where the adversary doesn't need to choose the interval set at the beginning and is provided with a decryption oracle. However, we only concentrate on IND-sI-CPA security.

3.5.3   Binary Tree Encryption and Forward Secure Encryption

The concept of binary tree encryption (BTE) was first proposed by Canetti, et al [64]. BTE is a relaxation of hierarchical identity-based encryption (HIBE) [122]. As in HIBE, a "master" public key PK is associated with a binary tree in BTE; each node ω in this tree has a corresponding secret key SK_ω. To encrypt a message "targeted" for some node, one uses both PK and the name of the target node; the resulting ciphertext can then be decrypted using the secret key of the target node. Moreover, as in HIBE the secret key of any node can be employed to derive the secret keys for the children of that node. The only difference between HIBE and BTE is that the latter insists on a *binary* tree, where each non-leaf node only has two child nodes.

Technically speaking, forward secure encryption (FSE) is an elegant application of BTE. Let the depth of a binary tree be d which implies it has $n = 2^d$ leaf nodes. In a FSE scheme, the lifetime of a system is divided into $n = 2^d$ time periods, each of which is associated with a unique leaf node of the tree. A user holding a private key for time period ω can open all the messages encrypted under the subsequent time periods, namely $\omega' \in [\omega, n]$. The private key D_ω in the FSE construction contains the node secret keys SK_ω for the leaf node ω as well as node secret keys for the right siblings of the nodes on the path from the root to node ω, where all these node secret keys come from the underlying BTE scheme. To encrypt a message for a certain period ω', one uses both PK and the name of respective leaf node ω' as in the BTE scheme; the resulting ciphertext can then be decrypted using node secret key $SK_{\omega'}$, which is also similar to the BTE scheme. As shown in Figure 3.11. (a), a private key D_2 containing the node secret keys SK_2, SK_c, and SK_b can be used to derive all the node secret keys for leaf nodes falling into the interval $[2, 8]$. Therefore, D_2 can be used to open all the messages encrypted under time periods in the interval $[2, 8]$.

Indeed, FSE can be viewed as a special case of interval encryption. As shown in Figure 3.11. (a), if we use ciphertext C_4 encrypted under leaf node 4 to represent the interval $[1, 4]$, then only the private key for time period $\omega \in [1, 4]$ can be used to open the message, e.g., D_2 could be used for the decryption of C_4 because SK_4 can be derived from SK_c, which belongs to D_2. However, D_5 cannot be used for decrypting C_4 as it is impossible to deduce SK_4 from any node secret keys included in D_5.

In the remainder of this section, we use a right direction arrow from a certain leaf node (or the corresponding index in the axis) to denote this particular private key distribution mode. A right direction arrow from a leaf node ω means that all the node

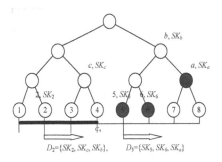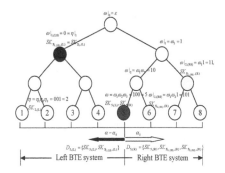

Figure 3.11: (a). The key distribution mode of forward secure encryption, C_4 represents the interval $[1, 4]$, and the private key for a user ω can be used to derive the node secret keys for all the nodes in the interval $[\omega, n]$ (b). Key distribution mode of interval encryption: we let $\omega = 5$ here. The respective private key D_5 contains left private key $D_{5,(L)}$ and right private key $D_{5,(R)}$. $D_{5,(L)} = \{SK_{5,(L)}, SK_{5|_{1,(LS)},(L)}\}$ which are derived from left master key $\alpha - \alpha_5$. Similarly, we have $D_{5,(R)} = \{SK_{5,(R)}, SK_{5|_{2,(RS)},(R)}, SK_{5|_{3,(RS)},(R)}\}$ derived from right master key α_5. Let the left bound η of an interval be 2 here, then $SK_{2,(L)}$ can be derived from $SK_{2|_{1,(L)}}$ which is equal to $SK_{5|_{1,(LS)},(L)}$ belonging to $D_{5,(L)}$.

secret keys of the leaf nodes in the interval $[\omega, n]$ are computable from its own private key. Therefore, this private key can be used to open all the message encrypted under these nodes. Besides, we also use a left direction arrow from a leaf node ω to denote an opposite decryption ability, namely the respective private key can be used to open all the messages encrypted under the leaf nodes in the interval $[1, \omega]$. It is feasible by simply assigning a user with the node secret keys for node ω as well as node secret keys for the left siblings of all the nodes on the path from the root to ω. Generally, the FSE construction is treated as a special interval encryption scheme in which the encryptor can set the interval form as $[1, j]$. The upper bound j depends on which leaf node the ciphertext corresponds to. Now, our goal is to realize an interval encryption scheme covering multiple intervals, each of which has two freely chosen bounds determined by the encryptor.

3.5.4 A Generic Transformation from BTE to Interval Encryption

Trivial Constructions: A trivial interval encryption scheme can be given directly from attribute-based encryption [129] if one treats $\log n$ bits to represent a number from 1 to n as attributes and builds an access tree allowing specific intervals. However, even the most efficient trivial methodology would inevitably result in an interval encryption construction with a ciphertext size of $\mathcal{O}(k \log n)$, where k is the number of intervals. As we have mentioned, our goal is to realize a BE system in which the ciphertext size is determined by the number of intervals k. *If all the messages are only encrypted under the bounds of each interval like in the FSE scheme*, then this goal is reachable. However, it is still a challenge for us to make sure that only those with an index within two bounds of each interval can open the message.

A Generic Transformation from BTE to Interval Encryption: There are some difficulties in the transformation from BTE to interval encryption. The first difficulty is how to differentiate the decryption ability of an index in and outside of an interval. Taking the interval [3, 6] shown in Figure 3.12.(a), for instance, we could easily find the required difference if we project two opposite direction arrows from each index in the axis, where the connotation of the arrows can be found in our exposition of the last paragraph in Section 3.5.3. The key observation of our transformation is that: *the two opposite direction arrows starting from index 5 can cross both bounds 3 and 6 , respectively, and therefore decrypt the corresponding partial ciphertext in two different manners* (We will show how to differentiate the partial decryption from two different directions, and how this will eventually lead to successful generation of the corresponding message encryption key in the sequel). *However, only one unique direction arrow from index 2 or 7 can cross the two bounds, i.e., only the right direction arrow from index 2 can cross 3 and 6 while only the left direction arrow from index 7 can cross 3 and 6. This implies that those outside of an interval can only decrypt the partial ciphertext in a unique manner.* The

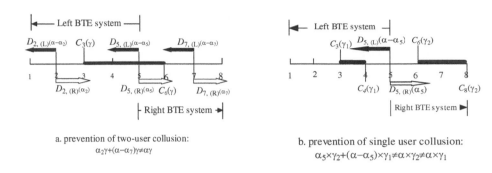

a. prevention of two-user collusion:
$\alpha_2\gamma+(\alpha-\alpha_7)\gamma\neq\alpha\gamma$

b. prevention of single user collusion:
$\alpha_5\times\gamma_2+(\alpha-\alpha_5)\times\gamma_1\neq\alpha\times\gamma_2\neq\alpha\times\gamma_1$

Figure 3.12: Collusion and its prevention

private key for right direction arrow is called right private key in the concrete construction and the one for left direction arrow is left private key.

The master key of the underlying BTE or HIBE scheme only contains *one group element*. The message encryption key for each interval corresponds to $\alpha \cdot \gamma$ in the exponent of a pairing, where α is the system master key and γ is randomly chosen by the encryptor. For each user ω, we choose a random number α_ω and split the master key α into two parts: one is the right master key α_ω, which serves as the root master key for the right BTE system, from which the right private key $D_{\omega,(R)}$ of ω is derived; the other one is the left master key $\alpha - \alpha_\omega$, i.e., the root master key for the left BTE system, from which the left private key $D_{\omega,(L)}$ of ω is derived. Note that the two private keys for ω can be distributed similarly to the FSE scheme as shown in Figure 3.11. (b). Consequently, a partial decryption using the user's right private key contains $\alpha_\omega \cdot \gamma$ in the exponent of a pairing while a partial decryption using the left private key will have $(\alpha - \alpha_\omega) \cdot \gamma$ in its exponent. Then, the message encryption key containing $\alpha \cdot \gamma$ in its exponent will be recovered since $\alpha = \alpha_\omega + \alpha - \alpha_\omega$ holds.

In this way, we can actually prevent a possible collusion attack called two-user collusion. For example (shown in Figure 3.12. (a), a user $\omega = 7$ with a left private key $D_{7,(L)}$ (which could decrypt the partial ciphertext C_3) and a user $\omega = 2$ with right private key $D_{2,(R)}$ (which can decrypt the partial ciphertext C_6) might collude to open the message for interval $[3, 6]$ (since they can also complete the partial decryption in two different manners) although neither of them is in this particular interval. In our system, the partial decryption from $D_{7,(L)}$ will contain $(\alpha - \alpha_7) \cdot \gamma$ while the partial decryption from $D_{2,(R)}$ contains $\alpha_2 \cdot \gamma$ in its exponents, and hence the collusion will fail since there's no way for them to obtain $\alpha \cdot \gamma$ in the final step.

Besides, the encryptor is required to use a unique random γ_j while generating the ciphertext for each interval NI_j. This aims to prevent another attack called a single-user collusion. This attack only occurs in the scenario with multiple intervals (where $k \geq 1$). For instance (shown in Figure 3.12. (b)), in an interval encryption system with two intervals $[3, 4] \bigcup [6, 8]$, the partial decryption on C_3 from the left private key $D_{5,(L)}$ contains $\alpha - \alpha_5$ and the partial decryption on C_8 from the right private key $D_{5,(R)}$ contains the other half randomness α_5, the message encryption key corresponding to $\alpha \cdot \gamma$ might be recovered if these two intervals use the same randomness. However, a unique randomness for each interval can guarantee that only a user within a certain interval can successfully open the message. For example (as shown in Figure 3.12.(b)), a random γ_1 is used in the ciphertext for interval $[C_3, C_4]$ and γ_2 is used in the encryption for interval $[C_6, C_8]$. The message encryption keys of these two intervals correspond to $\alpha \cdot \gamma_1$ and $\alpha \cdot \gamma_2$, respectively. A single-user collusion fails since the randomized partial decryption $(\alpha - \alpha_5)\gamma_1$ and $\alpha_5\gamma_2$ will not generate a meaningful encryption key in the final step (see Figure 3.12. b.).

The proposed methodology might not be generic, since it somehow relies on the property of bilinear mapping. Therefore, we only illustrate our methodology using concrete examples rather than providing a formal description of a generic interval encryption system in the following subsections.

3.5.5   Basic Construction: A Concrete Instantiation Based on HIBE

In this subsection, we will describe how the proposed methodology can be applied to the HIBE (viewed as a binary tree encryption scheme here) construction [35] to propose an interval encryption scheme. Note that there is an additional algorithm **DeckeyDer** $(D_\omega = \{D_{\omega,(L)}, D_{\omega,(R)}\}, \zeta, \eta)$ compared with the original definition of interval encryption in Section 3.5.2. This algorithm is a preliminary step for the decryption algorithm, and we treat it as an independent algorithm for clarity. Besides, there is an additional slight technical modification to the underlying BTE construction in the sense that we basically have two concrete instantiations of a hash function to guarantee that we could cover both the two bounds of each interval in the security proof.

Let $\mathbb{G}_1$ be a bilinear group of prime order p, and let g be a generator of $\mathbb{G}_1$. In addition, let $\hat{e} : \mathbb{G}_1 \times \mathbb{G}_1 \to \mathbb{G}_2$ denote the bilinear map. A security parameter, κ, will determine the size of the groups. Assume the system accommodates $n = 2^d$ users, where d is an integer.

Setup(n): Select a random $\alpha \in \mathbb{Z}_p$ and set $g_1 = g^\alpha$. Choose random elements g_2, $g_{3,(L)}, g_{3,(R)}, h_{1,(L)}, \cdots, h_{d,(L)}, h_{1,(R)}, \cdots, h_{d,(R)}$ from $\mathbb{G}_1$.

The public key is $PK = (g, g_1, g_2, g_{3,(L)}, g_{3,(R)}, h_{1,(L)}, \cdots, h_{d,(L)}, h_{1,(R)}, \cdots, h_{d,(R)})$. For a binary string $v = [v_1 v_2 \cdots v_j]$ where $j \in [1, d]$, define two publicly computable functions as: $F_{(L)}(v) = g_{3,(L)} \cdot \prod_{i=1}^{j} h_{i,(L)}^{v_i}$ and $F_{(R)}(v) = g_{3,(R)} \cdot \prod_{i=1}^{j} h_{i,(R)}^{v_i}$. The system master key is $SK_\varepsilon = g_2^\alpha$.

PvkGen(ω, SK_ε): For receiver $\omega = [\omega_1 \omega_2 \cdots \omega_d]$ which is associated with the ω-th leaf node (starting from left to right), the algorithm first chooses a random number α_ω. The right master key for ω is $SK_{\varepsilon,(R)} = g_2^{\alpha_\omega}$, and the left master key is $SK_{\varepsilon,(L)} = g_2^{\alpha - \alpha_\omega}$. The algorithm first generates two node secret keys $SK_{\omega,(R)} = [g_2^{\alpha_\omega} (F_{(R)}(\omega))^{r_\omega}, g^{r_\omega}]$ and $SK_{\omega,(L)} = [g_2^{\alpha - \alpha_\omega} (F_{(L)}(\omega))^{r_\omega}, g^{r_\omega}]$ for leaf node ω where r_ω is a random number from $\mathbb{Z}_p$. For all the nodes $\omega|_j, j = 1, \cdots, d$ on the path from the root to the leaf node ω, if it has a right sibling $\omega|_{j,\text{(RS)}}=[\omega_1\omega_2\cdots\omega_{j-1}1]$, the algorithm uses the right master key to generate the respective node secret key as $SK_{\omega|_{j,\text{(RS)}},\text{(R)}}=[g_2^{\alpha_\omega}\,(F_{\text{(R)}}(\omega|_{j,\text{(RS)}}))^{r_j},\,g^{r_j},$ $h_{j+1,\text{(R)}}^{r_j},\cdots,h_{d,\text{(R)}}^{r_j}]$ where r_j is also a random number; otherwise the algorithm uses the left master key to generate node secret key for its left sibling $\omega|_{j,\text{(LS)}}=[\omega_1\omega_2\cdots\omega_{j-1}0]$ as $SK_{\omega|_{j,\text{(LS)}},\text{(L)}}=[g_2^{\alpha-\alpha_\omega}\,(F_{\text{(L)}}(\omega|_{j,\text{(LS)}}))^{r_j},\,g^{r_j},h_{j+1,\text{(L)}}^{r_j},\cdots,h_{d,\text{(L)}}^{r_j}]$.

Output private key $D_\omega = \{D_{\omega,\text{(R)}}, D_{\omega,\text{(L)}}\}$, where

$$D_{\omega,\text{(R)}} = \{SK_{\omega,\text{(R)}}, SK_{\omega|_{j,\text{(RS)}},\text{(R)}}\}_{j\in[1,d]}, D_{\omega,\text{(L)}} = \{SK_{\omega,\text{(L)}}, SK_{\omega|_{j,\text{(LS)}},\text{(L)}}\}_{j\in[1,d]}.$$

Encrypt(S, PK): The encryptor first chooses a k-wise natural interval set $S - \bigcup_{j=1}^{k} NI_j$, where $NI_j = [l_j, r_j]$. For each interval, pick γ_j uniformly from $\mathbb{Z}_p$ at random. Let the binary name of the corresponding leaf nodes for the two bounds be $r_j = [r_{j1}\cdots r_{jd}]$ and $l_j = [l_{j1}\cdots l_{jd}]$.

Output the respective ciphertext $C_{l_j}=\{g^{\gamma_j}, (F_{\text{(L)}}(l_j))^{\gamma_j}\}$ and $C_{r_j}=\{g^{\gamma_j}, (F_{\text{(R)}}(r_j))^{\gamma_j}\}$. Set the message encryption key for each interval NI_j as $K_j=\hat{e}(g_1, g_2)^{\gamma_j} \in \mathbb{G}_2$. The collection of these partial ciphertexts constitute the header Hdr=$\{C_{l_j}, C_{r_j}\}_{j=1}^{k}$.

DeckeyDer ($D_\omega = \{D_{\omega,\text{(L)}}, D_{\omega,\text{(R)}}\}, \zeta, \eta$): This algorithm derives the node secret key $SK_{\eta,\text{(L)}}$ for the lower bound η, and $SK_{\zeta,\text{(R)}}$ for the upper bound ζ.

1. Let a natural number $\eta \le \omega$ denote the η-th leaf node, and thus η is on the left of ω in the binary tree. Assume the binary representation of η is $\eta = \eta_1\cdots\eta_d$. There must exist a node secret key $SK_{\eta|_{j},\text{(L)}}, j \in [1, d]$ which belongs to $D_{\omega,\text{(L)}}$ (as shown in Figure 3.11. b.). Run the derivation algorithm of the underlying BTE scheme iteratively, which means the following steps need to be executed iteratively for $i = j$ to $i = d - 1$:

   (a) Let $\eta|_i = \eta_1\cdots\eta_i$. Parse $SK_{\eta|_i,\text{(L)}}$ as $\left(g_2^{\alpha-\alpha_\omega}(F_{\text{(L)}}(\eta|_i))^{r_i}, g^{r_i}, h_{i+1,\text{(L)}}^{r_i}, \cdots,\right.$ $\left. h_{d,\text{(L)}}^{r_i}\right)=(a_0, a_1, b_{i+1}, \cdots, b_d)$.

   (b) Choose random $t \in \mathbb{Z}_p$, and output $SK_{\eta|_{i+1},\text{(L)}} = (a_0 \cdot b_{i+1}^{\eta_{i+1}} \cdot (F_{\text{(L)}}(\eta|_{i+1}))^t, a_1 \cdot g^t, b_{i+2} \cdot h_{i+2,\text{(L)}}^t, \cdots, b_d \cdot h_d^t)$ and set $i = i + 1$.

Finally, it will output a node secret key $SK_{\eta,(L)}=[g_2^{\alpha-\alpha_\omega}(F_{(L)}(\eta))^{r'},g^{r'}]$ for the lower bound η.

2. Let a natural number $\zeta \geq \omega$ denote the ζth leaf node. Assume the binary representation of ζ is $\zeta = \zeta_1 \cdots \zeta_d$. Therefore, there must exist a node secret key $SK_{\zeta|_j,(R)}$ which belongs to $D_{\omega,(R)}$. Run the derivation algorithm of the underlying BTE scheme iteratively, which means steps 1(a)-1(b) need to be executed iteratively.

Output a node secret key $SK_{\zeta,(R)}=[g_2^{\alpha_\omega}(F_{(R)}(\zeta))^{r''},g^{r''}]$ for the upper bound ζ.

Decrypt $(S,\omega,D_\omega,\text{Hdr},PK)$: If $\omega \in NI_j = [l_j,r_j], 1 \leq j \leq k$ which implies that $l_j \leq \omega \leq r_j$, then it runs **DeckeyDer** (D_ω,r_j,l_j) to generate decryption key $SK_{r_j,(R)}$ and $SK_{l_j,(L)}$. It obtains the corresponding secret key $SK_{r_j,(R)}=[g_2^{\alpha_\omega}(F_{(R)}(r_j))^{r''},g^{r''}]$ and the partial ciphertext for the upper bound $C_{r_j}=\{g^{\gamma_j},(F_{(R)}(r_j))^{\gamma_j}\}$. Compute $\frac{\hat{e}[g^{\gamma_j},g_2^{\alpha_\omega}(F_{(R)}(r_j))^{r''}]}{\hat{e}[g^{r''},(F_{(R)}(r_j))^{\gamma_j}]} = \hat{e}(g,g_2)^{\gamma_j\alpha_\omega}$. It also obtains the corresponding secret key $SK_{l_j,(L)}=[g_2^{\alpha-\alpha_\omega}(F_{(L)}(l_j))^{r'},g^{r'}]$ and the partial ciphertext for the lower bound $C_{l_j}=\{g^{\gamma_j},(F_{(L)}(l_j))^{\gamma_j}\}$. Compute $\frac{\hat{e}[g^{\gamma_j},g_2^{\alpha-\alpha_\omega}(F_{(L)}(l_j))^{r'}]}{\hat{e}[g^{r'},(F_{(L)}(l_j))^{\gamma_j}]} = \hat{e}(g,g_2)^{\gamma_j(\alpha-\alpha_\omega)}$. Finally, it computes $\hat{e}(g,g_2)^{\gamma_j\alpha_\omega} \cdot \hat{e}(g,g_2)^{\gamma_j(\alpha-\alpha_\omega)} = \hat{e}(g^\alpha,g_2)^{\gamma_j} = \hat{e}(g_1,g_2)^{\gamma_j}$.

3.5.6 Discussion on Efficiency and Security

In this construction, the public key size is $\mathcal{O}(\log n)$, and the private key only contains $\mathcal{O}(\log n)$ BTE node secret keys. Note that the private key in the DF transformation [101] of the NNL method or the HS construction contains $\mathcal{O}(\log^2 n)$ or $\mathcal{O}(\log^{1+\epsilon} n)$ node secret keys, respectively. It is important to point out that a widely used tool, updatable public storage in the FSE scheme [35] , can also be adopted in our proposed interval encryption system to limit the cost of private storage to $\mathcal{O}(\log n)$. The above efficiency parameters can be further improved if the random oracle is adopted, i.e., the public key size can be reduced to $\mathcal{O}(1)$ in this case.

The decryption cost is dominated by the derivation of the two node secret keys. The derivation cost can be reduced to $\mathcal{O}(\log n)$ by doing the following computation: in order to deduce the node secret key $SK_{\eta,(L)}=[g_2^{\alpha-\alpha_\omega} \cdot (F_{(L)}(\eta))^{r'}, g^{r'}]=(a'_0, a'_1)$ from $SK_{\eta|i,(L)}=\left(g_2^{\alpha-\alpha_\omega}(F_{(L)}(\eta|_i))^{r_i}, g^{r_i}, h_{i+1,(L)}^{r_i}, \cdots, h_{d,(L)}^{r_i}\right)=(a_0, a_1, b_{i+1}, \cdots, b_d)$, we can compute $a'_0 = a_0 \cdot \prod_{k=i+1}^{d} b_k^{\eta_k} \cdot (F_{(L)}(\eta))^t, a'_1 = a_1 \cdot g^t$ where we set $r' = r_i + t$. We can deduce the node secret key $SK_{\zeta,(R)}$ from $SK_{\zeta|i,(R)}$ in a similar way. The overall decryption time is then reduced to $\mathcal{O}(\log n)$ since the rest of the decryption procedure only requires a constant number of group operations.

Why is $\mathcal{O}(k)$ *better*: A system with a transmission overhead proportional to k is more efficient than the traditional systems, especially the system where communication load is linearly dependent on r such as the revocation system [240, 303]. To illustrate, we compare the performance of both systems in presence of different values for r as well as k. We assume that the total node number n is set to $2^{17} = 131072$ and let r increase from 1 to n. For a specific r, we randomly generate 1000 revoked sets, which correspond to 1000 different interval number k, and thus obtain an average interval number $\bar{k}$ as well as the average transmission overhead of the proposed scheme, which has been shown in Figure 3.13. From Figure 3.13, it is observed that, when the revocation set is small, the performance of the proposed scheme is very close to the tradition systems. However the difference will be scaled up along with the increase of r. If the revoked set number exceeds 50% of the total number, the communication load of the proposed scheme will decrease with increase of r. It is also observed that the proposed scheme can achieve the best performance if r is very large, which further demonstrates that the proposed scheme is suitable for cases when a small receiver set is employed. Compared with the BGW generalized construction [39] with a $\sqrt{n}$ size transmission overload which only serves as a better choice than the trivial solution and the traditional systems when $r > \sqrt{n}$, we have a benefit that our system keeps the advantage over the traditional constructions when r is a small number, namely $r \ll n$.

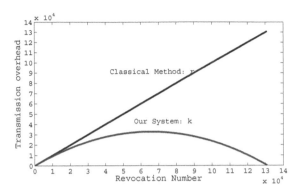

Figure 3.13: Comparison between k and r

From the above results, it is concluded that the proposed construction fits into more cases than the traditional systems which are dependent on r, and therefore constitutes a more favorable choice in practice.

The selective security of the proposed construction can be proven secure under the $(d+1)$-BDHE assumption, and it is stated as follows. We omit the concrete proof here.

Theorem 3.5.2. *If the Decisional $(d+1)$-BDHE assumption holds in $\mathbb{G}_1, \mathbb{G}_2$, the proposed interval encryption scheme is selective chosen plaintext secure.*

3.5.7 Extension Work

Inclusive Extended Interval Encryption: An inclusive extended interval encryption scheme deals with the scenario where the message is encrypted under a collection of intervals $S = \bigcup_{j=1}^{k}[l_j, r_j]$, and the private key D_ω of a user ω corresponds to an interval $[l_\omega, r_\omega]$. The decryption is successful if at least one interval $[l_j, r_j], j \in [1, k]$ exists such that $[l_\omega, r_\omega] \subseteq [l_j, r_j]$. To generate a private key corresponding to an interval $[l_\omega, r_\omega]$ (see Figure 3.14), we simply generate a left private key $D_{l_\omega,(\text{L})}$ corresponding to the lower bound l_ω using the left master key $\alpha - \alpha_\omega$ as in the basic construction. Similarly, we generate a right private key $D_{r_\omega,(\text{R})}$ for the upper bound r_ω using the right master key α_ω. The rest of the above algorithms have no significant differences from

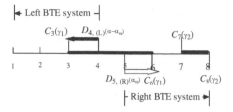

Figure 3.14: Extended Interval Encryption: The generation of a private key for an interval [4, 5]

those in the basic construction. Furthermore, it is easy to observe that a man holding a private key for an interval $[l_\omega, r_\omega]$ can delegate a private key for another interval $[l_{\omega'}, r_{\omega'}]$ using the **DeckeyDer** algorithm as long as $[l_\omega, r_\omega] \subseteq [l_{\omega'}, r_{\omega'}]$. This is a property somewhat close to a recently proposed concept called inclusive identity-based encryption (IBE) [40]. We consider this extended construction of important theoretical interest because few inclusive constructions exist [120] since the proposal of inclusive IBE.

Adaptively Secure Interval Encryption: An adaptively secure interval encryption is constructed from Lewko and Waters' HIBE construction [182]. The basic idea is to apply our proposed transformation method to Lewko and Waters' HIBE scheme.

Range Attribute-based Encryption: In a key-policy attribute-based encryption (ABE) [129], a private key might be associated with an access policy such as "Old man **AND** tall." A man holding this private key can open a message encrypted under an attribute set {"Old man," "tall"} since this attribute set satisfies the above access policy. In practice, the attributes in an attribute set might have certain range and the attributes in an access policy might be assigned with certain concrete evaluations. In the above example, the access policy might be denoted as a formula "Age: 60 **AND** Height: 180 (cm)". A man holding a private key associated with the above policy should be able to open a message encrypted under an attribute set {"Age: 50 to 100", "Height: 175 to 250 (cm)."} The reason for the successful decryption is that both evaluations of the two attributes fall into the range required in the attribute set and hence the access policy is satisfied. However, A man holding a private key associated with an access policy "Age: 49 **AND** Height: 180 (cm)" cannot decrypt this message since the evaluation "Age: 49" is not within the corresponding range "Age: 50 to 100" in the attribute set. A range ABE scheme is realizable from a traditional ABE scheme. However, the ciphertext will blow up with a $\log n$ factor as shown in our trivial example of constructing interval encryption from ABE. The proposed interval encryption scheme can be easily modified to a range ABE scheme with a constant ciphertext size.

Interval Encryption under Simpler Assumption: The proposed method is also applicable to those BTE or HIBE schemes, in which cases their master keys only contain a single group element such as [34, 62]. We can construct interval encryption schemes based on the decisional bilinear Diffie–Hellman assumption. The concrete steps are similar to the steps of Section3.5.5. The weakness of these constructions is that the ciphertext size will blow up with a $\log n$ factor compared with the basic construction while the private key size remains $\mathcal{O}(\log n)$.

Encryption Under a Graph: Consider the following application: A message might be encrypted under a digital map of a certain territory on earth (a close two-dimensional graph) and only those who hold a private key for a location in the territory can open the message. This notion might actually intrigue several interesting applications. For example, a launch order of a certain weapon might be encrypted under a map of a specific region and only those who have a private key corresponding to a location within this region can launch this weapon. Apparently, we can map all the points in a two-dimensional digital map to the points in a single dimensional axis. We can simply calculate $i = (y - 1)c + x$ where (x, y) is a point in a two-dimensional map with width c and height d. If we set $n = c * d$, then all the points can be mapped into an index $i \in [1, n]$. In other words, all the points within the territory of this digital map can be mapped into a collection of intervals. Therefore, the proposed interval encryption provides a solution for the above scenario. The number of intervals depends on the perimeter of this graph.

Future Work and Open Problems: The reason why the proposed construction can reduce the transmission cost is due to the difference between the points in and outside of a certain interval. How to reduce the transmission cost based the difference between points in and outside of a graph, especially to minimize the constant factor k during the multi-dimensional scenario is left as an important open problem. It is possible to borrow some idea from computational geometry to solve this problem. To propose a BTE or HIBE construction with improved efficiency or under a weaker assumption which is compatible with our framework is very interesting since this directly implies the improvement of interval encryption.

3.6 Fuzzy Identity-Based Signature Schemes

At present, while we are enjoying the facilities given by the modern technology, many security and privacy issues arise. One of them is reliable user authentication. In a generic cryptographic environment, the possession of the decryption key is sufficient to authenticate the user's identity. But since the decryption key is so long and complicated, it is difficult to memorize. As a result, it is stored somewhere. It relies on other authentication (e.g., password) to retrieve this key. By simple dictionary attacks, most passwords are easy to guess. Moreover, people tend to write down complex passwords on easily accessible locations. In addition, people often use the same password in many applications. Thus, if a single password is compromised, many doors are opened.

Considering the limitation of passwords, it is not suitable to use password based authentication in systems which require high security level. An alternative approach is to use biometrics (fingerprints, iris data, face, and voice). It is inherently more reliable than passwords, since people will never forget or lost their biometrics; they are extremely difficult to copy, distribute, or share; and the person being authenticated is

Figure 3.15: Two biometric scans of the same feature are rarely identical.

required to present at the time and point of the authentication. It is also difficult to forge biometrics. Thus, biometric authentication is a potential candidate to replace password based authentication.

Despite the advantages of biometric authentication, the following issue must be addressed: biometric data is *not exactly reproducible*, as shown in Figure 3.15 that two biometric scans of the same feature are rarely identical. Thus, traditional protocols can not guarantee the correctness when the parties use a shared secret derived from biometric data.

Furthermore, a common approach to biometric authentication is to capture the biometric templates of all users during the *enrollment phase* and to store the templates in a reference database. During the *authentication phase*, new scan result is compared with the stored templates. To store biometric templates in a database will incur a number of security and privacy risks, such as:

1. Impersonation. An attacker steals templates from a database and constructs artificial biometrics that could pass authentication.

2. Irrevokability. Once compromised, biometrics cannot be updated, reissued or destroyed.

3. Exposure of sensitive personal information.

The concept of fuzzy identity-based encryption (FIBE) was introduced by Sahai and Waters and further developed in a line of works, e.g., [16, 227]. In a nutshell a FIBE allows a user with the private key for identity ω to decrypt a ciphertext encrypted for identity ω' if and only if ω and ω' are within a certain distance judged by some metric.

In this section, we introduce a novel cryptographic primitive that is the signature analogue of a FIBE, we call it *fuzzy identity-based signature* (FIBS) [300]. [8] FIBS allows a user with identity ω to issue a signature which can be verified with identity ω' if ω and ω' are within a certain distance judged by some metric. Using FIBS, we construct a biometric authentication scheme and address the problems which lie in biometric authentication mentioned in the above paragraph. In our scheme, the server stores a FIBS signature and a public string **pub** which is shared by the users. When a user A is going to authenticate himself to user B, he will send his public string **pub** to the server and allow the server to retrieve the signature and reconstruct the shared secret. Our scheme meets the "robust to noise," "anti-impersonation," and "privacy protection" security requirements.

Another interesting application is attribute-based signature. In this application, a user can issue a signature on behalf of the group that has a certain set of attributes. For example, an IT company might want a C++ senior programmer whose age is above 50 to sign the technical report. In this scenario, it will sign to the identity {"C++","senior programmer","above 50"}. Any user who has an identity that contains all of these attributes can issue the signature.

Much work has focused on developing secure biometric authentication [92,164,165, 214,278]. Most recently, Dodis, Reyzin, and Smith [102] showed an approach different from ours to use biometric data to derive secure cryptographic keys which can be used for the purposes of authentication. Roughly speaking, they introduce two primitives: a *secure sketch* which allows recovery of a shared secret given a close approximation thereof, and a *fuzzy extractor* which extracts a uniformly string s from this shared secret in an error-tolerant manner. Both of the primitives work by constructing a "public"

---

[8]Since FIBE appeared in 2005, we are the first to propose the cryptographic primitive FIBS in 2007.

string **pub** which is stored by the server and transmitted to the user, **pub** encodes the re-
dundancy information needed for error-tolerant reconstruction. But the work of Dodis,
Reyzin, and Smith does not address the issue of malicious modification of **pub**. Indeed,
the adversary who maliciously alters the public string sent to a user may be able to learn
the user's biometric data. Boyen et al. [46] improved this result in a way that resists an
active adversary and provides mutual authentication and authenticated key exchange.

3.6.1 Definitions

Notations: From here on we use $\mathbb{Z}_q$ to denote the group $\{0, \ldots, q-1\}$ under addition
modulo q. For a group $\mathbb{G}$ of prime order we use $\mathbb{G}^*$ to denote the set $\mathbb{G}^* = \mathbb{G} - \{O\}$
where O is the identity element in the group $\mathbb{G}$. We use $\mathbb{Z}_q^*$ to denote the set of positive
integers.

Bilinear Pairings and Assumptions: Let us consider two multiplicative group $\mathbb{G}$ and
$\mathbb{G}_T$ of the same prime order p. A bilinear pairing is a map $e : \mathbb{G} \times \mathbb{G} \to \mathbb{G}_T$ with the
following properties:

1. Bilinear: $e(u^a, v^b) = e(u, v)^{ab}$, where $u, v \in \mathbb{G}$, and $a, b \in \mathbb{Z}_p^*$.

2. Nondegeneracy: there exist $u \in \mathbb{G}$ and $v \in \mathbb{G}$ such that $e(u, v) \neq 1$.

3. Computability: There is an efficient algorithm to compute $e(u, v)$ for all $u, v \in \mathbb{G}$.

Computational Diffie–Hellman (CDH) Assumption: We briefly review the computational Diffie–Hellman (CDH) Assumption. The readers can refer to previous literature [162] for more details.

The challenger chooses $a, b \in \mathbb{Z}_p$ at random and outputs $(g, A = g^a, B = g^b)$. The
adversary then attempts to output $g^{ab} \in \mathbb{G}$. An adversary, $\mathcal{B}$, has at least an ϵ advantage
if

$$\Pr[\mathcal{B}(g, g^a, g^b) = g^{ab}] \geq \epsilon$$

where the probability is over the randomly chosen a, b and the random bits consumed by $\mathcal{B}$.

Definition 3.6.1. The $(t, \epsilon) - CDH$ assumption holds if no t-time adversary has at least ϵ advantage in winning the above game.

Threshold Secret Sharing Schemes: A (n, t) threshold secret sharing scheme distributes a secret s among a set of $\mathcal{P} = \{R_1, .., R_n\}$ of n players by a dealer. Each player R_i will privately receive s_i as a share of the secret from the dealer. Then, those subsets with at least t players can recover the secret, while other subsets containing less than t players cannot gain any information about the secret.

Shamir's solution uses polynomial interpolation. Let $GF(q)$ be a finite field with $q \geq n$ elements, and let $s \in GF(q)$ be the secret to be shared. The dealer randomly picks a polynomial $f(x)$ of degree $t - 1$, and sets the constant of $f(x)$ as s. So $f(x)$ is of the following form $f(x) = s + \sum_{j=1}^{t-1} a_j x^j$.

If we assign every player R_i with a unique field element α_i, the dealer sends the secret share $s_i = f(\alpha_i)$ to R_i through a private channel. Now if the set of players $S \subset \mathcal{P}$ such that $|S| \geq t$, they could recover the secret $s = f(0)$ by using the following formula:

$$f(x) = \sum_{R_i \in S} \Delta_{\alpha_i, S}(x) f(\alpha_i) = \sum_{R_i \in S} \Delta_{\alpha_i, S}(x) s_i,$$

where

$$\Delta_{\alpha_i, S}(x) = \prod_{R_l \in S, l \neq i} \frac{x - \alpha_l}{\alpha_i - \alpha_l}.$$

On the other hand, it can be proved that if the subset $B \subset \mathcal{P}$ such that $|B| < t$, all players in B cannot get any information about the polynomial $f(x)$ even if they collude.

3.6.2   Security Models

Fuzzy Identity-Based Signature: The generic fuzzy identity-based signature (FIBS) scheme consists of the following algorithms:

- Setup($1^k, d$): The Setup algorithm is a probabilistic algorithm that takes a security parameter 1^k and an error tolerance parameter d as input. It generates the master key mk and public parameters $params$. Note that mk is kept secret.

- Extract(msk, ID): The private key extraction algorithm is a probabilistic algorithm that takes the master key mk and an identity ID as input. It outputs a private key associated with ID, denoted by D_{ID}.

- Sign($params, D_{ID}, M$): The signing algorithm is a probabilistic algorithm that takes the public parameters $params$, a private key D_{ID} associated with ID, and a message M as input. It outputs the signature σ.

- Verify($params, ID', M, \sigma$): The verification algorithm is a deterministic algorithm that takes the public parameters $params$, an identity ID' such that $|ID' \cap ID| \geq d$, the message M, and the corresponding signature σ as input. It returns a bit b, where $b = 1$ means that the signature is valid.

Security Model

Definition 3.6.2. (UF-FIBS-CMA). Let $\mathcal{A}$ be an adversary assumed to be a probabilistic Turing machine taking as input a security parameter k. Consider the following game in which $\mathcal{A}$ interacts with a challenger $\mathcal{C}$:

- **Setup:** The challenger $\mathcal{C}$ runs the setup phase of the algorithm and tells the adversary $\mathcal{A}$ the public parameters.

- **Phase 1:** $\mathcal{A}$ issues private key queries and signature queries for any identity γ_i adaptively.

- **Phase 2:** $\mathcal{A}$ declares the target identity α, where $|\alpha \cap \gamma_i| < d$ for all γ_i from Phase 1.

- **Phase 3:** $\mathcal{A}$ issues private key queries for many identities γ_j, where $|\gamma_j \cap \alpha| < d$ for all j. $\mathcal{A}$ issues signature queries for any identity.

- **Phase 4:** $\mathcal{A}$ outputs $(\alpha, \tilde{M}, \tilde{\sigma})$, where $\tilde{\sigma}$ is α's valid signature on the message $\tilde{M}$ and $\mathcal{A}$ does not make a signature query on $(\tilde{M}, \tilde{\sigma})$ for identity α .

We define $\mathcal{A}$'s success probability by

$$\mathsf{Succ}_{\mathrm{FIBS},\mathcal{A}}^{\mathrm{UF\text{-}FIBS\text{-}CMA}}(k) = \Pr[\mathsf{Verify}(\alpha, \tilde{M}, \tilde{\sigma}) = 1]$$

The fuzzy identity-based signature scheme FIBS is said to be UF-FIBS-CMA secure if $\mathsf{Succ}_{\mathrm{FIBS},\mathcal{A}}^{\mathrm{UF\text{-}FIBS\text{-}CMA}}(k)$ is negligible in the security parameter k.

3.6.3 Construction

Our scheme is based on the two level hierarchical signature in [48].

In what follows, we assume that groups $\mathbb{G}$ and $\mathbb{G}_T$ of prime order p such that a bilinear pairing $e : \mathbb{G} \times \mathbb{G} \to \mathbb{G}_T$ can be constructed, and g is a generator of $\mathbb{G}$.

Identities will be sets of n elements of $\mathbb{Z}_p^*$. We use the definition of Lagrange coefficient $\Delta_{i,S}(x)$ as in section 2.4.

Setup(n, d): To set up the system, first, choose $g_1 = g^y, g_2 \in \mathbb{G}$. Next, choose $t_1, \ldots, t_{n+1}$ uniformly at random from $\mathbb{G}$. Let N be the set $\{1, \ldots, n+1\}$ and we define a function, T, as:

$$T(x) = g_2^{x^n} \prod_{i=1}^{n+1} t_i^{\Delta_{i,N}(x)}.$$

Next, select a random integer $z' \in \mathbb{Z}_p$ and a random vector $\vec{z} = (z_1, \ldots, z_m) \in \mathbb{Z}_p^m$.

The public parameters of the system and the master key are given by

$$\mathbf{PP} = \Big(g_1, g_2, t_1, \ldots, t_{n+1}, v' = g^{z'}, v_1 = g^{z_1}, \ldots, v_m = g^{z_m},$$

$$A = e(g_1, g_2) \Big) \in \mathbb{G}^{n+m+4} \times \mathbb{G}_T$$

$$\mathbf{MK} = y.$$

Extract(PP, MK, ω): To generate the private key for the identity ω, first choose a random $d - 1$ degree polynomial q such that $q(0) = y$, and return

$$K_\omega = (\{D_i\}_{i \in \omega}, \{d_i\}_{i \in \omega}) \in \mathbb{G}^{2n},$$

where the elements are constructed as

$$D_i = g_2^{q(i)} T(i)^{r_i}, \quad d_i = g^{-r_i},$$

where r_i is a random number from $\mathbb{Z}_p$ defined for all $i \in \omega$.

Sign(PP, K_ω, M): To sign a message represented as a bit string $M = (\mu_1 \cdots \mu_m) \in \{0, 1\}^m$ for identity ω, using private key $K_\omega = (\{D_i\}_{i \in \omega}, \{d_i\}_{i \in \omega}) \in \mathbb{G}^{2n}$, select a random $s_i \in \mathbb{Z}_p$ for each i in ω , and output

$$S = \left(\left\{ D_i \cdot (v' \prod_{j=1}^{m} v_j^{\mu_j})^{s_i} \right\}_{i \in \omega}, \{d_i\}_{i \in \omega}, \{g^{-s_i}\}_{i \in \omega} \right)$$

$$= \left(\left\{ g_2^{q(i)} \cdot T(i)^{r_i} \cdot (v' \prod_{j=1}^{m} v_j^{\mu_j})^{s_i} \right\}_{i \in \omega}, \{g^{-r_i}\}_{i \in \omega}, \{g^{-s_i}\}_{i \in \omega} \right) \in \mathbb{G}^{3n}.$$

Verify(PP, ω', M, σ): To verify a signature $S = (\{S_1^{(i)}\}_{i \in \omega}, \{S_2^{(i)}\}_{i \in \omega}, \{S_3^{(i)}\}_{i \in \omega})$ with respect to an identity ω', where $|\omega' \cap \omega| \geq d$, and a message $M = (\mu_1, \ldots, \mu_m) \in$

$\{0,1\}^m$, choose an arbitrary $d-$element subset S of $\omega \cap \omega'$ and verify that

$$\prod_S \left(e\left(S_1^{(i)}, g\right) \cdot e\left(S_2^{(i)}, T(i)\right) \cdot e\left(S_3^{(i)}, v' \prod_{j=1}^m v_j^{\mu_j}\right) \right)^{\Delta_{i,S}(0)}$$

$$= \prod_S \left(e\left(g_2^{q(i)} \cdot T(i)^{r_i} \cdot \left(v' \prod_{j=1}^m v_j^{\mu_j}\right)^{s_i}, g\right) \cdot e\left(g^{-r_i}, T(i)\right) \right.$$
$$\left. \cdot e\left(g^{-s_i}, v' \prod_{j=1}^m v_j^{\mu_j}\right) \right)^{\Delta_{i,S}(0)}$$

$$= \prod_S \left(e\left(g_2^{q(i)}, g\right) \cdot e\left(T(i)^{r_i}, g\right) \cdot e\left(\left(v' \prod_{j=1}^m v_j^{\mu_j}\right)^{s_i}, g\right) \cdot e\left(g^{-r_i}, T(i)\right) \right.$$
$$\left. \cdot e\left(g^{-s_i}, v' \prod_{j=1}^m v_j^{\mu_j}\right) \right)^{\Delta_{i,S}(0)}$$

$$= \prod_S e\left(g_2^{q(i)}, g\right)^{\Delta_{i,S}(0)} \overset{?}{=} A.$$

If the equality holds, output **1**; otherwise, output **0**.

3.6.4   Security Proofs

We show security as in Theorem 5.1, the approach is based on that of [48, 129].

Theorem 3.6.3. *Let $\mathcal{A}$ be an adversary that makes at most $l \ll p$ signature queries and produces a successful forgery against our scheme with probability ϵ in time t. Then there exists an algorithm $\mathcal{B}$ that solves the CDH problem in $\mathbb{Z}_p$ with probability $\tilde{\epsilon} \geq \epsilon/(4p^n nl)$ in time $\tilde{t} \approx t$.*

Proof. The simulator $\mathcal{B}$ is given an instance $(g, g^a, g^b) \in \mathbb{G}^3$ of the CDH problem, and wants to produce g^{ab}. The simulation proceeds as follows:

Setup: $\mathcal{B}$ first selects a random identity α^*. Next, $\mathcal{B}$ chooses a random $k \in \{0, \ldots, m\}$, and random numbers $x', x_1, \ldots, x_m$ in the interval $\{0, \ldots, 2l - 1\}$. It also chooses additional random exponents $z', z_1, \ldots, z_m \in \mathbb{Z}_p$. It lets $g_1 = g^a, g_2 = g^b$. It then chooses a random n degree polynomial $f(x)$ and an n degree polynomial $u(x)$ such that $\forall x \; u(x) = -x^n$ if $x \in \alpha$. $\mathcal{B}$ sets $t_i = g_2^{u(i)} g^{f(i)}$ for i from 1 to $n+1$. Since t_i is chosen independently at random, we have $T(i) = g_2^{i^n} \prod_{j=1}^{n+1} (g_2^{u(j)} g^{f(j)})^{\Delta_{j,N}(i)} = g_2^{i^n + u(i)} g^{f(i)}$. The simulator gives the public parameters,

$$\mathbf{PP} = (g, g_1, g_2, t_1, \ldots, t_{n+1}, v' = g_2^{x'-2kl} g^{z'}, \{v_j = g_2^{x_j} g^{z_j}\}_{j=1,\ldots,m}, A = e(g_1, g_2)).$$

The corresponding master key, $\mathbf{MK} = a$, is unknown to $\mathcal{B}$.

To answer a private key query on identity γ that $|\gamma \cap \alpha^*| < d$, the simulator $\mathcal{B}$ proceeds as follows. We first define three sets Γ, Γ', S in the following manner:

$\Gamma = \gamma \cap \alpha$, Γ' be any set such that $\Gamma \subseteq \Gamma' \subseteq \gamma$ and, $|\Gamma'| = d-1$, and $S = \Gamma' \cup \{0\}$.

Then we define the private key K_γ for $i \in \Gamma'$ as $(\{D_i\}_{i \in \Gamma'} = \{g_2^{\lambda_i} T(i)^{r_i}\}_{i \in \Gamma'}$, $\{d_i\}_{i \in \Gamma'} = \{g^{r_i}\}_{i \in \Gamma'})$, where λ_i, r_i are chosen randomly in $\mathbb{Z}_p$. We define $d-1$ degree polynomial $q(x)$ as $q(i) = \lambda_i, q(0) = a$.

Next we compute the private key K_γ for $i \in \gamma - \Gamma'$ as follows:

$$D_i = \left(\prod_{j \in \Gamma'} g_2^{\lambda_j \Delta_{j,S}(i)} \right) \cdot \left(g_1^{\frac{-f(i)}{i^n+u(i)}} \left(g_2^{i^n+u(i)} g^{f(i)} \right)^{r_i'} \right)^{\Delta_{0,S}(i)}$$

$$d_i = \left(g_1^{\frac{-1}{i^n+u(i)}} g^{r_i'} \right)^{\Delta_{0,S}(i)}.$$

Since $i \notin \alpha$, $i^n + u(i)$ can not be zero. We claim that such construction is a valid response to this private key query. To see this, let $r_i = (r_i' - \frac{a}{i^n+u(i)})\Delta_{0,S}(i)$. Then we have that

$$
\begin{aligned}
D_i &= \left(\prod_{j \in \Gamma'} g_2^{\lambda_j \Delta_{j,S}(i)} \right) \cdot \left(g_1^{\frac{-f(i)}{i^n + u(i)}} \left(g_2^{i^n + u(i)} g^{f(i)} \right)^{r_i'} \right)^{\Delta_{0,S}(i)} \\
&= \left(\prod_{j \in \Gamma'} g_2^{\lambda_j \Delta_{j,S}(i)} \right) \cdot \left(g^{\frac{-a f(i)}{i^n + u(i)}} \left(g_2^{i^n + u(i)} g^{f(i)} \right)^{r_i'} \right)^{\Delta_{0,S}(i)} \\
&= \left(\prod_{j \in \Gamma'} g_2^{\lambda_j \Delta_{j,S}(i)} \right) \cdot \left(g_2^{a} \left(g_2^{i^n + u(i)} g^{f(i)} \right)^{\frac{-a}{i^n + u(i)}} \cdot \left(g_2^{i^n + u(i)} g^{f(i)} \right)^{r_i'} \right)^{\Delta_{0,S}(i)} \\
&= \left(\prod_{j \in \Gamma'} g_2^{\lambda_j \Delta_{j,S}(i)} \right) \cdot \left(g_2^{a} \left(g_2^{i^n + u(i)} g^{f(i)} \right)^{r_i' - \frac{a}{i^n + u(i)}} \right)^{\Delta_{0,S}(i)} \\
&= \left(\prod_{j \in \Gamma'} g_2^{\lambda_j \Delta_{j,S}(i)} \right) g_2^{a \Delta_{0,S}(i)} (T(i))^{r_i} \\
&= g_2^{q(i)} T_i^{r_i} \\
d_i &= \left(g_1^{\frac{-1}{i^n + u(i)}} g^{r_i'} \right)^{\Delta_{0,S}(i)} = \left(g^{r_i' - \frac{a}{i^n + u(i)}} \right)^{\Delta_{0,S}(i)}.
\end{aligned}
$$

It shows that D_i, d_i have the correct distribution. To answer the signature query on identity γ that $|\gamma \cap \alpha^*| < d$, $\mathcal{B}$ uses K_γ to create a signature on M exactly as in the actual scheme, and outputs the result.

To answer the signature query on identity α^* for some $M = (\mu_1 \cdots \mu_m)$, we define $F = -2kl + x' + \sum_{j=1}^{m} x_j \mu_j$ and $J = z' + \sum_{j=1}^{m} z_j \mu_j$. If $F \equiv 0 \pmod{p}$, the simulator aborts. Otherwise, $\mathcal{B}$ selects a random set Λ such that $\Lambda \subset \alpha^*$ and $|\Lambda| = d - 1$ and define $g^{q'(i)} = g^{\lambda_i'}$ for $i \in \Lambda$ where λ_i' is chosen randomly in $\mathbb{Z}_p$. Then it computes $g^{q'(i)} = (\prod_{j=1}^{d-1} g^{\lambda_j' \Delta_{j,\alpha^*}(i)}) g^{a \Delta_{0,\alpha^*}(i)}$ for $i \in \alpha^* - \Lambda$. $\mathcal{B}$ picks random r_i, s_i for $i \in \alpha^*$ and computes,

$$
\begin{aligned}
S_1^{(i)} &= (g^{q'(i)})^{-J/F} g^{f(i) r_i} (g^J g_2^F)^{s_i} \\
S_2^{(i)} &= g^{-r_i} \\
S_3^{(i)} &= (g^{q'(i)})^{1/F} g^{-s_i}.
\end{aligned}
$$

For $\tilde{s}_i = s_i - q'(i)/F$, we have

$$
\begin{aligned}
S_1^{(i)} &= (g^{q'(i)})^{-J/F} g^{f(i)r_i} (g^J g_2^F)^{s_i} \\
&= (g^{q'(i)})^{-J/F} g^{f(i)r_i} g^{q'(i)J/F} g_2^{q'(i)} (g^J g_2^F)^{s_i - q'(i)F} \\
&= g_2^{q'(i)} g^{f(i)r_i} (g^J g_2^F)^{\tilde{s}_i} = g_2^{q'(i)} T(i)^{r(i)} \left(v' \prod_{j=1}^m v_j^{\mu_j}\right)^{\tilde{s}_i} \\
S_3^{(i)} &= (g^{q'(i)})^{1/F} g^{-s_i} = (g^{q'(i)})^{1/F} g^{-q'(i)/F} g^{-\tilde{s}_i} = g^{-\tilde{s}_i}.
\end{aligned}
$$

It shows that $S_1^{(i)}$, $S_2^{(i)}$, and $S_3^{(i)}$ have the correct distribution. Eventually, $\mathcal{A}$ outputs a valid forgery $S^* = (\{S_1^{(i)*}\}_{i\in\alpha}, \{S_2^{(i)*}\}_{i\in\alpha}, \{S_3^{(i)*}\}_{i\in\alpha})$ on M^* where $M^* = (\mu_1^* \cdots \mu_m^*) \in \{0,1\}^m$ for identity α. Let $F^* = -2kl + x' + \sum_{j=1}^m x_j \mu_j^*$ and $J^* = z' + \sum_{j=1}^m z_j \mu_j^*$. If $\alpha \neq \alpha^*$ or $F^* \not\equiv 0 \pmod p$, $\mathcal{B}$ aborts. Otherwise, the forgery must be of the following form, for some $r_i^*, s_i^* \in \mathbb{Z}_p$,

$$
S_1^{(i)} = g_2^{q^*(i)} T(i)^{r_i^*} \left(v' \prod_{j=1}^m v_j^{\mu_j}\right)^{s_i^*} = g_2^{q^*(i)} g^{f(i)r_i^*} g^{J^* s_i^*}
$$

$$
S_2^{(i)} = g^{-r_i^*}
$$

$$
S_3^{(i)} = g^{-s_i^*}.
$$

We select a random set Λ' such that $\Lambda' \subset \alpha$ and $|\Lambda'| = d$, and compute as follows:

$$
\begin{aligned}
S_1^* &= \prod_{i\in\Lambda'} \left(S_1^{(i)}\right)^{\Delta_{i,\alpha}(i)} \\
&= \prod_{i\in\Lambda'} \left(g_2^{\Delta_{i,\alpha}(i)q^*(i)} T(i)^{\Delta_{i,\alpha}(i)r_i^*} \cdot \left(v' \prod_{j=1}^m v_j^{\mu_j}\right)^{\Delta_{i,\alpha}(i)s_i^*}\right) \\
&= \prod_{i\in\Lambda'} \left(g_2^{\Delta_{i,\alpha}(i)q^*(i)} g^{\Delta_{i,\alpha}(i)f(i)r_i^*} g^{\Delta_{i,\alpha}(i)J^* s_i^*}\right) \\
&= g^{ab} \prod_{i\in\Lambda'} \left(g^{\Delta_{i,\alpha}(i)f(i)r_i^*} g^{\Delta_{i,\alpha}(i)J^* s_i^*}\right)
\end{aligned}
$$

$$S_2^* = \prod_{i \in \Lambda'} \left(S_2^{(i)} \right)^{\Delta_{i,\alpha}(i)f(i)} = \prod_{i \in \Lambda'} g^{-\Delta_{i,\alpha}(i)f(i)r_i^*}$$

$$S_3^* = \prod_{i \in \Lambda'} \left(S_3^{(i)} \right)^{\Delta_{i,\alpha}(i)} = \prod_{i \in \Lambda'} g^{-\Delta_{i,\alpha}(i)s_i^*}.$$

$\mathcal{B}$ could solve the CDH instance by outputting $S_1^* \cdot S_2^* \cdot (S_3^*)^{J^*} = g^{ab}$.

$$\Pr[\text{the simulation not aborting}]$$

$$=\Pr[\alpha = \alpha^*] \cdot \Pr[F \not\equiv 0 \pmod{p}] \cdot \Pr[F^* \equiv 0 \pmod{p}]$$

$$=\frac{1}{p^n} \cdot \left(1 - \frac{1}{2l}\right) \cdot \frac{1}{2nl} \leq \frac{1}{4p^n nl}$$

Thus, $\tilde{\epsilon} \geq \epsilon \cdot \Pr[\text{the simulation not aborting}] \geq \epsilon \cdot \frac{1}{4p^n nl}$. $\qquad\square$

3.6.5 Applications to Biometric Authentication

An important application of FIBS deals with biometric authentication. A biometric authentication system is essentially a pattern-recognition system that recognizes a person based on a feature vector derived from a specific physiological or behavioral characteristic that the person possesses [228]. Since biometrics cannot be lost or forgotten like e.g., computer passwords, biometrics have the potential to offer higher security and more convenience for the users.

Our FIBS system provides an attractive solution to biometric authentication. We use our FIBS scheme (Setup, Extract, Sign, Verify) presented in Section 3.6.3 as the underlying algorithm. We adopt the biometric authentication model presented by Verbitskiy *et al.* It consists of two phases: An enrollment phase and an authentication / Verification phase. Figure 3.16 gives an illustration of our protocol. In Figure 3.16, the text above the arrow represents the sending message. For instance, 1. X means in step 1, Alice's biometric measurement data X is sent to Extract Box. Our protocol performs as follows:

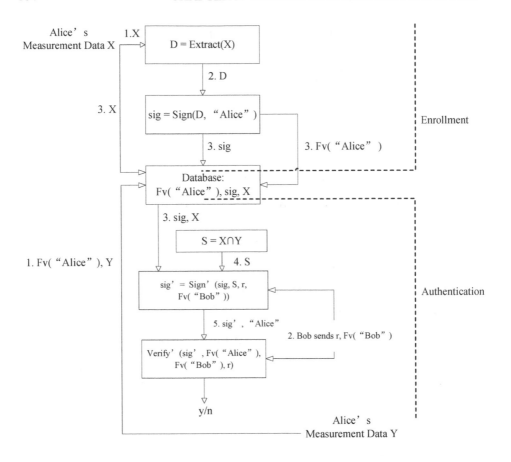

Figure 3.16: The proposed biometric authentication architecture.

Enrollment Phase:

1. First, Alice goes with her biometric data through an enrollment phase to a certification authority (CA). During this procedure the properties of her biometric data are measured with special equipment. We model the measurement data as a feature vector, which is a set of n elements of Z_p^*.

2. From the measurement data a private key D is derived using **Extract**.

3. Then the reference data stored in the database is obtained by applying **Sign** to D using "Alice" as the message. CA stores $F_v("Alice")$ together with the reference

data and the biometric measurement data. Here $F_v(M) = v' \prod_{j=1}^{m} v_j^{\mu_j}$, $M = (\mu_1, \ldots, \mu_m) \in \{0,1\}^m$, and $v', v_1, \ldots, v_m$ are public parameters.

4. CA then erases the private key D, the message "Alice," and all the intermediate data physically.

Authentication / Verification Phase:

1. When Alice wants to authenticate herself to Bob at a later point of time, a measurement that extracts her birometric data Y is taken. Alice sends $F_v(\text{"Alice"})$ and Y to CA.

2. Bob sends $F_v(\text{"Bob"})$ and a random number r to CA.

3. CA find the reference data $\sigma = \{\sigma_1^{(i)}, \sigma_2^{(i)}, \sigma_3^{(i)}\}$ and biometric measurement data X by searching $F_v(\text{"Alice"})$.

4. CA computes $S = X \cap Y$.

5. CA computes $\sigma^* = \{\sigma_1^{*(i)}, \sigma_2^{*(i)}, \sigma_3^{*(i)}, \sigma_4^{*(i)}\}_{i \in S} = \{(\sigma_1^{(i)} F_v(\text{"Bob"}))^{\Delta_{i,S}(0)r}$, $(\sigma_2^{(i)})^{\Delta_{i,S}(0)r}, (\sigma_3^{(i)})^{\Delta_{i,S}(0)r}, g^{\Delta_{i,S}(0)r}\}_{i \in S}$, and returns σ^* to Alice. Here $T(i)$ is the public function defined by our FIBS scheme and S's description is included in σ^*. Then, Alice sends σ^* and her identity "Alice" to Bob.

6. Taking σ^* and "Alice" as input parameters, Bob checks whether

$$\prod_{i \in S}(\sigma_1^{*(i)}, g) \cdot e(\sigma_2^{*(i)}, T(i)) \cdot e(\sigma_3^{*(i)}, F_v(\text{"Alice"}))e(\sigma_4^{*(i)}, F_v(\text{"Bob"}))) \stackrel{?}{=} A^r,$$

A is public parameters. If it holds, Alice passes the authentication. Otherwise, Alice fails the authentication.

Security Analysis: The security of our protocol is analyzed in terms of three important security measurements:

1. "Robust to noise": Biometric data is *not exactly reproducible*, as two biometric scans of the same feature are rarely identical.

2. "Anti-Impersonation": An attacker steals templates from a database and constructs artificial biometrics that pass authentication.

3. "Privacy protection": Exposure of sensitive personal information.

Analysis of "Robust to noise"   : With the proposed protocol, the Setup algorithm is run to set up the error tolerance parameter d according to the noise between two biometric scans. The reference data σ stored in the database is generated using d, such that the people with the biometric measurement ω' will be authenticated as ω if ω and ω' is within the distance d. Therefore, our protocol meets the "Robust to noise" requirements.

Analysis of "Anti-Impersonation":   This security requirement was recognized by several authors [33, 274, 275]. When an authentication system is used on a large scale, the reference database has to be made available to many different verifiers who cannot be trusted. Especially in the network environment, this is a serious threat. It was explicitly shown [211] that by using information stolen from a database, artificial biometrics can be constructed to impersonate people. Construction of artificial fingerprints is possible even if part of the template is available. It has been shown [141] that if minutiae templates of a fingerprint are available, it is still possible to construct artificial fingerprints that pass the authentication.

The proposed protocol satisfies "Anti-Impersonation" requirement because the underlying FIBS scheme is proved secure under UF-FIBS-CMA model. In UF-FIBS-CMA model, the adversary's probability to produce a valid signature of any message under a new private key is negligible, even though he can get any number of signatures and private keys. In our protocol, the reference data stored in the database is a FIBS signature. Therefore, if the adversary can construct an artificial template that is different from the template stored in the database, he can provide a successful forgery of the underlying FIBS scheme.

Analysis of "Privacy protection": It is shown [228] that several privacy concerns surround the use of biometrics for authentication. Because biometric identifiers are biological in origin, collectors may glean additional personal information from scanned biometric measurements. For instance, certain malformed fingers might be statistically correlated with certain genetic disorders. With the rapid advances in human genome research, the fear of inferring further information from biological measurements becomes serious.

Because the identity data "Alice" stored in the database is $F_v(\text{"}Alice\text{"})$ and the discrete logarithm problem is difficult in group $\mathbb{G}$, the probability to recover "Alice" is negligible. At the same time, the authentication request sent to CA only contains $F_v(\text{"}Alice\text{"})$. Therefore, the adversary cannot get any relationship between scanned biometric measurement and identity information stored in the database. Therefore, our protocol meet the "privacy protection" requirements.

Efficiency We now consider the efficiency of the scheme in terms of data size stored in the database and computation time in enrollment phase and authentication/verification phase.

Let $l_{\mathbb{G}}$ and $l_{Z_p^*}$ be the bit length of elements in group $\mathbb{G}$ and Z_p^* respectively. And the biometric measurement data contains n elements in Z_p^*. The signature consists of three group elements in $\mathbb{G}$ for every element in feature vector ω. That is, the length of the signature is $3nl_{\mathbb{G}}$. The length of $F_v(m)$ is $l_{\mathbb{G}}$. The length of biometric measurement data is $nl_{Z_p^*}$. Therefore, The total length of each data is $3nl_{\mathbb{G}} + nl_{Z_p^*} + l_{\mathbb{G}}$.

The computation time in enrollment phase is $m + n$ multiplication in $\mathbb{G}$ and $2n$ exponentiation computation in $\mathbb{G}$.

In our rudimentary Sign algorithm, the number of pairings to sign might be as large as the number of elements of the biometric measurement. However, it is possible to optimize this method. We now discuss methods to improve upon it.

We find out that

$$e(S_1^{(i)}, g), e(S_2^{(i)}, T(i)), e(S_3^{(i)}, v' \prod_{j=1}^{m} v_j^{\mu_j})$$

can be computed offline. Therefore, we compute and store

$$e(S_1^{(i)}, g) \cdot e(S_2^{(i)}, T(i)) \cdot e(S_3^{(i)}, v' \prod_{j=1}^{m} v_j^{\mu_j})$$

to replace the original signature stored in the database. There will be $3n$ more pairing computations in the enrollment phase. So the computation time for enrollment phase is $m + n$ multiplications in $\mathbb{G}$, $2n$ exponentiation computations in $\mathbb{G}$, $3n$ pairing computations. And the total length of each data will be $nl_{\mathbb{G}_T} + nl_{Z_p^*} + l_{\mathbb{G}}$. The computation time in the authentication/verification phase is drastically reduced to n exponentiation computations in $\mathbb{G}$, $2n$ exponentiation computations in $\mathbb{G}_T$, $2(d-1)$ multiplications in $\mathbb{G}_T$.

It is shown [97], the data length of iris biometric template is of 150 dimensions. Therefore, n is 150. And $l_{Z_p^*}$ is 1024 bits, while $l_{\mathbb{G}_T}$ and $l_{\mathbb{G}}$ are 512 bits in normal case. In summary, we have $nl_{\mathbb{G}_T} + nl_{Z_p^*} + l_{\mathbb{G}}$ bit length, which is nearly 28KB for each data stored in the database.

Furthermore, we observe that the computation for enrollment phase could be run concurrently, so the total computation time for Enrollment phase is roughly $3n$ pairing computation. We consider the benchmarks of the primitive cryptographic operations given by Jiang et al. [161], which is evaluated on Intel Core^{TM} 2 Duo 1.83 GHz Linux machine. The computation time for pairing is 2.75ms and multiplication is 0.75ms. So the enrollment phase takes less than 1.25s, and the authentication/verification phase takes less than 0.62s. Note that a pairing operation is roughly two to three times more expensive than exponentiation in $\mathbb{G}$ and $\mathbb{G}_T$.

We have demonstrated that the proposed protocol can function efficiently both in the authentication/verification phase and the Enrollment phase.

In the recent work, ABS has been a hot topic [185,220]. In 2012, two attribute-based signature schemes with constant size signatures are proposed [139]. Their security is proven in the selective-predicate and adaptive-message setting, in the standard model, under chosen message attacks, with respect to some algorithmic assumptions related to bilinear groups.

3.7 Notes

In this chapter, we have given a comprehensive description of the formal definitions and security models of attribute-based encryption cryptosystems. We also introduced some of our research work publicized recently in this field, namely a bounded ciphertext-policy encryption scheme published in the conference ASIACCS 2009, a multi-authority attribute-based encryption scheme published in the conference Indocrypt 2008 and ESORICS 2011, an interval encryption scheme published in the conference ACNS 2010 and a FIBS scheme published in the journal *Computers & Electrical Engineering*, which has become the attribute-based signature in fact. Readers can refer to [192, 198, 199, 201, 301], if they want to know more ABE schemes and ABS schemes.

There are also some interesting problems in attribute-based cryptography. For instance,

1. How to design an expressive CPABE system resisting malicious key leakage?

> *For the bounded ciphertext-policy attribute-based encryption, it is imperative for improving the efficiency of the proposed BCPABE scheme and designing a new kind of BCPABE excluding the assistance of dummy nodes. In a CPABE system, decryption keys are defined over*

*attributes shared by multiple users. Given a decryption key it may
not be possible to trace to the original key owner. Also, as a decryption privilege could be possessed by multiple users who own the same
set of attributes, malicious users would tempt to or be very willing to
leak their decryption privileges without the risk of being caught. This
problem severely limits the applications of CPABE. Although some attempts [142, 186, 187] have been made to address this problem, the
policy in these systems can only support a single* AND *gate with wild
card. To construct an expressive CPABE system resisting malicious
key leakage is still an open problem.*

2. How to design interval encryption schemes in the multi-dimensional scenario?

3. How to design attribute-based proxy re-encryption schemes with higher efficiency
   and stronger security?

   *Combining the proxy re-encryption technique with recently introduced
   attribute-based cryptosystem, attribute-based proxy re-encryption
   scheme is proposed. All the advantages of PRE scheme can be inher-
   ited into access control environment. The security model of ABPRE
   was defined for the first time and our scheme can be proved selective-
   structure chosen plaintext secure and master key secure in the standard
   model assuming that ADBDH problem and ADH problem are hard to
   solve. Moreover, another kind of key delegation algorithm is developed
   in the ABPRE scheme, providing more delegating capability to each
   valid user. The future work includes how to design a more delicate
   ABPRE scheme with higher efficiency and stronger security. Besides,
   it still remains an open problem to construct an ABPRE scheme that
   has a reduction based on a more standard and natural assumption.*

Chapter 4

Batch Cryptography

4.1 Introduction

From Chapter 1, we know the new direction of modern cryptography — *Batch Cryptography*, which can process the decryption, key agreement, and signature/verification in a batch way, instead of one by one. In this chapter, we give the formal definitions and concrete schemes as well as the related theories in mathematics. Batch cryptography focuses on the global efficiency and security. It includes batch decryption, batch key agreement and batch signature/verification. To the best of our knowledge, it is the first effort towards introducing batch cryptography.

The remainder of this chapter is organized as follows. A brief review on related work of batch signature/verification schemes as well as their security models and schemes are given in Section 4.2. Batch decryption and batch key agreement are introduced in Section 4.3. Batch RSA is considered as an important case of batch cryptography. New batch RSA schemes based on Diophantine equations are presented in Section 4.4. To implement these new schemes, we employ some related mathematical theory and techniques for solving these Diophantine equations in Section 4.5. Finally, a conclusion of the chapter is presented in Section 4.6.

4.2 Aggregate Signature and Batch Verification

Aggregate signature was first introduced by Boneh et al. [38] for batch verification, which was based on the short signature [41] with efficiently computable bilinear maps in 2003. An aggregate signature scheme is a special digital signature that supports aggregation which can be informally defined as follows: given b signatures on b distinct messages from b distinct users, it is possible to aggregate all these signatures $\sigma_1, \sigma_2, \ldots, \sigma_b$ into a single *compact* signature which convinces the verifier that the b users indeed signed the b original messages in a batch way (i.e., user i signed message m_i for $i = 1, 2, \ldots, b$). Moreover, the aggregation can be performed incrementally. That is, signatures σ_1, σ_2 can be aggregated into σ_{12} which can then be further aggregated with σ_3 to obtain σ_{123}.

$$\sigma_1 + \sigma_2 \to \sigma_{12} \text{ and } \sigma_{12} + \sigma_3 \to \sigma_{123}.$$

Lysyanskaya et al. [205] use certified trapdoor permutations which permits only sequential aggregation, i.e., the b-th signer must aggregate its own signature into the aggregate signature formed by the first $b - 1$ signers. Lu et al. [203] present the first aggregate signature scheme that is provably secure without random oracles. The signatures are sequentially constructed. However, unlike the scheme of Lysyanskaya et al. [205], a verifier need not know the order in which the aggregate signature was created. Additionally, these signatures are shorter than those of Lysyanskaya et al. [205] and can be verified more efficiently than those of Boneh et al. [38]. Ahn et al. [3] modify the stateful signatures of Hohenberger and Waters [146] by removing the chameleon hash and present a surprisingly efficient synchronized aggregate signature scheme which is secure under the computational Diffie–Hellman (CDH) assumption in the standard model.

In the batch verification, Naccache et al. [219] gave the first efficient batch verifier for DSA signatures in 1994; however, an interactive batch verifier presented in an
early version of their paper was broken by Lim and Lee [196]. In 1995, Laih and Yen
[302] proposed a new method for batch verification of DSA and RSA signatures, but
the RSA batch verifier was broken five years later by Boyd and Pavlovski [45]. In 1998,
Harn presented two batch verification techniques for DSA [137] and RSA [138] but
both were later broken [45, 152, 153]. In the same year, Bellare et al. [21] took the first
systematic look at batch verification and presented three generic methods for batching
modular exponentiations, called the random subset test, the small exponents test, and
the bucket test which are similar to the ideas from Laih and Yen [302]. Later, Cheon and
Lee [83] introduced two new methods called the sparse exponents test and the complex
exponents test, which they claim to be about twice as fast as the small exponents test.
In 2000, Boyd and Pavlovski [45] proposed some attacks against different batch verification schemes most of which were based on the small exponents test and related tests,
and repaired some broken schemes based on the small exponents test. In 2001, Hoshino
et al. [148] pointed out that the problem discovered by Boyd and Pavlovski [45] was
only critical for batch verification such as the zero-knowledge proofs. Other schemes
for batch verification based on bilinear maps were proposed [304, 308, 309] but all were
later broken by Cao et al. [67]. In 2006, a method was proposed for identifying invalid signatures in RSA-type batch signatures [180], but Stanek [262] showed that this
method is awed. Ferrara et al. [110] gave performance measurements for the schemes
herein, and also showed how to batch verify other types of signatures, such as group
and ring signatures. Law and Matt [178] pointed out that some identity-based signature schemes batch well, and gave methods for identifying invalid signatures in a batch.
Because the goals of batch verification and aggregate signature are slightly different, it
is important to clarify these distinct notions. Informally, batch verification's goal is to
verify b distinct signatures quickly, while the goal in aggregate signatures is to compress

these signatures.

4.2.1   Definitions

We follow [38] for formalizing the aggregate signature in the chosen-key security model. For the convenience of our presentation, the attacker is also called an adversary or a forger.

Definition 4.2.1. The adversary $\mathcal{A}$ is given a single public key, called challenge public key. $\mathcal{A}$'s goal is the existential forgery of an aggregate signature. $\mathcal{A}$ is granted the power to choose all public keys except the challenge public key and access to a signing oracle on the challenge public key. $\mathcal{A}$'s advantage, denoted $\text{Adv}_{\text{AggSig}\,\mathcal{A}}$, is defined to be the probability of success in the following game:

- **Setup**. $\mathcal{A}$ is provided with a public key pk_1, generated at random.

- **Queries**. $\mathcal{A}$ requests hash values of messages and signatures with pk_1 on messages of his choice.

- **Response**. Finally, $\mathcal{A}$ outputs $b-1$ additional public keys $pk_2, ..., pk_b$. These keys, along with the initial key pk_1, will be included in $\mathcal{A}$'s forged aggregate. $\mathcal{A}$ also outputs messages $m_1, m_2, ..., m_b$, and an aggregate signature σ, where σ is signed by the b users and each user signs on the corresponding message.

The aggregate signature σ is called nontrivial if $\mathcal{A}$ did not request a signature on m_1 under pk_1. We define that $\mathcal{A}$ wins if σ is nontrivial valid aggregate signature on messages $m_1, m_2, ..., m_b$ under keys $pk_1, pk_2, ..., pk_b$. The probability is over the coin tosses of the key-generation algorithm and of $\mathcal{A}$.

Definition 4.2.2. An adversary $\mathcal{A}$ $(t, q_H, q_S, b_{max}, \epsilon)$-breaks an aggregate signature scheme in the chosen-key model if: $\mathcal{A}$ runs in time at most t; $\mathcal{A}$ makes at most q_H queries to the hash function and at most q_S queries to the signing oracle; $Adv_{AggSig_{\mathcal{A}}}$

is at least ϵ; and the forged aggregate signature is signed by at most b_{max} users. An aggregate signature scheme is $(t, q_H, q_S, b_{max}, \epsilon)$-secure against existential forgery in the chosen-key model if no forger $(t, q_H, q_S, b_{max}, \epsilon)$-breaks it.

Recall that a digital signature scheme is a tuple of algorithms (**Gen**, **Sign**, **Verify**) that also is correct and secure. The correctness property states that for all **Gen**$(1^\ell \rightarrow (pk; sk)$, the algorithm **Verify**$(pk, m, Sign(sk, m)) = 1$.

Definition 4.2.3 (Camenisch et al. [60]). Let ℓ be the security parameter. Suppose (**Gen**, **Sign**, **Verify**) is a signature scheme and $(pk_1, sk_1), ..., (pk_b, sk_b)$ are independently generated according to **Gen** (1^ℓ). We define a probabilistic batch verification algorithm (**Batch**) as follows:

- **Batch**$((pk_1, m_1, \sigma_1), ..., (pk_b, m_b, \sigma_b)) = 1$ provided **Verify**$(pk_i, m_i, \sigma_i) = 1$ for all $i = 1, 2, ..., b$.

- **Batch**$((pk_1, m_1, \sigma_1), ..., (pk_b, m_b, \sigma_b)) = 0$ except with probability negligible in ϵ, taken over the randomness of **Batch**, provided **Verify**$(pk_i, m_i, \sigma_i) = 0$ for any $i = 1, 2, ..., b$.

Definition 4.2.4 (Gentry and Ramzan [121]). The security model for identity based aggregate signature (IBAS) is defined as follows:

- **Setup:** The adversary $\mathcal{A}$ is given public key pk of the Private Key Generation (**PKG**), an integer b_{max}, and any other needed parameters.

- **Queries:** Proceeding adaptively, $\mathcal{A}$ may choose identities ID_i and request the private key sk_i. Also, $\mathcal{A}$ may request an IBAS σ_S on $(pk, S, \{m_i\}_{i=1}^{k-1})$ where $S = \{ID_i\}_{i=1}^{k-1}$. We require that $\mathcal{A}$ has not to make a query $(pk, S', \{m_i\}_{i=1}^{k-1})$ where $ID_i \in S \cap S'$ and $m_i' \neq m_i$.

- **Response:** For some $(pk, \{ID_i\}_{i=1}^{b}, \{m_i\}_{i=1}^{b})$ with $b \leq b_{max}$, $\mathcal{A}$ outputs an IBAS σ_l.

We call that the signature is nontrivial, where for some i, $1 \leq i \leq b$, $\mathcal{A}$ did not request the private key for ID_i and did not request a signature including the pair (ID_i, m_i). We define that $\mathcal{A}$ wins if σ_b is a nontrivial valid signature on $(pk, \{ID_i\}_{i=1}^b, \{m_i\}_{i=1}^b)$

Definition 4.2.5. An IBAS adversary $\mathcal{A}$ $(t, \epsilon, b_{max}, q_H, q_E, q_S)$-breaks an IBAS scheme in the above model if: for the integer b_{max} as above, $\mathcal{A}$ runs in time at most t; $\mathcal{A}$ makes at most q_H hash function queries, q_E private key extraction queries, and at most q_S signing oracle queries; and $\text{Adv}_{\text{IBAS}\mathcal{A}}$ is at least ϵ.   An IBAS scheme is $(t, \epsilon, b_{max}, q_H, q_E, q_S)$-secured against existential forgery if no adversary $(t, \epsilon, b_{max}, q_H, q_E, q_S)$-breaks it.

4.2.2   Aggregate Signature

Aggregation is obtained by combining the resulting b signatures $\{\sigma_1, \sigma_2, ..., \sigma_b\}$ into one aggregate signature σ:

$$\sigma \leftarrow \prod_{i=1}^b \sigma_i.$$

We introduce aggregate signature scheme given by Boneh et al.'s [38] as follows:

- **Setup:** Let $e : \mathbb{G}_1 \times \mathbb{G}_2 \rightarrow \mathbb{G}_T$, where g generates the group $\mathbb{G}_2$ of prime order q.

- **Gen:** Chooses a random $sk \in \mathbb{Z}_q$ and outputs $pk = g^{sk}$.

- **Sig:** A signature on message m is $\sigma = H(m)^{sk}$, where H is a hash function $H : \{0,1\}^* \rightarrow \mathbb{G}_1$.

- **Ver:** To verify signature σ on message m, one checks that $e(\sigma, g) \overset{?}{=} e(H(m), pk)$.

Batch verification is very simple. What we need to do are the following two steps: checking that the m_i's are mutually distinct and ensuring that

$$e(g, \sigma) = \prod_{i=1}^{b} e(pk_i, h_i) \tag{4.1}$$

where $h_i = h(m_i)$. Equality (4.1) holds because

$$e(g, \sigma) = e(g, \prod_{i=1}^{b} h_i^{sk_i}) = \prod_{i=1}^{b} e(g, h_i)^{sk_i} = \prod_{i=1}^{b} e(g^{sk_i}, h_i) = \prod_{i=1}^{b} e(pk_i, h_i). \tag{4.2}$$

In the special case when all b signatures are issued by the same public key pk_0 (same signer), the verification is faster *only for 2 pairings*. One needs to verify that

$$e(g, \sigma) = e(\prod_{i=1}^{b} h(m_i), pk_0),$$

where $m_1, m_2, ..., m_b$ are the signed messages.

However, Camenisch et al. [60] pointed out that the aggregate scheme in [38] is not a batch verification scheme since, for any $a \neq 1 \in \mathbb{G}_1$, the two invalid message-signature pairs $P_1 = (m_1, a \cdot H(m_1)^{sk})$ and $P_2 = (m_2, a^{-1} \cdot H(m_2)^{sk})$ will verify under Definition 4.2.3.

4.2.3 Identity-Based Aggregate Signature

In 2006, Gentry and Ramzan [121] gave the first identity-based aggregate signature (IBAS) as follows:

- **Setup:** It generates groups $\mathbb{G}_1$ and $\mathbb{G}_2$ of prime order q and an admissible pairing $e : \mathbb{G}_1 \times \mathbb{G}_1 \rightarrow \mathbb{G}_2$; chooses an arbitrary generator $P \in \mathbb{G}_1$. In addition, it chooses three cryptographic hash functions $H_1, H_2 : \{0, 1\}^* \rightarrow \mathbb{G}_1$, and $H_3 : \{0, 1\}^* \rightarrow \mathbb{Z}_q^*$. After setting the system parameters, **PKG** picks a random number

$sk_{PKG} \in \mathbb{Z}_q^*$ as its secret key and sets its public key as $P_{pub} = sk_{PKG} \cdot P$. The system parameters are $params = (\mathbb{G}_1, \mathbb{G}_2, q, e, P, P_{pub}, H_1, H_2, H_3)$. The sytem's secret is sk_{PKG}.

- **Extract:**

$$sk_i^0 = sk_{PKG} \cdot Q_{ID_i}^0 \text{ and } sk_i^1 = sk_{PKG} \cdot Q_{ID_i}^1$$

where $Q_{ID_i}^0 = H_1(ID_i \| 0)$ and $Q_{ID_i}^1 = H_1(ID_i \| 1)$.

- **Individual Signing:**
  (1) Choose a unique string ω and compute $P_\omega = H_2(\omega) \in \mathbb{G}_1$;
  (2) Compute $c_i = H_3(m_i, ID_i, \omega) \in \mathbb{Z}_q^*$;
  (3) Generate a random number r_i, where $r_i \in \mathbb{Z}_q^*$;
  (4) Compute signature $Sig_i = (\omega, U_i, V_i)$ for message m_i, where

$$U_i = r_i \cdot P \quad \text{and} \quad V_i = r_i \cdot P_\omega + sk_i^0 + c_i \cdot sk_i^1. \tag{4.3}$$

- **Aggregate:** Anyone can aggregate a collection of individual signatures that use the same string ω. Signatures $Sig_i = (U_i, V_i)$ where $0 \le i \le N$ can be aggregated to $Sig = (U, V)$ where $U = \sum_{i=0}^b U_i$ and $V = \sum_{i=0}^b V_i$.

- **Verification:**

$$e(V, P) = e(P_\omega, U)e\left(\sum_{i=1}^b Q_{ID_i}^0 + \sum_{i=1}^b c_i \cdot Q_{ID_i}^1, P_{pub}\right) \tag{4.4}$$

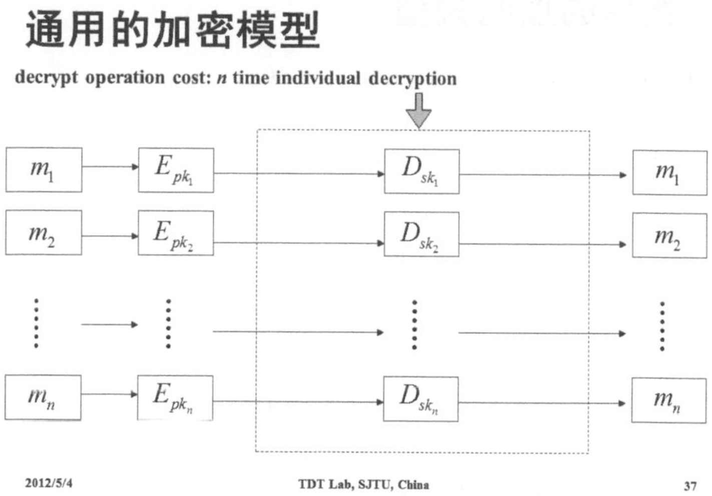

通用的加密模型

Figure 4.1: Individual decryption model

where $c_i = H_3(m_i, ID_i, \omega)$. The batch verification equality (4.4) holds because:

$$
\begin{aligned}
e(V, P) &= e\left(\sum_{i=1}^{b} V_i, P\right) = e\left(\sum_{i=1}^{b} r_i P_\omega + \sum_{i=1}^{b} sk_i^0 + \sum_{i=1}^{b} c_i sk_i^1, P\right) \\
&= \prod_{i=0}^{i} e\left(r_i P_\omega, P\right) \prod_{i=0}^{i} e\left(sk_{PKG} Q_{ID_i}^0, P\right) \prod_{i=0}^{i} e\left(c_i sk_{PKG} Q_{ID_i}^1, P\right) \\
&= \prod_{i=0}^{i} e\left(P_\omega, r_i P\right) \prod_{i=0}^{i} e\left(Q_{ID_i}^0, sk_{PKG} P\right) \prod_{i=0}^{i} e\left(c_i Q_{ID_i}^1, sk_{PKG} P\right) \\
&= e\left(P_\omega, \sum_{i=1}^{b} U_i\right) e\left(\sum_{i=1}^{b} Q_{ID_i}^0, P_{pub}\right) e\left(\sum_{i=1}^{b} c_i Q_{ID_i}^1, P_{pub}\right) \\
&= e(P_\omega, U) e\left(\sum_{i=1}^{b} Q_{ID_i}^0 + \sum_{i=1}^{b} c_i Q_{ID_i}^1, P_{pub}\right)
\end{aligned}
$$

4.3 Batch Decryption and Batch Key Agreement

In the above sections, we focus on signature aggregation and batch verification. From now on, we focus our attention to batch decryption and key agreement schemes. In batch decryption, batch RSA based on the RSA scheme is a frequently used example to explain this concept. Batch RSA was first proposed by Fiat [111] in 1989 and later developed by Boneh and Shacham [43] in 2002. It is designed to speed up RSA decryp-

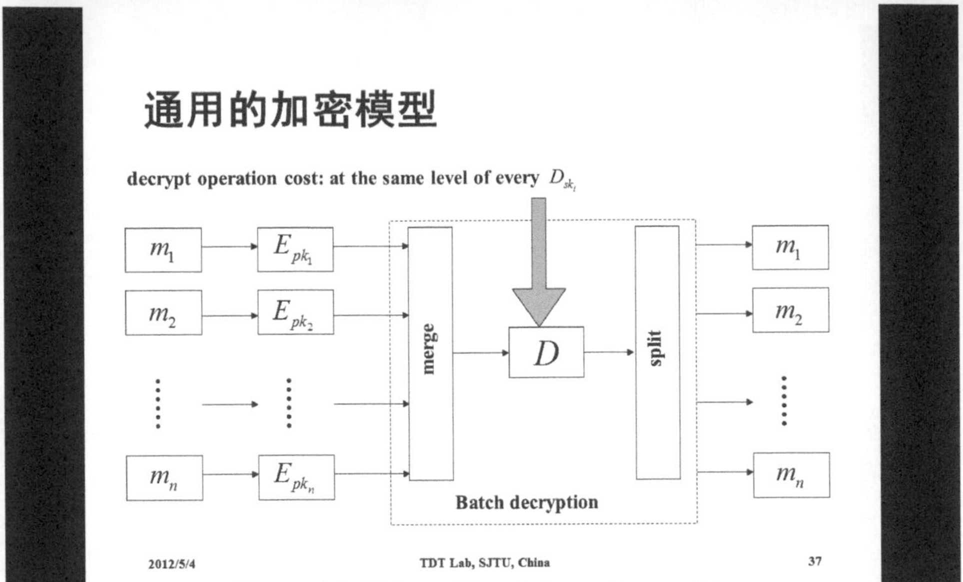

通用的加密模型

Figure 4.2: Universal batch decryption model

tion and signing algorithms. Batch RSA has many applications. Since it was proposed, it has attracted many researchers and industry engineers. Shacham and Boneh [242] improved performance of SSL based on batch RSA. Yacobi and Beller [298] applied the mathematical ideas of batch RSA to DH-based schemes. They proposed a batch Diffie–Hellman key agreement scheme. Figure 4.1 gives an individual decryption model where messages are encrypted by the different public keys and sent to decryption module individually and the decryption operations are performed one by one. Thus, the total cost is n times individual decryption in the dotted line box (we assume that it is a server) in Figure 4.1.

We are trying to find a way to improve the efficiency by a batch model instead of this individual decryption model. Figure 4.2 gives a universal batch decryption model where the messages are encrypted by the different public keys and sent to decryption module individually. The decryption module first merges these ciphertexts, decrypts using its secret key *only once*, and splits to individual plaintexts.

Batch RSA can do a number of RSA decryptions for approximately the cost of a single decryption.

For the integrally introducing batch RSA, we first give the RSA scheme and then give Fiat's batch RSA and Boneh and Shacham's batch RSA.

4.3.1   Review of RSA

RSA [233] is the most widely deployed public key cryptosystem. It is used for securing web traffic, e-mail, and some wireless devices. Since RSA is based on arithmetic
modulo large integer numbers it might be slow in constrained environments. For example, 1024-bit RSA decryption on a small handheld device such as the PalmPilot III can
take as long as 30 seconds. Similarly, on a heavily loaded web server, RSA decryption
significantly reduces the number of SSL requests per second that the server can handle.
Typically, one improves performance of RSA using special-purpose hardware. Current
RSA coprocessors can perform as many as 10,000 RSA decryptions per second (using
a 1024-bit modulus) and even faster processors are coming out.

The RSA encryption scheme can be summarized as follows. The public key is the
pair (N, e), where N is the product of two large primes p and q, and e is chosen to be
coprime to Euler's function $\phi(N)$. The private key is d where $d \equiv e^{-1} \pmod{\phi(N)}$.

To encrypt a message m, one computes the ciphertext c as follows:

$$c \equiv m^e \pmod{N} \tag{4.5}$$

To decrypt the ciphertext c, one computes

$$m \equiv c^d \pmod{N} \equiv c^{1/e} \pmod{N}. \tag{4.6}$$

Thus, every decryption consists of one full-sized modular exponentiation.

Later, Quisquater and Couvreur [230] suggested the use of the Chinese Remainder Theorem (CRT) in order to make the decryption run slightly faster. It is standard
practice to employ the Chinese Remainder Theorem for RSA decryption. Rather than
compute $\sigma \leftarrow c^d \bmod N$, the decryptor evaluates:

$$m_p \leftarrow c_p^{d_p} \bmod p \tag{4.7}$$

and

$$m_q \leftarrow c_q^{d_q} \mod q \qquad\qquad (4.8)$$

Here, $d_p \equiv d \pmod{p-1}$ and $d_q \equiv d \pmod{q-1}$. Then the decryptor uses the CRT to calculate m from m_p and m_q. This is approximately four times faster than evaluating $c^d \pmod{N}$ directly.

Generally speaking, a small value of e is chosen, which is coprime to $\phi(N)$, and d will be $\mathcal{O}(\phi(N))$. In fact, if d was too small (less than exponential in the security parameter), it would allow the cryptanalyst to attack the scheme. Weiner [296] gives attacks on the short private key d of RSA.

Problem 4.3.1. How to cut down the cost of decryption with a batch of ciphertext?

Fiat's batch RSA gives us the first answer to this problem.

4.3.2   Batch RSA

Fiat's batch RSA [111] can be shown in Figure 4.3. Let $e_1, e_2, \ldots, e_b$ be b different

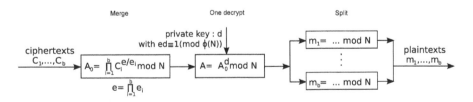

Figure 4.3: Batch RSA decryption

encryption exponents, coprime to $\phi(N)$ and to each other where N is defined as in the RSA scheme. Let

$$e = \prod_{i=1}^{b} e_i \qquad\qquad (4.9)$$

be $\mathcal{O}$ bits long. Given ciphertext $c_1, c_2, \ldots, c_b$ related to plaintext $m_1, m_2, \ldots, m_b$, our goal is to generate the b roots (decryptions):

$$c_1^{1/e_1}, c_2^{1/e_2}, \ldots, c_b^{1/e_b} \quad \text{mod } N. \tag{4.10}$$

Let T be a binary tree with leaves labeled $e_1, e_2, \ldots, e_b$. Let d_i denote the depth of the leaf labeled e_i, T should be constructed so that $W = \sum_{i=1}^{b} d_i \log e_i$ is minimized similar to the Huffman code tree construction. For the main result of $\mathcal{O}(\log^2 N)$ multiplications per RSA operation, Fiat's scheme could simply assume that T is a full binary tree. In practice, there is some advantage in using a tree that minimizes the sum of weight times path length because the work performed is proportional to the sum W above. Note that $W = \mathcal{O}(\log b \log e)$. Finally, Fiat shows that the number of multiplications required to compute the b roots above is $\mathcal{O}(W + \log N)$.

In Fiat's scheme, the first goal is to generate the product:

$$A_0 = c_1^{e/e_1} \cdot c_2^{e/e_2} \cdot \ldots \cdot c_b^{e/e_b}. \tag{4.11}$$

Fiat also shows that this requires $\mathcal{O}(W)$ multiplications. Use the binary tree T as a guide, working from the leaves to the root. Every internal node takes the recursive result from the left branch (L) raises it to the power E_R where E_R is the product of the labels associated with leaves on the right branch. Similarly, each node takes the result from the right branch (R) and raises it to the power E_L, which is the product of the labels on the left branch. Each node saves the intermediate results L^{E_R} and R^{E_L} (required later). The result associated with this node is $L^{E_R} \cdot R^{E_L}$. The product A_0 is simply the result associated with the root.

Second, it extracts the e-th root of the product A_0:

$$A = A_0^{1/e} = c_1^{1/e_1} \cdot c_2^{1/e_2} \cdot \ldots \cdot c_b^{1/e_b} \quad \text{mod } N \tag{4.12}$$

This involves $\mathcal{O}(\log N)$ modular multiplications which is equivalent to one RSA decryption.

Third, the factors of A are the roots. The next goal is to break the product A into two subproducts, the breakup is implied by the structure of the binary tree T used to generate the product C. The scheme repeats this recursively to break up the product into its b factors.

Let $e_1, e_2, \ldots, e_k$ be the labels associated with the left branch of the root of the binary tree T. Define an exponent X by means of the Chinese Remainder Theorem (CRT):

$$
\begin{cases}
X \equiv 0 \pmod{e_1} \\[4pt]
X \equiv 0 \pmod{e_2} \\[4pt]
\cdots \\[4pt]
X \equiv 0 \pmod{e_k} \\[4pt]
X \equiv 1 \pmod{e_{k+1}} \\[4pt]
\cdots \\[4pt]
X \equiv 1 \pmod{e_b}
\end{cases}
\tag{4.13}
$$

There is a unique solution for X modulo $e = \prod_{i=1}^{b} e_i$ by CRT. By (4.13), let

$$
X = \left(\prod_{i=1}^{k} e_i \right) \cdot X_1
\tag{4.14}
$$

and

$$
X - 1 = \left(\prod_{i=k+1}^{b} e_i \right) \cdot X_2,
\tag{4.15}
$$

where X_1, X_2 are positive integers. Let $P_1 = \prod_{i=1}^{k} e_i$ and $P_2 = \prod_{i=k+1}^{b} e_i$, then $X = P_1 \cdot X_1$ and $X - 1 = P_2 \cdot X_2$. Compute

$$
C_1 = c_1^{P_1/e_1} \cdot c_2^{P_1/e_2} \cdot \ \cdots \ \cdot c_k^{P_1/e_k} \quad \bmod N
\tag{4.16}
$$

and

$$C_2 = c_{k+1}^{P_2/e_{k+1}} \cdot c_{k+2}^{P_2/e_{k+2}} \cdot \ \cdots \ \cdot c_b^{P_2/e_b} \quad \bmod N \tag{4.17}$$

Note that C_1 and C_2 have been computed, as the left and right branch values of the root, during the tree-based computation of A_0.

Raise A to the X-th power modulo N, it follows from

$$
\begin{aligned}
A^X &= \left(\prod_{i=1}^{b} c_i^{1/e_i} \right)^X \\
&= \left(\prod_{i-1}^{k} c_i^{1/e_i} \right)^{P_1 X_1} \cdot \left(\prod_{i=k+1}^{b} c_i^{1/e_i} \right)^{P_2 X_2} \cdot \prod_{i=k+1}^{b} c_i^{1/e_i} \\
&= C_1^{X_1} \cdot C_2^{X_2} \cdot \prod_{i=k+1}^{b} c_i^{1/e_i},
\end{aligned}
$$

that $Q_2 = \prod_{k+1}^{b} c_i^{1/e_i}$, and so $Q_1 = \prod_{i=1}^{k} c_i^{1/e_i}$ since $Q_1 = A/Q_2$. After that, it recursively breaks up the two products Q_1 and Q_2, respectively, until each c_i^{1/e_i} is obtained. Thus, the overall number of multiplications is $\mathcal{O}(W)$. The number of modular divisions required is $\mathcal{O}(b)$.

To summarize this method, we introduce the following theorem.

Theorem 4.3.2 (Fiat [111]). *Let $e_1, e_2, \ldots, e_b$ be b different encryption exponents, relatively prime to $\phi(N)$ and to each other. Given the ciphertexts $c_1, c_2, \ldots, c_b$, we can generate the b roots*

$$c_1^{1/e_1}, c_2^{1/e_2}, \ldots, c_b^{1/e_b} \quad \bmod N$$

in $\mathcal{O}(\log b (\sum_{i=1}^{b} \log e_i) + \log N)$ modular multiplications and $\mathcal{O}(b)$ modular divisions.

Fiat's batch RSA gave an example when using small public exponents e_1 and e_2, it decrypts two ciphertexts for approximately the price of one decryption.

Suppose c_1 is a ciphertext obtained by encrypting some plaintext m_1 using the public key $(N, 3)$, and c_2 is a ciphertext for some m_2 using $(N, 5)$. To decrypt, they must compute $c_1^{1/3}$ and $c_2^{1/5}$ (mod N). Fiat observed that by setting

$$A = (c_1^5 \cdot c_2^3)^{1/15},$$ (4.18)

one obtains

$$c_1^{1/3} = \frac{A^{10}}{c_1^3 \cdot c_2^2}$$ (4.19)

and

$$c_2^{1/5} = \frac{A^6}{c_1^2 \cdot c_2}.$$ (4.20)

Hence, at the cost of computing a single 15th root and some additional arithmetic, it is able to decrypt both c_1 and c_2. Computing a 15th root is considered to take the same time as a single RSA decryption in this case.

Boneh and Shacham [43] generalized (4.19) and (4.20) to

$$m_i \equiv c_i^{1/e_i} \equiv \frac{A^{\alpha_i}}{c_i^{(\alpha_i-1)/e_i} \cdot \prod_{j \neq i} c_j^{\alpha_i/e_j}} \quad (\text{mod } N)$$ (4.21)

where α_i satisfies

$$\alpha_i \equiv \begin{cases} 1 & (\text{mod } e_i), \\ 0 & (\text{mod } e_j), \quad \text{for } j \neq i, \end{cases}$$ (4.22)

where $1 \leq i, j \leq b$.

This method requires b different modular inversions whereas Fiat's tree-based method requires $2b$ modular inversions, but fewer auxiliary multiplications. Thus, Boneh and Shacham give an improvement to Fiat's scheme. However, in Boneh and Shacham's scheme, it is still required to solve the systems of linear congruences Equations (4.22) with b modular inversions. In addition, it is not easy to implement in hardware using these different public keys $(e_1, e_2, ..., e_b)$ as the modular equipped in advance in the chip based systems. Thus, we leave the following problems:

Problem 4.3.3. For each b, how to obtain all α_i's satisfying Equation (4.22) without any modular operation?

Problem 4.3.4. Are there any other methods to split the plaintext directly like Equation (4.21)?

Once Problem 4.3.3 is solved, we can directly obtain m_i satisfying Equation (4.21) using only one modular N, which allows us to implement batch RSA in hardware easily. If the answer to Problem 4.3.4 is positive, we can get more methods to split plaintext to individual. We will discuss it in Section 4.4.

4.3.3 Batch Key Agreement

In this section, we consider a star-like network to present the batch key agreement. For example, in a personal communication system network, the central server, through the ports, agrees upon a session key with each of the users. Each port holds a secret key that is used to establish a secret channel in the network.

Yacobi and Beller [298] proposed a solution, in which each of the users picks a random number, encrypts it with the public key of the corresponding port, and sends it to the counterpart port. The central server decrypts the received ciphertext, and uses the random numbers or their hash values as session keys. Yacobi and Beller [298] summarized the difference between RSA-based and DH-based key exchange. The basic DH scheme was first proposed in [99], which makes the key exchanged individually and needs b operations when the number of the users is b. Yacobi and Beller [298] presented a batch RSA key agreement as follows:

- First, the central server with b ports communicates with the users. The central server generates a public/private key pair for each port. The modulus is a composite N, which is the product of two large primes p and q. Each port's public key e_i is chosen to be prime to Euler's function $\phi(N)$ and pairwise coprime. The private key is d_i, where $d_i \equiv e_i^{-1} \pmod{\phi(N)}$.

- Second, when b users want to agree on session keys individually with the central server, each user needs to find an available port and encrypts a chosen random number using the port's public key e_i.

- Third, when the central server receives the ciphertexts from each port i, it takes one decryption similar to the batch RSA decryption in the above section and gets the random numbers that could finally generate the session keys.

4.4   Batch RSA's Implementation Based on Diophantine Equations

In this section, we introduce our batch RSA implementation based on Diophantine Equations. Our implementation of batch RSA reduces $b - 1$ modular inversion computation compared with Fiat's batch RSA and Boneh and Shacham's implementations. Furthermore, our implementation is more suitable for hardware circuits. If we apply this algorithm to a hardware circuit for batch RSA decryption, our implementation requires only one circuit module to do modular operations for N, while each of the other two implementations requires b different modules to do modular operations, which is difficult to be written into chips in advance and varies for every decryption.

4.4.1   Implementation Based on Plus-Type Equations

We find that the solutions of the Diophantine equation

$$\sum_{1 \leq i \leq n} \frac{1}{x_i} + \prod_{1 \leq i \leq n} \frac{1}{x_i} = 1, \quad 2 \leq x_1 < x_2 < \cdots < x_n \tag{4.23}$$

can be used to implement batch RSA so that the plaintexts can be constructed directly and the whole decryption only uses modular N. We call Equation (4.23) a *Plus-Type Diophantine Equation* or *Plus-Type Equation*. For convenience, throughout this book,

Equation (4.23) with n unknown variables is denoted as Equation $(4.23)_n$.

Our new implementation of batch RSA contains three algorithms: KeyGen, Encrypt, and Decrypt. The essence of the new implementation is in the decryption algorithm.

- KeyGen: This algorithm follows Fiat's batch RSA key generation algorithm except choosing the public keys $e_1, e_2, \ldots, e_b$ from solutions of Equation $(4.23)_n$, that is, $\{e_1, e_2, \ldots, e_b\} \subseteq \{x_1, x_2, \ldots, x_n\}$.

- Encrypt: This algorithm is same as in Fiat's batch RSA.

- Decrypt: This algorithm consists of three steps:

  - "Merge": The algorithm evaluates $A_0 = \prod_{1 \leq i \leq b} c_i^{e/e_i} \mod N$. This computation can be finished by Fiat's optimized algorithm [111].

  - "One decrypt": The algorithm evaluates one full-scale modular exponentiation $A \equiv A_0^d \mod N$.

  - "Split": The essence of our implementation is

  $$\alpha_i = \frac{\prod_{1 \leq j \leq n} x_j}{e_i}, \tag{4.24}$$

  where $1 \leq i \leq b$, and it outputs[1]

  $$m_i \equiv \frac{c_i^{(\alpha_i+1)/e_i} \cdot \prod_{j \neq i} c_j^{\alpha_i/e_j}}{A^{\alpha_i}} \pmod{N}. \tag{4.25}$$

Remark 4.4.1. In our decrypt algorithm, after the plaintext $(m_1, m_2, \ldots, m_{b-1})$ are decrypted, the final m_b can be calculated by

$$m_b \equiv \frac{A}{\prod_{j=1}^{b-1} m_j} \pmod{N}, \tag{4.26}$$

---

[1]Equation (4.25) is slightly different from Equation (4.22) since we choose the Plus-Type Equation.

where the computation can be reduced.

Remark 4.4.2. If the public keys $e_1, e_2, ..., e_b$ are chosen arbitrarily (not from the solution $(x_1, x_2, ..., x_n)$ of Equation $(4.23)_n$), Equation (4.25) also satisfies when α_i satisfies

$$\alpha_i \equiv \begin{cases} -1 \pmod{e_i}, \\ 0 \pmod{e_j}, \quad \text{for } j \neq i. \end{cases} \tag{4.27}$$

Theorem 4.4.3 (Correctness). *The above scheme is correct.*

Proof. From $A \equiv A_0^d \mod N$, we have

$$A \equiv \prod_{1 \leq i \leq b} c_i^{1/e_i} \pmod{N}.$$

Since $(x_1, \ldots, x_n)$ is a solution of Equation $(4.23)_n$, we have

$$\frac{\prod_{1 \leq k \leq n} x_k}{x_i} \equiv -1 \pmod{x_i}$$

and $\frac{\prod_{1 \leq k \leq n} x_k}{x_i} \equiv 0 \pmod{x_j}$ (for all i, j such that $1 \leq i \leq n, 1 \leq j \leq n, j \neq i$). Hence $e_i | \alpha_i + 1$ and $e_j | \alpha_i$ (for all i, j such that $1 \leq i \leq b, 1 \leq j \leq b, j \neq i$). In Equation (4.25), all the modular exponentiations are integer exponentiations. Therefore, it follows from a direct computation that $m_i \equiv c_i^{1/e_i} \pmod{N}$, where $1 \leq i \leq n$. $\square$

For a simple example, we choose a solution $(2, 3, 7)$ of Equation $(4.23)_3$ and take 3 and 7 as the public keys e_1 and e_2, respectively. Thus,

$$A = (c_1^7 \cdot c_2^3)^{1/21}, \tag{4.28}$$

$$\alpha_1 = \frac{\prod_{1 \leq j \leq 3} x_j}{e_1} = \frac{2 \cdot 3 \cdot 7}{3} = 14, \tag{4.29}$$

and

$$\alpha_2 = \frac{\prod_{1 \le j \le 3} x_j}{e_2} = \frac{2 \cdot 3 \cdot 7}{7} = 6. \tag{4.30}$$

Thus, we can get m_i as follows:

$$m_1 \equiv \frac{c_1^5 \cdot c_2^2}{A^{14}} \pmod{N}, \tag{4.31}$$

and

$$m_2 \equiv \frac{c_1^2 \cdot c_2}{A^6} \pmod{N}. \tag{4.32}$$

In fact, m_b can be obtained through Remark 4.4.1. As shown in the above example, if we compute m_2 first and then compute $m_1 = \frac{A}{m_2}$, we can finish the split (decryption) efficiently. Therefore, we give a new method to compute α_i without any modular operation and a new method to split the plaintext. And so, the answer to Problem 4.3.4 is "YES". As far as Problem 4.3.3 is concerned, for arbitrary positive integer b, we are able to find many α_i's. Furthermore, if b is not greater than 8, we are able to find all α_i's. We will introduce our work towards these problems in detail in Section 4.5.

Here, we would like to point out the differences among Fiat's RSA, Boneh and Shacham's Batch RSA and our implementation. The advantage of our implementation is that there is only one modular, $\mod N$, while both Fiat's batch RSA and Boneh and Shacham's implementation require at least b modulars.

Fiat's batch RSA: In KeyGen, $\{e_1, \ldots, e_b\}$ are randomly picked. In Decrypt, Fiat's tree-based method requires $2b + 1$ modulars. k is chosen so that $1 \le k < b$. Let $P_1 = \prod_{1 \le i \le k} e_i$, $P_2 = \prod_{k < i \le b} e_i$, $X_1 = P_1^{-1} \mod P_2$, $X_2 = \frac{P_1 X_1 - 1}{P_2}$. Then compute

$$\frac{A^{P_1 X_1}}{(\prod_{1 \le i \le k} c_i^{X_1/e_i})^{P_1} (\prod_{k < i \le b} c_i^{X_2/e_2})^{P_2}} \pmod{N} = \prod_{k < i \le b} m_i \pmod{N},$$

and

$$\frac{A}{\prod_{k<i\leq b} m_i} \pmod{N} = \prod_{1\leq i\leq k} m_i \pmod{N}.$$

This process is done recursively until $m_1, m_2, \ldots, m_b$ are calculated.

Boneh and Shacham's improvement: In Decrypt, α_i is obtained from Equations (4.22). Then $m_1, \ldots, m_{b-1}$ are evaluated by Equation (4.21). m_b is calculated by $m_b = \frac{A}{\prod_{1\leq j<b} m_j} \mod N$. The evaluation of $\alpha_1, \alpha_2, \ldots, \alpha_b$ requires b modulars.

Our implementation: $\alpha_1, \alpha_2, \ldots, \alpha_b$ are evaluated directly from the solution of Equation (4.23). Therefore, $b-1$ modulars are reduced. Furthermore, we get the plaintext m_i from Equation (4.21).

4.4.2   A Concrete Example Based on Plus-Type Equations

We give a concrete example to explain our algorithm to show its advantage.

KeyGen: Let the batch size $b = 7$. Randomly choose two 512-bit primes p and q:

$p =$ 10117548898104292384286149905619010120991278203895691114024386173114298

44196706850901843445909220802938862195711111823138805089269233175589739

8024486601967,

$q =$ 11942104782310447106884510125435655207765533763318483151065949254099976

07148379656268155866005262065749842699695077677106764786988072219334757

0651658382099.

Let $N = p \cdot q$.

$N = $12082482908131106460128885334610470733722100002825862245001246171456821

    80012763601040492827413417501769408868451698340800241737759849184798713

    84311467341679006113447482380675139998093264657381057047502767534877118

    98885364204757448057797883429393903920166667955622009797159159329896972

    336035744801291741098733.

We also randomly choose a solution $(x_1, x_2, ..., x_8)$ of Equation $(4.23)_8$ where

$$(x_1, x_2, ..., x_8) = (2, 5, 7, 11, 17, 157, 961, 4398619).$$

Choose the public keys as a subset of $\{x_1, x_2, ..., x_8\}$, for example,

$$e_1 = 5, e_2 = 7, e_3 = 11, e_4 = 17, e_5 = 157, e_6 = 961, e_7 = 4398619.$$

We compute

$$x = \prod_{1 \leq i \leq 8} x_i = 8687184244716670.$$

Thus, the public key is $(N, b = 7, e_1, \ldots, e_7, x)$. We compute the private key $d \equiv (\prod_{1 \leq i \leq 7} e_i)^{-1} \pmod{\phi(N)}$. d is as follows:

$d = $51876628765295878339434317579819334450607160465137205859602663251881894

    76248150332865697383721278328392362980310483479730993621055438208436678

    79229731524489835689222861536030650222353925195560713659815910775865522

    82418304145301163771498370785315888078449179919532145099665908400332933

    9007006290015363771 65451.

Encrypt: In this algorithm, we randomly choose a set of plaintexts as follows:

$m_1 =$4221633548392243042282864630522663863695555118703675833821894036905077171838777670539200939753400187558122765253611872592118414072653271581340495571506863788161897605670123187057436117651128827996382232338704024966061230507065541604612461999235606847271280370184105753033610196263742970851861312460607203291 3,

$m_2 =$1151533947837365379940910016280409200162737684752451945445689787972275572450525649782084136951971592156312409505014881084133776507758249194091017451835595363479534246194440683517296729380028180824767812284337195884903045805855691532524657250014881831387932259391677502088344734860780936543318050974074778512 12,

$m_3 =$4756419614981397000736388413173719088476313830179405877240364272308047312035300434644530039033892729163849746746554561004845842905132968056260466897769229999703867702384986954748583611402515591693169465909304907907485507778362215229654756766785989010660597936510888289677276864253387812967153352739183525981 3,

$m_4 =$4707611369554423725312206467113642205595821086075262302567832558108181584726955718832078797363240141336812574782622687307406041247418256240938544009923625721389868585352204679423469820466914319036593211022030949477538608235973279788444073132098629466136717240729350202564079397599873540789572436535089064620 50,

$m_5 =$3056804421400053909810480533457398075865795888037808223347639426486659611473267935711445240332623905770776659086150428617242444099500783232312046769108166931072005148517411753793343069196447770511541342922801797320718860211955245691146753456943490615147871871859543250963054245752591013194621357895629296339 6,

$m_6 =$ 41377747175811034232018876246074059499644813172580074927939214929882998

79343270705904474700195302890675803041195770969714258620664937410522441

78736605455167725304146829924000744594614850114240963709159146374669699

62099112312462790622604288246922576918641763721175549148057922321723822

247230264707138113937042,

and

$m_7 =$ 18771896161902354107492151420608960655175748847557766591096333115544603

88639115772459886403263920691549973938090162854541237354779139974823373

22150711283782344174258728142522160188013242155927394712955764457748723

37172809149112059845297173025509381745188281479782761341318653942719134

67100812393204054489176 7.

Since the ciphertext c_i is computed as follows $c_i \equiv m_i^{e_i} \pmod{N}$, where $1 \leq i \leq 7$. The ciphertexts are

$c_1 =$ 32162445244401743657170446510623926600937713705351525278004836278347467

78927818565931306900626346594961086912725141809256522635855273218400519

72284664046732942258138795733271680027669332436406173493987951365152414

60797459530585405018120324737308623849797695612736566101235396092740354

722520913907481538932914,

$c_2 =$ 11126467052899354989487810154732066615105697509420849586673823566312709

37106371291521967449468618459618164119164667029310985402655564125542964

94010110454544233209099018186584811824903310123116178433727720898855237

28984076826348484900130452142401788332892321945373838606441449511625255

94292269601993575417058 76,

$c_3 =$ 42702738253874326011706347202093584148640737529891335287560057328232353

12936031281751279979434019357164808476168319559847561796277935971011335

12500689235545118116809962307591329406164127984458850739739240726187291

08244500104462787099197514522083354004527717032842340463375898989280266

03903546435635766975589 4,

$c_4 =$ 12040562557445113732473911652223993486907469290295457468128810943319836
898652398813786228195028822201946321471521227160539917722201991567652 30
27510298197878920424163371390730684468025560664928791033866500650711867
64062267173825842140914417574375291482513842223759443700367430229129300
51810122901328794776191 17,

$c_5 =$ 666992272967985743433970599682280235635250170724122647764727400690596 8
10673854955408065485554029221240872924799602606228672828557689341973222
51563295263105644219421197195443848039478669501756964029791476750411047
04900146183224578433038704943628899192782802758784232421703243761756962
89463428899640082290844 9,

$c_6 =$ 358428109076265458629697920690411048239716849764148792509010863697928 21
45580427757242397757516407715510081504624842581515104846891405331533253
61167679156090524981249642173885098604795315049563957046069182482249951
05623301480564801725462585729948675579745223831303862071442959713127023
481367052128284782369023,

and

$c_7 =$ 471556714510312154098441247160426411655724947766101869428412309392811 62
67241186266669860841735961194282501443823932882005051795195976808037886
40329386789302925870414294037418371443238801520123232628229674559133870
46744561955317811032439098967200824031961905675407869670714450791407371
569282383073692093702050.

Decrypt: In this algorithm, we first merge these ciphertexts to evaluate A_0 by Fiat's method.

$A_0 =$5235986644401950336921913868924412987643513613186688702380709736255942485204585255108074606984294095293919326326201521808678582533044952634349945106581854769334225343690855354241718422562683767770434276839098284531602132751452520137852437744111403924748204787935806406418492128763349311895521745079888137070,

After that, we take one decryption which does a full-scale modular exponentiation $A \equiv A_0^d \pmod{N}$:

$A =$97020381281501165266847566088201413180464241138685997123851463155123151439767358689202218874711897855236400747921160125364634061010391848918619864291769257801862202109448200560836407848792167375605028317938474390697318487470321735651171108807814620809108832763769829503607711597977756914765803133937106665598.

Finally, from Equation (4.24), we split the plaintext using the following α_i.

$$\begin{array}{llll}
\alpha_1 = & 1737436848943334, & \alpha_2 = & 1241026320673810, \\
\alpha_3 = & 789744022246970, & \alpha_4 = & 511010837924510, \\
\alpha_5 = & 55332383724310, & \alpha_6 = & 9039733865470, \\
\alpha_7 = & 1974979930.
\end{array}$$

We get the plaintext from Equation (4.25).

$$m_1' \equiv \frac{c_1^{(\alpha_1+1)/e_1} \prod_{2 \le j \le 7} c_j^{\alpha_1/e_j}}{A^{\alpha_1}}$$

$$\equiv c_1^{347487369788667} \cdot c_2^{248205264134762} \cdot c_3^{157948804449394}$$

$$\cdot c_4^{102202167584902} \cdot c_5^{11066476744862} \cdot c_6^{1807946773094}$$

$$\cdot c_7^{394995986} \cdot A^{-1737436848943334} \quad (\text{mod } N)$$

We can find that $m_1' = m_1$, which shows the correctness of our implementation. Note that $m_2, m_3, \ldots, m_7$ can be computed in the same way from Equation (4.25) and Equation (4.26). In fact, we compute $m_2, m_3, \ldots, m_7$ first and then compute m_1 as Remark 4.4.1, we can finish the split (decryption) efficiently.

4.4.3 Implementation Based on Minus-Type Equations

We find that the solutions of the following Diophantine Equation

$$\sum_{1 \le i \le n} \frac{1}{x_i} - \prod_{1 \le i \le n} \frac{1}{x_i} = 1, \quad 2 \le x_1 < x_2 < \cdots < x_n \tag{4.33}$$

can also be used to implement batch RSA so that α_i, which satisfies Equation (4.22), can be constructed directly and the whole decryption only uses modular N. We call Equation (4.33) the *Minus-Type Diophantine Equation* or *Minus-Type Equation*. For convenience, throughout this book, Equation (4.33) with n unknown variables is denoted as Equation $(4.33)_n$.

Our new implementation of batch RSA also contains three algorithms: KeyGen, Encrypt, and Decrypt. The essence of the implementation is in the decryption algorithm. Algorithm KeyGen and Encrypt are the same as our implementation based on the Plus-Type Equations.

Decrypt: This algorithm takes a public key $(N, b, e_1, e_2, \ldots, e_b, x = \prod_{1 \le i \le n} x_i)$, a private key d, and ciphertexts $c_1, c_2, \ldots, c_b$ as input. It evaluates $A_0 \equiv \prod_{1 \le i \le b} c_i^{e/e_i}$

$(\text{mod } N)$ by the methods proposed by Fiat, and $A \equiv A_0^d \pmod{N}$. Obviously

$$A \equiv \prod_{1 \leq i \leq b} m_i \pmod{N}.$$

Then it evaluates

$$\alpha_i = \frac{x}{e_i}, \quad 1 \leq i \leq b,$$

and it outputs

$$m_i \equiv \frac{A^{\alpha_i}}{C_i^{(\alpha_i - 1)/e_i} \cdot \prod_{j \neq i} C_j^{\alpha_i/e_j}} \pmod{N}, \quad 1 \leq i \leq b. \tag{4.34}$$

Remark 4.4.4. In our decrypt algorithm, after the plaintext $(m_1, m_2, \ldots, m_{b-1})$ are decrypted, the final m_b can be calculated by

$$m_b = \frac{A}{\prod_{1 \leq j < b} m_j} \mod N$$

where computation can be reduced.

In our implementation, $b - 1$ modulars are reduced, compared with Fiat's batch RSA and Boneh and Shacham's improvement. Since $(x_1, x_2, \ldots, x_n)$ is a solution of Equation $(4.33)_n$ and $\{e_1, e_2, \ldots, e_b\} \subseteq \{x_1, x_2, \ldots, x_n\}$, $\alpha_1, \alpha_2, \ldots, \alpha_b$ are evaluated by $\alpha_i = \frac{\prod_{1 \leq j \leq n} x_j}{e_i}$ without modular inversion. Therefore, there is only one kind of modular operations, which is "mod N." If we make a hardware circuit for batch RSA decryption, then our implementation requires one circuit module doing modular operations, and other two implementations both require b modules doing modular operations. Our implementation is more suitable for hardware circuit.

Theorem 4.4.5 (Correctness). *The above scheme is correct.*

Proof. See the proof of Theorem 4.4.3. □

For a simple example, we choose a solution $(2, 3, 11, 13)$ of Equation $(4.33)_4$ and take 3,11, and 13 as the public keys e_1, e_2, and e_3, respectively.

First, we have

$$x = 2 \cdot 3 \cdot 11 \cdot 13 = 858,$$

and

$$e = e_1 \cdot e_2 \cdot e_3 = 429,$$

and

$$A \equiv (c_1^{429/3} \cdot c_2^{429/11} \cdot c_3^{429/13})^{1/429} \equiv (c_1^{143} \cdot c_2^{39} \cdot c_3^{33})^{1/429} \pmod{N}. \quad (4.35)$$

Second, we can get α_i:

$$\alpha_1 = \frac{x}{e_1} = \frac{858}{3} = 286,$$

$$\alpha_2 = \frac{x}{e_2} = \frac{858}{11} = 78,$$

and

$$\alpha_3 = \frac{x}{e_3} = \frac{858}{13} = 66.$$

Third, we can get m_i:

$$m_1 \equiv \frac{A^{286}}{c_1^{95} \cdot c_2^{26} \cdot c_3^{22}} \pmod{N}, \quad (4.36)$$

m_2 and m_3 can be obtained in the same way.

4.4.4   A Concrete Example Based on Minus-Type Equations

We explain the implementation by a concrete example.

KeyGen: Let the batch size $b = 4$. We choose $p = 107$ and $q = 83$. Let $N = p \cdot q = 8881$. From the solution $\{x_1 = 2, x_2 = 3, x_3 = 11, x_4 = 17, x_5 = 59\}$ of

Equation (4.33))$_5$, we choose the public key as follows: $e_1 = 3, e_2 = 11, e_3 = 17$, and $e_4 = 59$. Let $e = \prod_{1 \leq i \leq b} e_i = 33099$ and $x = \prod_{1 \leq i \leq 5} x_i = 66198$. The private key is $d \equiv e^{-1} \pmod{\phi(N)} \equiv 4083$.

Encrypt: Suppose the plaintexts are $m_1 = 1000, m_2 = 2000, m_3 = 3000$, and $m_4 = 4000$. We compute the ciphertexts $c_1 = 8281, c_2 = 1316, c_3 = 4824$, and $c_4 = 169$.

Decrypt: The algorithm evaluates the following values. We have $A_0 = 7441$ and take one decrypt operation $A = A_0{}^{4083} \mod 8881 = 4982$. We can obtain α_i as follows:

$$\alpha_1 = \frac{66198}{3} = 22066, \quad \alpha_2 = \frac{66198}{11} = 6018,$$

$$\alpha_3 = \frac{66198}{17} = 3894, \quad \alpha_4 = \frac{66198}{59} = 1122.$$

Then the algorithm outputs:

$$m_1' \equiv \frac{4982^{22066}}{8281^{(22066-1)/3} \times 1316^{22066/11} \times 4824^{22066/17} \times 169^{22066/59}} \pmod{8881} = 1000,$$

where it shows that $m_1' = m_1$. Note that m_2', m_3', and m_4' can be computed in the same way:

$$m_2' \equiv \frac{4982^{6018}}{8281^{6018/3} \times 1316^{(6018-1)/11} \times 4824^{6018/17} \times 169^{6018/59}} \pmod{8881} = 2000$$

$$m_3' \equiv \frac{4982^{3894}}{8281^{3894/3} \times 1316^{3894/11} \times 4824^{(3894-1)/17} \times 169^{3894/59}} \pmod{8881} = 3000$$

$$m_4' \equiv \frac{4982^{1122}}{8281^{1122/3} \times 1316^{1122/11} \times 4824^{1122/17} \times 169^{(1122-1)/59}} \pmod{8881} = 4000.$$

Obviously, $m_2' = m_2, m_3' = m_3$, and $m_4' = m_4$.

4.5   Solving the Diophantine Equations

4.5.1   Plus-Type Equations

Algorithms for Solving the Plus-Type Equations

Equation (4.23) is a special case of the problem of expressing 1 as the sum of distinct unit fractions. Equation (4.23) also provides solutions to Znám's problem: find all sequences $\{N_1, \ldots, N_k\}$ of integers ≥ 2 with the property that for each i, N_i properly divides $1+\prod_{1 \leq j \leq k, j \neq i} N_j$. In the past forty years, many mathematical researchers have worked on this equation [50, 51, 53, 57, 58, 74–76, 109, 169, 263, 264, 267, 268, 270]. In recent years, applications of Equation (4.23) are also found in the graph theory [52, 58], differential geometry [49, 51], and computation theory [103].

However, it is impossible to solve the Plus-Type Equation when $n \geq 8$ by hand. See Table 4.1 for the known possible value intervals for each variable ($n = 8$).

Table 4.1: The known possible values

x_i	possible values
x_1	2
x_2	[3, 5]
x_3	[7, 13]
x_4	[11, 71]
x_5	[17, 3559]
x_6	[67, 3.5×10^6]
x_7	[551, 1.0×10^{13}]
x_8	[8.6×10^4, 1.1×10^{26}]

Although the following results are known, Equation $(4.23)_8$ has not been solved completely until our work was published. Table 4.2 gives the number of solutions for $n \leq 8$ in previous works. We have presented an algorithm to solve Equation (4.23) by elliptic curve method. We found all of the 122 solutions for $n = 8$ in November, 2008. The computation was carried out on one personal computer with 2.3 GHz CPU

Table 4.2: The number of solutions for $n \leq 8$ for Equation $(4.23)_n$

Equations	Number of solutions	Year	Contributor
$n \leq 6$	15	1964	Ke and Sun [169]
$n = 7$	18	1978	Janak and Skula [158]
$n = 7$	26	1987	Cao, Liu and Zhang [75] [76]
$n = 7$	26	1988	Brenton and Hill [51]
$n = 8$	119	1998	Brenton and Vasiliu [53]

and lasted 21 days. We have provided the algorithm and all 122 solutions. Although we have not solved Equation (4.23) for all the solutions when $n = 9$, we have generated solutions based on known solutions [70]. We have found 411 solutions for $n = 9$, and 2318 solutions for $n = 10$. Since our method is universal, it could be used to find a number of solutions for $n > 10$.

Now, we introduce the algorithm to solve Equation (4.23). Firstly, we state some lemmas about Equation (4.23). Then we present the outline and the pseudo codes of the algorithm. Finally we propose a new theorem.

Lemma 4.5.1. *Suppose that* $(x_1, x_2, \ldots, x_n)$ *is a solution of Equation* $(4.23)_n$. *Then* $x_1, x_2, \ldots, x_n$ *are pairwise coprime.*

Proof. See [51]. $\square$

Lemma 4.5.2 (Sun and Cao [268]). *Let* $(x_1^{(j)}, x_2^{(j)}, \ldots, x_{n-1}^{(j)})$ *be the j-th solution of Equation* $(4.23)_{n-1}$, $1 \leq j \leq \Omega(n-1)$. *Put*

$$k_j(n) = \left(x_1^{(j)} \cdot x_2^{(j)} \cdot \cdots \cdot x_{n-1}^{(j)} \right)^2 + 1, 1 \leq j \leq \Omega(n-1).$$

Then

$$\Omega(n+1) \geq \Omega(n) + \sum_{j=1}^{\Omega(n-1)} \left(\frac{d(k_j(n))}{2} - 1 \right)$$

where $d(k_j(n))$ *denotes the number of different positive factors of* $k_j(n)$.

Proof. The proof consists of two steps:

Step 1: Let $(x_1, x_2, \ldots, x_n)$ be any solution of Equation $(4.23)_n$. Put $x_{n+1} = \prod_{1 \leq i \leq n} x_i + 1$. By a direct computation, $(x_1, x_2, \ldots, x_{n+1})$ is a solution of Equation $(4.23)_{n+1}$. Hence, $\Omega(n+1) \geq \Omega(n)$.

Step 2: Let $(x_1^{(j)}, x_2^{(j)}, \ldots, x_{n-1}^{(j)})$ be as above. Let k be one positive factor of $k_j(n)$, such that $1 < k < \sqrt{k_j(n)}$. Put

$$x_n^{(j)} = \prod_{1 \leq i \leq n-1} x_i + k, x_{n+1}^{(j)} = \prod_{1 \leq i \leq n-1} x_i + \frac{k_j(n)}{k}.$$

By a direct computation, it is proved that $(x_1^{(j)}, x_2^{(j)}, \ldots, x_{n+1}^{(j)})$ is a solution of Equation $(4.23)_{n+1}$. For each j, $1 \leq j \leq \Omega(n-1)$, the number of k satisfying $1 < k < \sqrt{k_j(n)}$ is $\frac{d(k_j(n))}{2} - 1$. Hence, we can generate $\sum_{j=1}^{\Omega(n-1)} \left(\frac{d(k_j(n))}{2} - 1 \right)$ solutions of Equation $(4.23)_{n+1}$. These solutions are all different from the solutions generated in Step 1, since $x_n^{(j)} \neq \prod_{1 \leq i \leq n-1} x_i + 1$. Therefore, $\Omega(n+1) \geq \Omega(n) + \sum_{j=1}^{\Omega(n-1)} \left(\frac{d(k_j(n))}{2} - 1 \right)$. □

Lemma 4.5.3. *Suppose* $2 \leq x_1 < \cdots < x_{n-2}$ *and* $x_1, x_2, \ldots, x_{n-2}$ *are pairwise coprime. Put*

$$A = \prod_{1 \leq i \leq n-2} x_i, B = A - \sum_{1 \leq i \leq n-2} \frac{A}{x_i}.$$

Let P *and* Q *be positive integers such that* $PQ = A^2 + B, P < Q$. *Put*

$$x_{n-1} = \frac{A+P}{B}, x_n = \frac{A+Q}{B}.$$

If $B|A + P$ *and* $B|A + Q$, $(x_1, x_2, \ldots, x_n)$ *is a solution to Equation* $(4.23)_n$.

Proof. This lemma is proved by a direct computation. □

Lemma 4.5.4. *Put* $x_1 = A$, *where* A *is an arbitrary positive integer. For* $1 \leq i < n$, *let* x_{i+1} *be the minimum positive integer such that* $x_{i+1} > x_i$ *and* x_{i+1} *is coprime to* all $x_1, x_2, \ldots, x_i$. If

$$\frac{1}{A} + \sum_{1 < i \leq n} \frac{1}{x_i} + \frac{1}{A} \prod_{1 < i \leq n} \frac{1}{x_i} < 1,$$

there is no solution of Equation $(4.23)_n$ with $x_1 = A$.

Proof. Suppose there is one solution of Equation $(4.23)_n$ with $x_1 = A$. For $1 \leq i < n$, x_{i+1} is the minimum positive integer such that $x_{i+1} > x_i$ and x_{i+1} is coprime to $x_1, x_2, \ldots, x_i$

For all solutions $(A, x_2^*, x_3^*, \ldots, x_n^*)$ of Equation $(4.23)_n$, we have

$$\sum_{1 \leq i \leq n} \frac{1}{x_i} + \prod_{1 \leq i \leq n} \frac{1}{x_i} \geq \sum_{1 \leq i \leq n} \frac{1}{x_i^*} + \prod_{1 \leq i \leq n} \frac{1}{x_i^*} = 1$$

It is a contradiction. □

Algorithm Outline–Direct Finding Algorithms

Let $(x_1, x_2, \ldots, x_n)$ be a solution of Equation $(4.23)_n$.

- First we calculate the lower bound of x_1, L_1, and the upper bound of x_1, U_1.

  Let x_1 be A, and x_{i+1} be the minimum positive integer such that $x_{i+1} > x_i$, and x_{i+1} is coprime to $x_1, x_2, \ldots, x_i$, $1 \leq i < n$. If $\sum_{1 \leq i \leq n} \frac{1}{x_i} + \prod_{1 \leq i \leq n} \frac{1}{x_i} \geq 1$, then $A \in [L_1, U_1]$ by Lemma 4.5.17.

  For example, for Equation $(4.23)_8$, we fix $x_1 = 2$ and then we get $x_2 = 3$, $x_3 = 5$, $x_4 = 7$, $x_5 = 11$, $x_6 = 13$, $x_7 = 17$, and $x_8 = 19$. From

  $$\frac{1}{2} + \frac{1}{3} + \frac{1}{5} + \frac{1}{7} + \frac{1}{11} + \frac{1}{13} + \frac{1}{17} + \frac{1}{19} + \frac{1}{2 \cdot 3 \cdot 5 \cdot 7 \cdot 11 \cdot 13 \cdot 17 \cdot 19} > 1$$

  we have $2 \in [L_1, U_1]$.

- Then for all values of x_1 over the range $[L_1, U_1]$, we calculate the lower bound of

x_2, L_2, and the upper bound of x_2, U_2. In this way we fix $(x_1, x_2, \ldots, x_i)$, then we calculate L_{i+1} and U_{i+1}, $1 \leq i < n - 1$.

– From

$$\frac{1}{x_1} + \cdots + \frac{1}{x_i} < \sum_{1 \leq j \leq n} \frac{1}{x_j} + \prod_{1 \leq j \leq n} \frac{1}{x_j} = 1,$$

we have

$$x_i > \frac{\prod_{1 \leq k \leq i-1} x_k}{\prod_{1 \leq k \leq i-1} x_k - \sum_{1 \leq k \leq i-1} \prod_{1 \leq j \leq i-1, j \neq k} x_j}.$$

Hence,

$$L_i = \max\left\{ x_{i-1} + 1, \left\lceil \frac{\prod_{1 \leq k \leq i-1} x_k}{\prod_{1 \leq k \leq i-1} x_k - \sum_{1 \leq k \leq i-1} \prod_{1 \leq j \leq i-1, j \neq k} x_j} \right\rceil \right\}.$$

– It follows from

$$\sum_{1 \leq k < i} \frac{1}{x_k} + (n - i + 1) \frac{1}{x_i} > \sum_{1 \leq i \leq n} \frac{1}{x_i} + \prod_{1 \leq i \leq n} \frac{1}{x_i} = 1$$

that

$$x_i < \frac{\left(\prod_{1 \leq k \leq i-1} x_k \right)(n - i + 1)}{\prod_{1 \leq k \leq i-1} x_k - \sum_{1 \leq k \leq i-1} \prod_{1 \leq j \leq i-1, j \neq k} x_j}.$$

Therefore,

$$U_i = \left\lfloor \frac{\left(\prod_{1 \leq k \leq i-1} x_k \right)(n - i + 1)}{\prod_{1 \leq k \leq i-1} x_k - \sum_{1 \leq k \leq i-1} \prod_{1 \leq j \leq i-1, j \neq k} x_j} \right\rfloor.$$

– When $(x_1, x_2, \ldots, x_{i-1})$ are fixed, put $x_i = U_i$ and let $x_j (i < j \leq n)$ be the minimum positive integer such that x_j is coprime to $x_1, \ldots, x_{j-1}$ and $x_j > x_{j-1}$. Then $\sum_{1 \leq i \leq n} \frac{1}{x_i} + \prod_{1 \leq i \leq n} \frac{1}{x_i} \geq 1$. We can further reduce U_i based on this fact.

- After all the lower/upper bounds of $(x_1, x_2, \ldots, x_{n-2})$ are calculated, for all values of x_i over the range $[L_i, U_i]$ $(1 \leq i \leq n-2)$, put $A = \prod_{1 \leq i \leq n-2} x_i$, $B = A - \sum_{1 \leq i \leq n-2} \frac{A}{x_i}$. Factorize $A^2 + B$. For all positive integers P and Q such that $PQ = A^2 + B$, $P < Q$, put $x_{n-1} = \frac{A+P}{B}$, $x_n = \frac{A+Q}{B}$. If both x_{n-1} and x_n are integers, $(x_1, x_2, \ldots, x_n)$ is a solution of Equation (4.23)$_n$.

There are many methods to factorize $A^2 + B$, such as general number field sieve, multiple polynomial quadratic sieve, elliptic curve method, and quantum algorithm. The general number field sieve [56, 88] is the most efficient algorithm known for factoring integers larger than 100 digits. Its complexity is $L_N[\frac{1}{3}, c]$, where $L_N[c_1, c_2] = \mathcal{O}(e^{(c_2+o(1))(\log N)^{c_1}(\log \log N)^{1-c_1}})$, and N is the integer to be factorized. The complexity of multiple polynomial quadratic sieve is $L_N[\frac{1}{2}, 1]$. The elliptic curve method [103] is the most suitable method for finding small factors. Its running time is dominated by the size of the smallest factor, while the complexity of the first two methods is both determined by the size of the number to be factored.

In our algorithm, integers are factorized by the elliptic curve method. Other methods also work well.

Pseudo Codes

We present the pseudo codes of algorithm for further explanation.

Algorithm 1 Search1

Procedure Search1$(depth)$

Input: $depth$

Output: All solutions of Equation (4.23)$_n$.

1:    **if** $depth = n$ **then**

2:      $x_n \leftarrow \dfrac{\prod_{1 \leq k \leq n-1} x_k + 1}{\prod_{1 \leq k \leq n-1} x_k - \sum_{1 \leq k \leq n-1} \prod_{1 \leq i \leq n-1, i \neq k} x_i}$

3:     **if** x_n is an integer **then**

4:         output $x_1, x_2, \ldots, x_n$

5:     **end if**

6: **end if**

7: **if** $depth \leq n$ **then**

8:     $L_{depth} \leftarrow \max \Big\{ x_{depth-1} + 1,$
$$\left\lceil \frac{\prod_{1 \leq k \leq depth-1} x_k}{\prod_{1 \leq k \leq depth-1} x_k - \sum_{1 \leq k \leq depth-1} \prod_{1 \leq j \leq depth-1, j \neq k} x_j} \right\rceil \Big\}$$

9:     $U_{depth} \leftarrow \left\lfloor \dfrac{(\prod_{1 \leq k \leq depth-1} x_k)(n-depth+1)}{\prod_{1 \leq k \leq depth-1} x_k - \sum_{1 \leq k \leq depth-1} \prod_{1 \leq j \leq depth-1, j \neq k} x_j} \right\rfloor$

10:    **call** ReduceU$(depth)$

11:    **for** all $x_{depth} \in [L_{depth}, U_{depth}]$ **do**

12:        **if** $(x_1, \ldots, x_{depth})$ are all relatively prime **then**

13:            **Call** Search1$(depth + 1)$

14:        **end if**

15:    **end for**

16: **end if**

17: **if** $depth = n - 1$ **then**

18:     $L_{depth} \leftarrow \max \Big\{ x_{depth-1} + 1,$
$$\left\lceil \frac{\prod_{1 \leq k \leq depth-1} x_k}{\prod_{1 \leq k \leq depth-1} x_k - \sum_{1 \leq k \leq depth-1} \prod_{1 \leq j \leq depth-1, j \neq k} x_j} \right\rceil \Big\}$$

19:     $U_{depth} \leftarrow \left\lfloor \dfrac{(\prod_{1 \leq k \leq depth-1} x_k)(n-depth+1)}{\prod_{1 \leq k \leq depth-1} x_k - \sum_{1 \leq k \leq depth-1} \prod_{1 \leq j \leq depth-1, j \neq k} x_j} \right\rfloor$

20:    **call** ReduceU$(depth)$

21:    **if** $U_{depth} - L_{depth} < 120000$ **then**

22:        **for** all $x_{depth} \in [L_{depth}, U_{depth}]$ **do**

23:            **if** $(x_1, \ldots, x_{depth})$ are all relatively prime **then**

24:                **call** Search1$(depth + 1)$

25:            **end if**

26:        **end for**

27:    **else**

28:         $A \leftarrow \prod_{1 \leq i \leq n-2} x_i$

29:         $B \leftarrow A - \sum_{1 \leq i \leq n-2} \frac{A}{x_i}$

30:         $C \leftarrow A^2 + B$

31:         **for** all positive factors P of C **do**

32:             $Q \leftarrow \frac{C}{P}$

33:             **if** $P < Q$ **then**

34:                 $x_{n-1} \leftarrow \frac{A+P}{B}$

35:                 $x_n \leftarrow \frac{A+Q}{B}$

36:                 **if** $x_{n-1} > x_{n-2}$ and both x_{n-1} and x_n are integers **then**

37:                     output $x_1, x_2, \ldots, x_n$

38:                 **end if**

39:             **end if**

40:         **end for**

41:     **end if**

42: **end if**

Function 1 $reduceU(depth)$

Input: $depth$

Output: Nothing.

Step 1. $x_{depth} \leftarrow U_{depth}$

Step 2. $x_{depth+1} \leftarrow$ the minimum value such that $(x_1, \ldots, x_{depth+1})$ are all
            relatively prime.

            $x_{depth+2} \leftarrow$ the minimum value such that $(x_1, \ldots, x_{depth+2})$ are all
            relatively prime.

            $\ldots$

            $x_n \leftarrow$ the minimum value such that $(x_1, \ldots, x_n)$ are all relatively prime.

Step 3. If $\sum_{1 \leq i \leq n} \frac{1}{x_i} + \prod_{1 \leq i \leq n} \frac{1}{x_i} < 1$, then

$$U_{depth} \leftarrow U_{depth} - 1$$

Goto Step 1.

End if

Remark 4.5.5. The complete solutions of Equation (4.23) can be found by running Algorithm Search1$(depth = 1)$.

Remark 4.5.6. Algorithm *search* can be further optimized. The computation amount of decomposing an integer is about 60,000 times as the computation amount of computing x_n based on $(x_1, \ldots, x_{n-1})$. Hence, we factorize $A^2 + B$ to get x_{n-1}, x_n only if $U_{n-1} - L_{n-1} > 120,000$.

Remark 4.5.7. We apply the computer program in [4] to factorization. We would like to thank D. Alpern again.

The computer program examines all possible tuple $(x_1, \ldots, x_6)$, and executes the factorization to find x_7 and x_8 for each tuple. The maximum value to be factorized is about 10^{26}, which can be factorized efficiently by elliptic curve method. Elliptic curve method is a randomized method. It may fail to factorize by treating a composite number as a prime. We make sure that this method is always successful when the computer continues searching for the solutions.

Theorem 4.5.8. *Equation* (4.23)$_8$ *has only 122 solutions. There is only one solution such that* $x_1, \ldots, x_8$ *are all primes:* $2, 3, 11, 23, 31, 47059, 2217342227, 1729101023519.$

All these 122 solutions can be found in Appendix A.

New Solutions Generated from Known Solutions

We propose a new method, which can generate solutions of Equation (4.23)$_{n+3}$ from solutions of Equation (4.23)$_n$ [68–70]. We call it *Gap-three extension* algorithm. See the following pseudo codes.

Algorithm 2 Gap-three Extension

Procedure GapThreeExtension

Input: $(x_1, x_2, \ldots, x_n)$, one solution of Equation (4.23)$_n$;

$\qquad$ LOW, lower bound of t; $HIGH$, upper bound of t.

Output: Solutions of Equation (4.23)$_{n+3}$

1: $\quad J \leftarrow \prod_{1 < i < n} x_i$

2: $\quad t \leftarrow 1$

3: $\quad$ **while** $t < LOW$ **do**

4: $\qquad$ $t \leftarrow t + 4$

5: $\quad$ **end while**

6: $\quad$ **while** $t + 4 \leq HIGH$ **do**

7: $\qquad$ **if** t is not coprime to any of $x_1, x_2, \ldots, x_n$ **then**

8: $\qquad\qquad$ **Continue**

9: $\qquad$ **end if**

10: $\qquad$ Factorize $J^2 + t$

11: $\qquad$ **for** each odd positive factor k of $J^2 + t$ **do**

12: $\qquad\qquad$ **if** $k \neq 1 \pmod 4$ or $k = 1$ or k is not coprime with any of

$\qquad\qquad\qquad$ $(x_1, x_2, \ldots, x_n)$ **then**

13: $\qquad\qquad\qquad$ **continue**

14: $\qquad\qquad$ **end if**

15: $\qquad\qquad$ $x_{n+1} \leftarrow J + k$

16: $\qquad\qquad$ $x_{n+2} \leftarrow J + \frac{J^2 + t}{k}$

17: $\qquad\qquad$ **if** $x_{n+1} \geq x_{n+2}$, or x_{n+1} and x_{n+2} are not coprime **then**

18: $\qquad\qquad\qquad$ **continue**

19: $\qquad\qquad$ **end if**

20: $\qquad\qquad$ **if** $t \nmid [(J^2 + J \cdot k)^2 + k]$ **then**

21:                      **continue**

22:                      **end if**

23:                      $x_{n+3} \leftarrow \frac{J \cdot x_{n+1} \cdot x_{n+2} + 1}{t}$

24:                      **if** $x_{n+2} >= x_{n+3}$, or x_{n+3} is not coprime with any of

                              $(x_1, x_2, \ldots, x_{n+2})$ **then**

25:                          **continue**

26:                      **end if**

27:                      **return** $(x_1, x_2, \ldots, x_{n+3})$

28:              **end for**

29:              $t \leftarrow t + 4$

30:      **end while**

The correctness of this method is based on the following lemma.

Lemma 4.5.9 (Cao [70]). *Let* $(x_1, x_2, \ldots, x_{n-2})$ *be a solution of Equation* $(4.23)_{n-2}$. *Put* $N = \prod_{1 \le i \le n-2} x_i$. *Then*

$$
\begin{aligned}
(x_1, x_2, \ldots, x_{n-2}, x_{n-1} = N + k, x_n &= N + \frac{N^2 + t}{k}, x_{n+1} \\
&= \frac{1}{t}[N(N+k)(N + \frac{N^2 + t}{k}) + 1])
\end{aligned}
$$

is a solution of Equation $(4.23)_{n+1}$, *where* x_{n-1}, x_n, x_{n+1} *are integers satisfying* $x_{n-1} < x_n < x_{n+1}$ *and* t, k *are positive integers. Moreover,* N, k, t *are pairwise coprime. If* $2|N$, $k \equiv t \equiv 1 \pmod 4$.

Proof. The proof consists of the following steps:

Step 1: By a direct computation, it is easy to check that $(x_1, x_2, \ldots, x_{n+1})$ is a solution of Equation $(4.23)_{n+1}$.

Step 2: Set $y = x_{n-1}$, $z = x_n$, $w = x_{n+1}$. Then from

$$
1 = \sum_{1 \le i \le n+1} \frac{1}{x_i} + \prod_{1 \le i \le n+1} \frac{1}{x_i},
$$

we have

$$1 = 1 - \frac{1}{N} + \frac{1}{y} + \frac{1}{z} + \frac{1}{w} + \frac{1}{N \cdot y \cdot z \cdot w}.$$

Therefore,

$$\frac{k}{N \cdot (N + k)} = \frac{1}{z} + \frac{1}{w} + \frac{1}{N \cdot (N + k) \cdot z \cdot w}.$$

Hence,

$$z \cdot w \cdot k = N \cdot (N + k) \cdot w + N \cdot (N + k) \cdot z + 1.$$

Let d denote $\gcd(N, k)$. Then $z \cdot w \cdot k \equiv 0 \pmod{d}$, and $N \cdot (N + k) \cdot w + N \cdot (N + k) \cdot z + 1 \equiv 1 \pmod{d}$. Therefore, $d | 1$. Hence, N and k are coprime.

Step 3: Let d' denote $\gcd(k, t)$. From $w \in \mathbb{Z}$, we have $t | [N \cdot (N+k)(N + \frac{N^2 + t}{k}) + 1]$. Hence, $d' | [N \cdot (N + k)(N + \frac{N^2 + t}{k}) + 1]$. From $d' | k$, we get $d' | [N^2(N + \frac{N^2 + t}{k}) + 1]$. Since $\frac{N^2 + t}{k} \in \mathbb{Z}$, we have $k | (N^2 + t)$. Therefore $d' | N^2$. It follows that $d' | 1$. Hence, k and t are coprime.

Step 4: Let d'' denote $\gcd(N, t)$. From

$$w = \frac{1}{t} \left[N \cdot (N + k)\left(N + \frac{N^2 + t}{k}\right) + 1 \right],$$

it follows that

$$w \cdot t = N \cdot (N + k)\left(N + \frac{N^2 + t}{k}\right) + 1.$$

Since $d'' | (wt)$ and $d'' | N$, we have $d'' | 1$. Therefore, N and t are coprime.

It follows from Steps 2, 3, and 4 that N, k, and t are pairwise coprime.

Step 5: Assume that $2 | N$. Since $t | [N \cdot (N + k) \cdot (N + \frac{N^2 + t}{k}) + 1]$ and $2 | N$, we have $2 \nmid t$. From $k | (N^2 + t)$, we have $2 \nmid k$. From $(k, t) = 1$, we get $(\frac{-t}{k}) = 1$. Then from $t | [N \cdot (N + k) \cdot (N + \frac{N^2 + t}{k}) + 1]$, we have $t | [N \cdot (N + k) \cdot (nk + N^2 + t) + k]$, i.e., $t | [N^2 \cdot (N + k)^2 + k]$. Therefore $(\frac{-k}{t}) = 1$. We have,

$$1 = \left(\frac{-t}{k}\right)\left(\frac{-k}{t}\right) = (-1)^{\frac{k-1}{2} + \frac{t-1}{2}} \left(\frac{t}{k}\right)\left(\frac{k}{t}\right) = (-1)^{\frac{k-1}{2} + \frac{t-1}{2} + \frac{k-1}{2} \cdot \frac{t-1}{2}}.$$

Thus,

$$\frac{k-1}{2} + \frac{t-1}{2} + \frac{k-1}{2} \cdot \frac{t-1}{2} \equiv 0 \pmod 2.$$

$$k \equiv t \equiv 1 \pmod 4.$$

$\square$

Now we propose the following new theorem.

Theorem 4.5.10. *Let $\Omega(k)$ denote the number of solutions of Equation $(4.23)_k$. We have*

$$\Omega(n+3) \geq \Omega(n+2) + A(n+1) + B(n) + \Gamma(n+2),$$

where $A(n+1) = \sum_{j=1}^{\Omega(n+1)} \left(\frac{d(k_j(n+2))}{2} - 1 \right)$, $d(k_j(n+2))$ is defined in Lemma 4.5.2, $B(n)$ denotes the number of solutions found by the Algorithm 2 based on n, restricted by $t > 1$ and $k > 1$, $\Gamma(n+2)$ denotes the number of solutions generated as follows: $(x_1, x_2, \ldots, x_{n+2})$ be a solution of Equation $(4.23)_{n+2}$ satisfying that $(x_1, x_2, \ldots, x_{n+1})$ is not a solution of Equation $(4.23)_{n+1}$, and $C = (\prod_{1 \leq i \leq n+1} x_i)^2 + 1$. Let k be a factor of C, such that $1 < k < \sqrt{C}$. Put

$$x'_{n+2} = \prod_{1 \leq i \leq n+1} x_i + k, \quad x'_{n+3} = \prod_{1 \leq i \leq n+1} x_i + \frac{C}{k}.$$

Then

$$(x_1, x_2, \ldots, x_{n+1}, x'_{n+2}, x'_{n+3})$$

is a solution of Equation $(4.23)_{n+3}$.

Proof. Let $S(k)$ be the set of all solutions of Equation $(4.23)_k$. In fact there are at least four different methods to generate solutions of Equation $(4.23)_{n+3}$ listed in Table 4.6. Methods 3 and 4 are from [70]. They are used firstly to generate solutions of Equation (4.23) when $n = 9, 10$ in Theorems 4.5.11 and 4.5.12.

Table 4.6: Four methods to generate solutions of Equation $(4.23)_{n+3}$

Methods	Number of solutions	Brief illustration
Method 1: Gap-one extension	$\Omega(n+2)$	$x_1, x_2, \ldots, x_{n+2},$ $x_{n+3} = (\prod_{1 \le i \le n+2} x_i) + 1.$ $(x_1, x_2, \ldots, x_{n+2}) \in S(n+2).$
Method 2: Gap-two extension	$A(n+1)$	$x_1, x_2, \ldots, x_{n+1}, x_{n+2} = J + k, x_{n+3} = J + \frac{J^2+1}{k}.$ $(x_1, x_2, \ldots, x_{n+1}) \in S(n+1).$ $J = \prod_{1 \le i \le n+1} x_i, \ k | (J^2 + 1), 1 < k < \sqrt{J^2 + 1}.$
Method 3: Gap-three extension	$B(n)$	$x_1, x_2, \ldots, x_n, x_{n+1} = N + k, x_{n+2} = N + \frac{N^2+t}{k},$ $x_{n+3} = \frac{1}{t}(\prod_{1 \le i \le n+2} x_i + 1), (x_1, x_2, \ldots, x_n) \in S(n).$ $N = \prod_{1 \le i \le n} x_i, k > 1, t > 1,$ $x_{n+1} \in \mathbb{Z}, x_{n+2} \in \mathbb{Z}, x_{n+3} \in \mathbb{Z}.$
Method 4: (Another gap-one extension)	$\Gamma(n+2)$	$x_1, x_2, \ldots, x_{n+1}, x'_{n+2} = J + k, x'_{n+3} = J + \frac{J^2+1}{k}.$ $(x_1, x_2, \ldots, x_{n+2}) \in S(n+2),$ $(x_1, x_2, \ldots, x_{n+1}) \notin S(n+1),$ $J = \prod_{1 \le i \le n+1} x_i, k | (J^2 + 1), 1 < k < \sqrt{J^2 + 1}.$

We prove that the solutions generated by the four methods are all different. The proof consists of the following three steps.

Step 1: It follows from Lemma 4.5.2 that the solutions generated by Method 2 are different from Method 1.

Step 2: The solutions generated by Method 3 are all different from Method 1. Because $t > 1$, $x_{n+3} < \prod_{1 \le i \le n+2} x_i + 1$ in Method 3, while $x_{n+3} = \prod_{1 \le i \le n+2} x_i + 1$ in Method 1.

The solutions generated by Method 3 are also different from Method 2. Since $k > 1$, $(x_1, x_2, \ldots, x_{n+1}) \notin S(n+1)$ in Method 3, while $(x_1, x_2, \ldots, x_{n+1}) \in S(n+1)$ in Method 2.

Step 3: The solutions generated by Method 4 are all different from Methods 2 and 3, because $(x_1, x_2, \ldots, x_{n+1}) \notin S(n+1)$, $(x_1, x_2, \ldots, x_n) \notin S(n)$ in Method 4, but $(x_1, x_2, \ldots, x_{n+1}) \in S(n+1)$ in Method 2, $(x_1, x_2, \ldots, x_n) \in S(n)$ in Method 3.

The solutions generated by Method 4 are also different from Method 1. Since $k > 1$, $x_{n+3} < \prod_{1 \le i \le n+2} x_i + 1$ in Method 4, while $x_{n+3} = \prod_{1 \le i \le n+2} x_i + 1$ in Method 1.

Therefore, we have $\Omega(n+3) \geq \Omega(n+2) + A(n+1) + B(n) + \Gamma(n+2)$.   □

The number of solutions for $n = 9, 10$

Theorem 4.5.11. *There are at least 411 solutions of Equation* (4.23)$_9$.

We found 411 solutions of Equation (4.23)$_9$. The number of solutions found by each method is listed in the following Table 4.7

Table 4.7: Solutions found in Equation (4.23)$_9$.

Solutions found at least	Method
122	Method 1, $\Omega(8)$
205	Method 2, $A(7)$
47	Method 3, $B(6)$
14	Method 4, $\Gamma(8)$
23	Direct finding

Because of space limitations, these solutions are listed in `http://tdt.sjtu.edu.cn/9plus.html` and solutions found by different methods are also listed separately there.

Theorem 4.5.12. *There are at least 2318 solutions of Equation* (4.23)$_{10}$.

We also found 2318 solutions of Equation (4.23)$_{10}$. The number of solutions found by each method is listed in Table 4.8

Table 4.8: Solutions found in Equation (4.23)$_{10}$.

Solutions found at least	Method
411	Method 1, $\Omega(9)$
1623	Method 2, $A(8)$
284	Method 3, $B(7)$

These solutions are listed in `http://tdt.sjtu.edu.cn/10plus.html`. Solutions found by different methods are also listed separately there.

4.5.2 Minus-Type Equations

Similar to the Plus-Type Equations, we propose a fast algorithm to solve Equation (4.33) in this section.

Algorithms for Solving the Minus-Type Equations

Traditional algorithms to solve Equation (4.33) have high computational complexity. Our algorithm is based on factorizing integers using the elliptic curve method. The elliptic curve method is a fast, sub-exponential running time algorithm. We found all 550 solutions for $n = 8$, and we list them in this section. But the computation amount for $n = 9$ is beyond our computational capacity. We cannot solve Equation (4.33) for $n = 9$ directly. Therefore, we propose new methods to generate solutions from known solutions of Equation (4.23) with less variables. We have found 1547 solutions for $n = 9$ and 18984 solutions for $n = 10$. Due to space limitations, these solutions are published at webpage `http://tdt.sjtu.edu.cn/9minus.html` and `http://tdt.sjtu.edu.cn/10minus.html` . These solutions can be used in the new batch RSA implementation.

Table 4.9: Research results about Equation (4.33)

Results	Researchers	Year
$n \geq 9 \to A(n+1) \geq \Omega(n) + \Omega(n-1) + 6$	Sun and Cao [266]	1985
$2 \nmid n \geq 9 \to A(n+1) \geq \Omega(n) + \Omega(n-1) + 10$	Sun and Cao [266]	1985
found the complete solutions for $n \leq 6$	Sun and Cao [265]	1986
$n \geq 9 \to A(n+1) \geq \Omega(n) + \Omega(n-1) + 10$	Sun and Cao [267]	1986
$2 \mid n \geq 12 \to A(n+1) \geq \Omega(n) + \Omega(n-1) + 14$	Sun and Cao [267]	1986
$n \geq 10 \to A(n+1) \geq \Omega(n) + \Omega(n-1) + 16$	Cao, Liu and Zhang [75]	1987
$2 \mid n \geq 12 \to A(n+1) \geq \Omega(n) + \Omega(n-1) + 18$	Cao, Liu and Zhang [75]	1987
$n \geq 10 \to A(n+1) \geq \Omega(n) + \Omega(n-1) + 34$	Cao [69]	1988
$2 \mid n \geq 12 \to A(n+1) \geq \Omega(n) + \Omega(n-1) + 46$	Cao [69]	1988

Equation (4.33) also helps to find Giuga numbers. A number K is a Giuga number if and only if $p|(\frac{K}{p}-1)$ for all prime divisors p of K, or $\sum_{p|K}\frac{1}{p}-\prod_{p|K}\frac{1}{p}\in\mathbb{N}$ [124]. Giuga numbers are named after Giuga [124], who formulated a conjecture on primality in 1950: N is prime if and only if $\sum_{k=1}^{N-1}k^{N-1}\equiv 1\pmod{N}$. In 1996, Borwein et al. [44] published 11 Giuga numbers. Then Hogan and Mangilin found the 12th Giuga number [58], and Girgensohn found another one [58] in 2000. Sloane presented 13 Giuga numbers [261]. We prove that there are only 12 Giuga numbers with less than eight distinct prime factors, through the complete solutions of Equation (4.33) for $n \leq 8$, which are listed in Appendix B and Appendix C. The basic idea is that the equation

$$\sum_{p|K}\frac{1}{p}-\prod_{p|K}\frac{1}{p}=1 \tag{4.37}$$

is a special case of Equation (4.33). $\prod_{1\leq i\leq n}x_i$ is a Giuga number if and only if $(x_1, x_2, \ldots, x_n)$ is a solution of Equation $(4.33)_n$ and $x_1, x_2, \ldots, x_n$ are all primes.

In the following section, we propose a fast algorithm to solve Equation (4.33) and a method to generate new solutions of Equation (4.33) from known solutions of Equation (4.23). We first state some useful lemmas.

Lemma 4.5.13. *If* $(x_1, x_2, \ldots, x_n)$ *is a solution of Equation* (4.33), *then* $x_1, x_2, \ldots, x_n$ *are pairwise coprime.*

Proof. For any $i, j, 1 \leq i < j \leq n$, put $d = \gcd(x_i, x_j)$. We have

$$\prod_{1\leq i\leq n}x_i[\sum_{1\leq i\leq n}\frac{1}{x_i}-\prod_{1\leq i\leq n}\frac{1}{x_i}]=\prod_{1\leq i\leq n}x_i, d|\prod_{1\leq i\leq n}x_i.$$

Therefore,

$$d|\prod_{1\leq i\leq n}x_i[\sum_{1\leq i\leq n}\frac{1}{x_i}-\prod_{1\leq i\leq n}\frac{1}{x_i}].$$

Hence, $d|1$, and so $d = 1$. This completes the proof. $\square$

Lemma 4.5.14. *If* $(x_1, x_2, \ldots, x_n)$ *is a solution of Equation (4.33), then*

$$\frac{\prod_{1 \le j \le n} x_j}{x_i} \equiv 1 \pmod{x_i}, 1 \le i \le n.$$

Proof. By a direct computation, it is easy to prove the lemma. □

Lemma 4.5.15 (Sun and Cao [266]). *Let* $(x_1^{(j)}, (x_2^{(j)}, \ldots, x_{n-1}^{(j)})$ *be the j-th solution of Equation* $(4.23)_{n-1}$, $1 \le j \le \Omega(n-1)$. *Put* $l_j(n) = (\prod_{1 \le i \le n-1} x_i^{(j)})^2 - 1$. *We have*

$$A(n+1) \ge \Omega(n) + \sum_{1 \le j < \Omega(n-1)} (\frac{d(l_j(n))}{2} - 1),$$

where $d(l_j(n))$ *denote the number of different positive factors of* $l_j(n)$.

Proof. The proof consists of two steps (Cao [70]) :

Step 1: Let k be a positive factor of $l_j(n)$ satisfying $1 < k < \sqrt{l_j(n)}$. Put

$$x_n = \prod_{1 \le i \le n-1} x_i^{(j)} + k, x_{n+1} = \prod_{1 \le i \le n-1} x_i^{(j)} + \frac{l_j(n)}{k}.$$

Direct computation shows that $(x_1, x_2, \ldots, x_{n+1})$ is a solution of Equation $(4.33)_{n+1}$. The number of k such that $1 < k < \sqrt{l_j(n)}$ is $\frac{d(l_j(n))}{2} - 1$. Therefore, we can construct $\sum_{1 \le j \le \Omega(n-1)} (\frac{d(l_j(n))}{2} - 1)$ solutions of Equation $(4.33)_{n+1}$, and all of the solutions satisfy $x_{n+1} \ne \prod_{1 \le i \le n} x_i - 1$.

Step 2: Let $(x_1, x_2, \ldots, x_n)$ be a solution of Equation $(4.23)_n$. Put $x_{n+1} = \prod_{1 \le i \le n} x_i - 1$. Then $(x_1, x_2, \ldots, x_{n+1})$ is a solution of Equation $(4.33)_{n+1}$. Since $x_{n+1} = \prod_{1 \le i \le n} x_i - 1$, this solution is different from any other solutions generated in step 1. Therefore, $A(n+1) \ge \Omega(n) + \sum_{1 \le j \le \Omega(n-1)} (\frac{d(l_j(n))}{2} - 1)$. □

Lemma 4.5.16. *For any solution* $(x_1, x_2, \ldots, x_n)$ *of Equation* $(4.23)_n$, *put*

$$j = \prod_{1 \le i \le n} x_i, x_{n+1} = j + k, x_{n+2} = j + \frac{j^2 + t}{k}, x_{n+3} = \frac{1}{t} [j(j+k)(j + \frac{j^2 + t}{k}) - 1],$$

*where k, t are positive integers. If $x_{n+1} < x_{n+2} < x_{n+3}$ and $x_{n+1}, x_{n+2}, x_{n+3} \in \mathbb{Z}$,
then $(x_1, x_2, \ldots, x_{n+3})$ is a solution of Equation (4.33)$_{n+3}$, and*

$$k|(j^2 + t), t|[j(j + k)(j + \frac{j^2 + t}{k}) - 1], \gcd(j, k) = \gcd(k, t) = \gcd(j, t) = 1.$$

*Moreover, $2|j$ implies (i) the Jacobi symbol $(\frac{k}{t}) = 1$; (ii) $k \equiv 1 \pmod 4$ or $k \equiv t \equiv 3$
(mod 4).*

Proof. Assume that $(x_1, x_2, \ldots, x_n)$ is a solution of Equation (4.23)$_n$. $j = \prod_{1 \le i \le n} x_i$,
$x_{n+1} = j + k$, $x_{n+2} = j + \frac{j^2 + t}{k}$, $x_{n+3} = \frac{1}{t}[j(j + k)(j + \frac{j^2 + t}{k}) - 1]$, where k, and t
are positive integers. Assume that $x_{n+1} < x_{n+2} < x_{n+3}$ and $x_{n+1}, x_{n+2}, x_{n+3} \in \mathbb{Z}$.
Direct computation shows that $(x_1, x_2, \ldots, x_{n+3})$ is a solution of Equation (4.33)$_{n+3}$.
Put $y = x_{n+1}, z = x_{n+2}, w = x_{n+3}$. We have

$$\sum_{1 \le i \le n+3} \frac{1}{x_i} - \prod_{1 \le i \le n+3} \frac{1}{x_i} = 1,$$

$$1 - \frac{1}{j} + \frac{1}{y} + \frac{1}{z} + \frac{1}{w} - \frac{1}{jyzw} = 1.$$

Therefore, $zwj + ywj + yzj - 1 = yzw$.

Let d denote $\gcd(j, k)$, we have $d|y$ since $y = j + k$. Therefore $d|yzw$, and so $d|1$.

Let d' denote $\gcd(k, t)$. Since $t|[j(j + k)(j + \frac{j^2 + t}{k}) - 1]$, we have

$$d'|[j(j + k)(j + \frac{j^2 + t}{k}) - 1].$$

From $d'|k$, we get
$$d'|[j^2(j + \frac{j^2 + t}{k}) - 1], \quad k|(j^2 + t).$$

Therefore $d'|j^2$. We have $d'|1$.

Let d'' denote $\gcd(j, t)$. From $wt = j(j + k)(j + \frac{j^2 + t}{k}) - 1$, and $d''|j$, $d''|wt$, it follows that $d'' = 1$.

Hence, $\gcd(j, k) = \gcd(k, t) = \gcd(j, t) = 1$.

Assume that $2|j$. Since $t|[j(j + k)(j + \frac{j^2+t}{k}) - 1]$, we have $2 \nmid t$. From $k|(j^2 + t)$ we have $2 \nmid k$. Then from $k|(j^2 + t)$ we have

$$(\frac{-t}{k}) = 1, \quad j(j + k)(j + \frac{j^2 + t}{k}) - 1 \equiv 0 \pmod{t}.$$

Hence $j^2(j + k)^2 \equiv k \pmod{t}$. We have

$$(\frac{k}{t}) = 1, \ 1 = (\frac{-t}{k})(\frac{k}{t}) = (-1)^{\frac{k-1}{2}}(\frac{t}{k})(\frac{k}{l}) = (-1)^{\frac{k-1}{2} + \frac{(k-1)(t-1)}{4}}.$$

Hence, $k \equiv 1 \pmod 4$ or $k \equiv t \equiv 3 \pmod 4$. $\qquad \square$

Lemma 4.5.17. *A is any positive integer. Put $x_1 = A$. For $1 < i \leq n$, let x_i be the minimum positive integer satisfying $x_i > x_{i-1}$ and x_i is coprime to all $x_1, x_2, \ldots, x_{i-1}$. If $\sum_{1 \leq i \leq n} \frac{1}{x_i} - \prod_{1 \leq i \leq n} \frac{1}{x_i} < 1$, there is no solution of Equation (4.33)$_n$ with $x_1 = A$.*

Proof. Suppose there is one solution of Equation (4.33)$_n$ with $x_1 = A$, which is $(A, x_2^*, \ldots, x_n^*)$. For $1 < i \leq n$, since x_i is the minimum positive integer satisfying $x_i > x_{i-1}$ and x_i is coprime to all $x_1, x_2, \ldots, x_{i-1}$. We have

$$\sum_{1 \leq i \leq n} \frac{1}{x_i} - \prod_{1 \leq i \leq n} \frac{1}{x_i} \geq \frac{1}{x_1^*} + \sum_{2 \leq i \leq n} \frac{1}{x_i} - \frac{1}{x_1^*} \prod_{2 \leq i \leq n} \frac{1}{x_i} \geq$$
$$\frac{1}{x_1^*} + \frac{1}{x_2^*} + \sum_{3 \leq i \leq n} \frac{1}{x_i} - \frac{1}{x_1^* x_2^*} \prod_{3 \leq i \leq n} \frac{1}{x_i} \geq \cdots \geq \sum_{1 \leq i \leq n} \frac{1}{x_i^*} - \prod_{1 \leq i \leq n} \frac{1}{x_i^*} = 1.$$

This completes the proof. $\qquad \square$

Lemma 4.5.18. *Suppose $2 \leq x_1 < \cdots < x_{n-2}$ and $x_1, x_2, \ldots, x_{n-2}$ are pairwise coprime. Put*

$$A = \prod_{1 \leq i \leq n-2} x_i, B = A - \sum_{1 \leq i \leq n-2} \frac{A}{x_i}.$$

Let P and Q be positive integers such that $PQ = A^2 - B, P < Q$. Put

$$x_{n-1} = \frac{A + P}{B}, x_n = \frac{A + Q}{B}.$$

If $x_{n-2} < x_{n-1}$, $x_{n-1} \in \mathbb{Z}$ and $x_n \in \mathbb{Z}$, then $(x_1, x_2, \ldots, x_n)$ is a solution of Equation (4.33)$_n$.

Proof. By a direct computation, it is easy to prove the lemma. $\square$

Algorithm Outline–Direct Finding Algorithms

Let $(x_1, x_2, \ldots, x_n)$ be a solution of Equation (4.33)$_n$.

- First, we calculate L_1 (the lower bound of x_1) and U_1 (the upper bound of x_1).

  Let x_1 be $\mathcal{A}$, and let x_{i+1} be the minimum positive integer such that $x_{i+1} > x_i$ and x_{i+1} are coprimes to any of $x_1, x_2, \ldots, x_i$, $1 \leq i < n$. If $\sum_{1 \leq i \leq n} \frac{1}{x_i} - \prod_{1 \leq i \leq n} \frac{1}{x_i} \geq 1$, then $\mathcal{A} \in [L_1, U_1]$ by Lemma 4.5.17.

  For example, for Equation (4.33)$_8$, we fix $\mathcal{A} = 2$, then we get $x_1 = 2$, $x_2 = 3$, $x_3 = 5$, $x_4 = 7$, $x_5 = 11$, $x_6 = 13$, $x_7 = 17$, and $x_8 = 19$.

  $$\frac{1}{2} + \frac{1}{3} + \frac{1}{5} + \frac{1}{7} + \frac{1}{11} + \frac{1}{13} + \frac{1}{17} + \frac{1}{19} - \frac{1}{2 \cdot 3 \cdot 5 \cdot 7 \cdot 11 \cdot 13 \cdot 17 \cdot 19} > 1.$$

  We fix $\mathcal{A} = 3$, then we get $x_1 = 3, x_2 = 4, x_3 = 5, x_4 = 7, x_5 = 11, x_6 = 13, x_7 = 17$, and $x_8 = 19$.

  $$\frac{1}{3} + \frac{1}{4} + \frac{1}{5} + \frac{1}{7} + \frac{1}{11} + \frac{1}{13} + \frac{1}{17} + \frac{1}{19} - \frac{1}{3 \cdot 4 \cdot 5 \cdot 7 \cdot 11 \cdot 13 \cdot 17 \cdot 19} > 1.$$

  We fix $\mathcal{A} = 4$, then we get $x_1 = 4, x_2 = 5, x_3 = 7, x_4 = 9, x_5 = 11, x_6 = $

$13, x_7 = 17$, and $x_8 = 19$.

$$\frac{1}{4} + \frac{1}{5} + \frac{1}{7} + \frac{1}{9} + \frac{1}{11} + \frac{1}{13} + \frac{1}{17} + \frac{1}{19} - \frac{1}{4 \cdot 5 \cdot 7 \cdot 9 \cdot 11 \cdot 13 \cdot 17 \cdot 19} < 1.$$

Hence, $L_1 = 2, U_1 = 3$.

- Then for all values of $x_1, x_2, \ldots, x_i, x_j \in [L_j, U_j], 1 \le j \le i$, we calculate L_{i+1} and $U_{i+1}, 1 \le i < n$ 1.

  – From

  $$\frac{1}{x_1} + \cdots + \frac{1}{x_i} < \sum_{1 \le j \le n} \frac{1}{x_j} - \prod_{1 \le j \le n} \frac{1}{x_j} = 1,$$

  we have

  $$x_i > \frac{\prod_{1 \le k \le i-1} x_k}{\prod_{1 \le k \le i-1} x_k - \sum_{1 \le k \le i-1} \prod_{1 \le j \le i-1, j \ne k} x_j}.$$

  Hence,

  $$L_i = \max \left\{ x_{i-1}+1, \left\lceil \frac{\prod_{1 \le k \le i-1} x_k}{\prod_{1 \le k \le i-1} x_k - \sum_{1 \le k \le i-1} \prod_{1 \le j \le i-1, j \ne k} x_j} \right\rceil \right\}.$$

  – It follows from

  $$\sum_{1 \le k < i} \frac{1}{x_k} + (n - i + 1)\frac{1}{x_i} > \sum_{1 \le i \le n} \frac{1}{x_i} - \prod_{1 \le i \le n} \frac{1}{x_i} = 1$$

  that

  $$x_i < \frac{\left(\prod_{1 \le k \le i-1} x_k \right)(n - i + 1)}{\prod_{1 \le k \le i-1} x_k - \sum_{1 \le k \le i-1} \prod_{1 \le j \le i-1, j \ne k} x_j}.$$

  Therefore,

  $$U_i = \left\lfloor \frac{\left(\prod_{1 \le k \le i-1} x_k \right)(n - i + 1)}{\prod_{1 \le k \le i-1} x_k - \sum_{1 \le k \le i-1} \prod_{1 \le j \le i-1, j \ne k} x_j} \right\rfloor.$$

– When $(x_1, x_2, \ldots, x_{i-1})$ are fixed, we set $x_i = U_i$ and we let $x_j (i < j \leq n)$ be the minimum positive integer such that x_j is coprime to any of $x_1, x_2, \ldots, x_{j-1}$ and $x_j > x_{j-1}$. Then

$$\sum_{1 \leq i \leq n} \frac{1}{x_i} - \prod_{1 \leq i \leq n} \frac{1}{x_i} \geq 1.$$

We can reduce U_i based on this fact.

- When $(x_1, x_2, \ldots, x_{n-2})$ are fixed, calculate x_{n-1}, x_n by Lemma 4.5.18. Put

$$A = \prod_{1 \leq i \leq n-2} x_i, \quad B = A - \sum_{1 \leq i \leq n-2} \frac{A}{x_i}.$$

Factorize $A^2 - B$. For all positive integers P and Q such that $PQ = A^2 - B, P < Q$, put

$$x_{n-1} = \frac{A + P}{B}, \quad x_n = \frac{A + Q}{B}.$$

If $x_{n-2} < x_{n-1}, x_{n-1} \in \mathbb{Z}$ and $x_n \in \mathbb{Z}$, then $(x_1, x_2, \ldots, x_n)$ is a solution of Equation $(4.33)_n$.

Pseudo Codes

The pseudo codes of the above algorithms are listed as follows:

Algorithm 3 Search2

Procedure Search2$(depth)$

Input: $depth$

Output: All solutions of Equation $(4.33)_n$

1:   **if** $depth = n$ **then**

2:       $x_n \leftarrow \dfrac{\prod_{1 \leq k \leq n-1} x_k - 1}{\prod_{1 \leq k \leq n-1} x_k - \sum_{1 \leq k \leq n-1} \prod_{1 \leq i \leq n-1, i \neq k} x_i}$

3:       **if** x_n is an integer **then**

4:        output $x_1, x_2, \ldots, x_n$

5:      **end if**

6: **else**

7:      **if** $depth < n - 1$ **then**

8:          $L_{depth} \leftarrow \max \Big\{ x_{depth-1} + 1,$
$$\left\lceil \frac{\prod_{1 \leq k \leq depth-1} x_k}{\prod_{1 \leq k \leq depth-1} x_k - \sum_{1 \leq k \leq depth-1} \prod_{1 \leq j \leq depth-1, j \neq k} x_j} \right\rceil \Big\}$$

9:          $U_{depth} \leftarrow \left\lfloor \frac{(\prod_{1 \leq k \leq depth-1} x_k)(n - depth + 1)}{\prod_{1 \leq k \leq depth-1} x_k - \sum_{1 \leq k \leq depth-1} \prod_{1 \leq j \leq depth-1, j \neq k} x_j} \right\rfloor$

10:        **call** ReduceUpperBound($depth$).

11:        **for** all $x_{depth} \in [L_{depth}, U_{depth}]$ **do**

12:            **if** $(x_1, \ldots, x_{depth})$ are all relatively prime **then**

13:                **call** Search2($depth + 1$)

14:            **end if**

15:        **end for**

16:      **else**

17:          **if** $depth = n - 1$ **then**

18:              $L_{depth} \leftarrow \max \Big\{ x_{depth-1} + 1,$
$$\left\lceil \frac{\prod_{1 \leq k \leq depth-1} x_k}{\prod_{1 \leq k \leq depth-1} x_k - \sum_{1 \leq k \leq depth-1} \prod_{1 \leq j \leq depth-1, j \neq k} x_j} \right\rceil \Big\}$$

19:              $U_{depth} \leftarrow \left\lfloor \frac{(\prod_{1 \leq k \leq depth-1} x_k)(n - depth + 1)}{\prod_{1 \leq k \leq depth-1} x_k - \sum_{1 \leq k \leq depth-1} \prod_{1 \leq j \leq depth-1, j \neq k} x_j} \right\rfloor$

20:            **call** ReduceUpperBound($depth$)

21:            **if** $U_{depth} - L_{depth} < 120000$ (see the remark)**, then**

22:                **for** all $x_{depth} \in [L_{depth}, U_{depth}]$ **do**

23:                    **if** $(x_1, \ldots, x_{depth})$ are all relatively prime **then**

24:                        **call** Search2($depth + 1$)

25:                    **end if**

26:                **end for**

27:            **else**

28:                $A \leftarrow \prod_{1 \leq i \leq n-2} x_i$

29:          $B \leftarrow A - \sum_{1 \leq i \leq n-2} \frac{A}{x_i}$

30:          $C \leftarrow A^2 - B$

31:              **for** all positive factors P of C (see the remark) **do**

32:                  $Q \leftarrow \frac{C}{P}$

33:                  **If** $P < Q$ **then**

34:                      $x_{n-1} \leftarrow \frac{A+P}{B}, x_n \leftarrow \frac{A+Q}{B}$

35:                          **if** $x_{n-1} > x_{n-2}$ and both x_{n-1} and x_n are integers

                            then

36:                              **return** $x_1, x_2, \ldots, x_n$

37:                          **end if**

38:                      **end if**

39:                  **end for**

40:              **end if**

41:          **end if**

42:      **end if**

43: **end if**

Algorithm 4 Reduce Upper Bound

Procedure: ReduceUpperBound($depth$)

Input: $depth$

Output: Nothing

1:    **while** TRUE **do**

2:        $x_{depth} \leftarrow U_{depth}$

3:        **for** $i = depth + 1$ to n **do**

4:            $x_i \leftarrow$ the minimum value such that $(x_1, x_2, \ldots, x_i)$ are all relatively

              prime

5:        **end for**

6:      **if** $\sum_{1 \leq i \leq n} \frac{1}{x_i} - \prod_{1 \leq i \leq n} \frac{1}{x_i} < 1$ **then**

7:          $U_{depth} \leftarrow U_{depth} - 1$

8:      **else**

9:          **return**

10:      **end if**

11:   **end while**

Remark 4.5.19. The complete solutions of Equation (4.33) are found by running Algorithm Search2($depth = 1$).

Remark 4.5.20. Algorithm *Search2* can be further optimized. The computation amount of factoring an integer is about 60,000 times as that of computing x_n from $(x_1, x_2, \ldots, x_{n-1})$. Hence, we factorize $A^2 - B$ to get x_{n-1}, x_n only if $U_{n-1} - L_{n-1} > 120,000$.

Remark 4.5.21. The integer factorization algorithm is based on Alpern's [4] contribution.

New Solutions Generated from Known Solutions

The new solutions of Minus-type Equation (4.33) can be extracted from the solutions of Plus-type Equation (4.23). The following four methods generate new solutions from known solutions of Equation (4.23).

- Method 1: Gap-one extension. For each solution $(x_1, x_2, \ldots, x_n)$ of Equation (4.23)$_n$, we set $x_{n+1} = \prod_{1 \leq i \leq n} x_i - 1$, then $(x_1, x_2, \ldots, x_{n+1})$ is a solution of Equation (4.33)$_{n+1}$.

- Method 2: Gap-two extension. For each solution $(x_1, x_2, \ldots, x_n)$ of Equation (4.23)$_n$, we set $J = \prod_{1 \leq i \leq n} x_i$, $C = J^2 - 1$. Then we factor C. For each positive factor k of C, we set

$$x_{n+1} = J + k, \quad x_{n+2} = J + \frac{C}{k}.$$

If $x_n < x_{n+1} < x_{n+2}$, then $(x_1, x_2, \ldots, x_{n+2})$ is a solution of Equation $(4.33)_{n+2}$.

- Method 3: Gap-three extension. For each solution $(x_1, x_2, \ldots, x_n)$ of Equation $(4.23)_n$, put $J = \prod_{1 \le i \le n} x_i$. For each odd integer $t > 0$, one factors $J^2 + t$. For each positive factor k of $J^2 + t$, put

$$x_{n+1} = J + k, \ x_{n+2} = J + \frac{J^2 + t}{k}, \ x_{n+3} = \frac{1}{t}[J \cdot x_{n+1} \cdot x_{n+2} - 1].$$

Then $(x_1, x_2, \ldots, x_{n+3})$ is probably a solution of Equation $(4.33)_{n+3}$.

- Method 4: Gap-two extension$^+$. Randomly choose $(x_1, x_2, \ldots, x_n)$, such that $x_1, x_2, \ldots, x_n$ are pairwise coprime, $2 \le x_1 < \cdots < x_n$. Let

$$A = \prod_{1 \le i \le n} x_i, \ B = A - \sum_{1 \le i \le n} \frac{A}{x_i}.$$

For all P, Q such that $1 \le P < Q$, $PQ = A^2 - B$, let

$$x_{n+1} = \frac{A + P}{B}, \ x_{n+2} = \frac{A + Q}{B}.$$

If $x_n < x_{n+1}$, $x_{n+1} \in \mathbb{Z}$ and $x_{n+2} \in \mathbb{Z}$, then $(x_1, x_2, \ldots, x_{n+2})$ is a solution of Equation $(4.33)_{n+2}$.

The number of solutions for $n = 8, 9, 10$

All the solutions of Minus-Type Equations for $n \le 7$ are listed in Appendix B. We also have the following results:

Theorem 4.5.22. *The number of the complete solutions of Equation* (4.33)$_8$ *is 550 (listed in Appendix C).*

Corollary 4.5.23. *There are only 12 Giuga numbers with no more than eight distinct prime factors.*

Proof. A number K is a Giuga number if and only if $\sum_{p|K} \frac{1}{p} - \prod_{p|K} \frac{1}{p} \in \mathbb{N}$ [124]. If n has no more than 8 distinct prime factors, then K is a Giuga number if and only if $\sum_{p|K} \frac{1}{p} - \prod_{p|K} \frac{1}{p} = 1$, since

$$\sum_{p|K} \frac{1}{p} - \prod_{p|K} \frac{1}{p} \le \frac{1}{2} + \frac{1}{3} + \frac{1}{5} + \frac{1}{7} + \frac{1}{11} + \frac{1}{13} + \frac{1}{17} + \frac{1}{19} < 2.$$

Each solution of Equation (4.33)$_n$ in which $x_1, x_2, \ldots, x_n$ are all primes corresponding to a Giuga number $\prod_{1 \le i \le n} x_i$. Therefore, this corollary can be verified by all solutions of Equation (4.33)$_n$ for $n \le 8$, which are listed in Appendices B and C. Giuga numbers with no more than eight distinct prime factors are:

$2 \times 3 \times 5$,

$2 \times 3 \times 7 \times 41$,

$2 \times 3 \times 11 \times 13$,

$2 \times 3 \times 11 \times 17 \times 59$,

$2 \times 3 \times 7 \times 43 \times 3041 \times 4447$,

$2 \times 3 \times 11 \times 23 \times 31 \times 47057$,

$2 \times 3 \times 7 \times 59 \times 163 \times 1381 \times 775807$,

$2 \times 3 \times 7 \times 71 \times 103 \times 61559 \times 29133437$,

$2 \times 3 \times 7 \times 71 \times 103 \times 67213 \times 713863$,

$2 \times 3 \times 7 \times 43 \times 1831 \times 138683 \times 2861051 \times 1456230512169437$,

$2 \times 3 \times 11 \times 23 \times 31 \times 47059 \times 2259696349 \times 110725121051$,

$2 \times 3 \times 11 \times 23 \times 31 \times 47137 \times 28282147 \times 3892535183$. $\qquad\square$

Theorem 4.5.24. *Equation* $(4.33)_9$ *has at least 1547 solutions. These solutions are available at* `http://tdt.sjtu.edu.cn/9minus.html`

Proof. The number of solutions found by each method are listed in Table 4.12.

Table 4.12: Numbers of solutions of Equation $(4.33)_9$ found

Numbers of solutions found	Method
122	Method 1
1342	Method 2
51	Method 4
32	Direct finding

$\square$

Theorem 4.5.25. *Equation* $(4.33)_{10}$ *has at least 18984 solutions. These solutions are available at* `http://tdt.sjtu.edu.cn/10minus.html`

Proof. From the known 411 solutions of Equation $(4.23)_9$, we can find 411 solutions of Equation $(4.33)_{10}$ by the Method 1 (Gap-one extension). The others are from the Method 2 (Gap-two extension). $\square$

4.6   Notes

In this chapter, we have introduced a new direction of modern cryptography—*Batch Cryptography*, which includes the batch encryption/decryption, batch key agreement, and batch signature/verification. We implemented batch RSA schemes based on the Diophantine Equations (Plus-Type Equations and Minus-Type Equations), and gave the methods to solve these Diophantine equations[2].

There are some interesting problems in batch cryptography which are still open. For instance,

---

[2] We have finished two manuscripts about efficiently implementing batch RSA and solving the Diophantine Equations (Plus-Type Equations and Minus-Type Equations) in 2008. The readers who are interested in this area can refer to [71, 72].

1. How to design aggregate signature and batch verification schemes without extra requirements, such that they are secure in the standard model?

   Researches in the batch cryptography are focusing on the aggregate signature and batch verification, which are secure in the random oracle model. In order to make them secure in the standard model, it is usual to introduce extra requirements. However, it remains open to construct a practical aggregation scheme in the standard model without extra requirements such as timing, interactive restrictions, or requiring each user to be able to prove knowledge of his/her secret key. It is also open to explore other relaxations of the full aggregation model.

2. How to design batch encryption schemes based on other assumptions such as quadratic residue, discrete logarithm, and pairing based assumptions?

   Almost all existing researches are focusing on the batch RSA. To avoid putting all eggs in one basket, it is better to design new feasible batch cryptographic algorithms based on quadratic residues or discrete logarithm problem. Besides, there is no formal security proof for the batch RSA. This is still an open problem.

3. How to design batch key exchange/agreement protocols?

   So far, the batch key exchange/agreement has attracted little attention. In the special case where there are many sends and only one receiver, each sender sends the ciphertext of a random number to the receiver under the receiver's public key. Then, the receiver is able to batch decrypt them and establish a secure channel with each sender. To design batch key exchange/agreement protocols becomes more complicated if the attackers exist and the property of entity authentication is required.

4. How to combine batch technique with other cryptographic algorithms?

   In this chapter, we only considered batch algorithms without com-
   bining them with other algorithms, such as proxy re-encryption/re-
   signature and attribute-based encryption/signature. It must be inter-
   esting to construct cryptographic algorithms using batch techniques
   to improve the efficiency as well as the security.

5. How to find all the solutions of the Plus-Type Equations and Minus-Type Equations with $n = 9, 10$, which can be used as the public keys in batch RSA? Furthermore, how to find the solutions to these equations when the right side of these equations is 2, not 1?

   We have given the new implementation of batch RSA. It requires only
   one modular, N. This implementation is more suitable to construct
   hardware circuits for batch RSA decryption. This implementation is
   based on the Diophantine Equations $\sum_{1 \leq i \leq N} \frac{1}{x_i} \pm \prod_{1 \leq i \leq N} \frac{1}{x_i} = 1$,
   $2 \leq x_1 < x_2 < \cdots < x_N$. All solutions to these equations for $N \leq 8$
   are known, but the complete solutions for $N \geq 9$ are unknown.

Chapter 5

Noncommutative Cryptography

5.1 Introduction

As a new direction in modern cryptography, noncommutative cryptography differs much from those discussed in previous chapters. In this chapter, we will discuss this subject through exploring the following questions.

Question 1: What is noncommutative cryptography?

An immediate reply: It is an abbreviation of the cryptography based on noncommutative algebraic structures, such as

- noncommutative groups or semigroups,

- noncommutative rings,

- even arbitrary sets over which some noncommutative operations could be well-defined.

Question 2: Why do we need noncommutative cryptography?

Before answering this question, let us pay attention to the following issues:

- In 1994, Shor [257] proposed an efficient quantum algorithm for solving integer factoring problem (IFP[1]) and the discrete logarithm problem (DLP). In 2003, Shor's algorithm was also extended to solve the discrete logarithm problem over elliptic curves (ECDLP) [229]. Shor's algorithm can be regarded as a special case of Kitaev's framework [172] for solving the so-called hidden subgroup (or subfield) problems (HSP). Now, we are aware of efficient quantum algorithms for HSP over arbitrary *commutative* groups. But there are evidences to suggest that HSP over *noncommutative* groups might be much harder: the progress in quantum algorithms for HSP over some noncommutative groups (e.g., symmetric groups) is very limited, even negative in some cases [234].

- In 2002, Stinson et al. [207] observed that most unbroken public-key cryptosystems used today were based on commutative algebraic structures, such as

  - $\mathbb{Z}_N$ (where N is a large composite integer), $\mathbb{Z}_p$ (where p is a large prime), and $\mathbb{F}_q$ (where $q = p^m$ is a power of some prime p and m is a positive integer), over which the well-known cryptosystems such as RSA encryption, ElGamal encryption, and Schnorr signature, are defined respectively;

  - $\mathbb{E}(\mathbb{F}_q)$ (i.e., the elliptic curves over the finite filed F_q) over which the elliptic curve cryptosytems and the pairing-based cryptosystems are defined;

  - $\mathbb{C}_K$ (i.e., the ideal class group over some algebraic (closed) field K) over which the (real or imaginary) quadratic field cryptosystems are defined.

---

[1] Shor's factoring algorithm uses two quantum registers. By introducing a more quantum register, we show that the measured numbers in the second and third quantum registers, which are of the same pre-measurement state, should be equal if Shor's complexity analysis is sound. This seems to contradict the Shor's argument that there are r possible observed values for the second register. Someone argues that the three quantum registers are entangled. If so, it is a peculiar entanglement which has not yet been mentioned. In our opinion, Shor's algorithm is not doubtless from a theoretical point of view. See Zhengjun Cao and Zhenfu Cao: On Shor's factoring algorithm with three quantum registers (under review).

For clarity, we would like to refer to these cryptosystems as *commutative cryptography*. The theoretical foundations for the above cryptosystems are based on the intractability of problems closer to number theory than group theory [207]. (A less rigid illustration is given in Figure 5.1.)

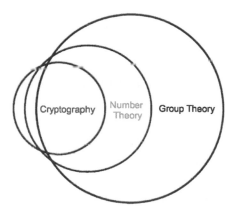

Figure 5.1: Cryptography, Number Theory, and Group Theory

From the above facts, we see that the classical commutative cryptography might be vulnerable to quantum computing. Moreover, cryptographers Goldreich and Lee [179] advised that *not to put all eggs in one basket*. Therefore, it seems applicable to introduce noncommutativity into cryptography.

Question 3: How about the state of arts of noncommutative cryptography?

The beginning of using noncommutative groups in cryptography goes back to Magyarik and Wagner [279] who proposed an approach to design public-key cryptosystems based on the undecidable word problem over groups and semigroups. This work has not attracted much attention until recently: In 2006, Birget et al. [28] pointed out that Magyarik and Wagner's scheme is actually not based on word problem, instead on premise problem, which is generally easier than the former. Now, when we look backward, it seems that Magarik and Wagner's pioneering idea comes *a bit earlier*: The glory of the

public-key cryptography community in 1980s was dominated by RSA, ElGamal, ECC

etc. However, there is no silver bullet in this world. In 1994, Shor's efficient quantum

algorithm for solving IFP and DLP emerged. Although practical quantum computer

might be at least decades away, its potential powerful capability has already casted

distrust in current cryptographic methods [179]. Since then many attempts have been

made for developing alternative public-key cryptosystems based on different platforms.

Therefore, noncommutative cryptography takes advantage of a turn of events and gets

on.

In 1999, Anshel et al. [6] proposed an elegant algebraic key establishment proto-

col. The foundation of their method lies in the difficulty of solving equations over

algebraic structures, in particular Garside groups [6]. In their pioneering paper, they

also suggested that braid groups might be a good alternative. Shortly afterward, Ko et

al. [175] published a fully fledged encryption scheme using braid groups. Since then

the subject has met with a quick success (see Section 5.2). However, from 2001 to

2003, repeated cryptanalytic successes also diminished the initial optimism on the sub-

ject [93]. It seems that more researches are still needed to reach a definite conclusion

on cryptographic potential of braid groups. Garber [117] made a comprehensive survey

of braid-based cryptography in 2007.

In 2001, Paeng et al. [224] proposed another new public-key cryptosystem built on

finite non-abelian groups. Paeng's method is based on the discrete logarithm problem

in the inner automorphism group defined by the conjugate action. Paeng's systems

were later improved and referred to as the so-called MOR system. In 2002, Magliveras

et al. [207] also developed new approaches to design public key cryptosystems using

one-way functions and trapdoors in finite groups. Their method originated from group

theory. They introduced two public-key cryptosystems, MST1 and MST2, based on the

difficulty of computing certain factorizations in finite groups. Later, Vasco et al. [276]

demonstrated that the factorization concept used in MST1 and MST2 admits a uniform description of several cryptographic primitives. Also, it turned out that a generalization of MST2 can serve as a unified framework for several public-key cryptosystems, including the ElGamal system [107], the braid group based system [175] and the MOR system.

In 2002, Grigoriev and Ponomarenko firstly constructed [131] homomorphic cryptosystems over non-abelian groups. Shortly afterward, based on the difficulty of the membership problem for groups of integer matrices, their method was extended to arbitrary finite groups [132, 133].

In 2004, inspired by Anshel et al.'s idea in the algebraic key exchange, Eick and Kahrobaei [106] proposed a new cryptosystem over *polycyclic groups*[2]. Further, Mahalanobis [208] in 2006 generalized the Diffie–Hellman key exchange protocol from a cyclic group to a finitely presented non-abelian *nilpotent group* of class 2.

In 2005, Shpilrain and Ushakov [259] suggested that Thompson's group might be a good platform for constructing public-key cryptosystems. In their contribution, the basic assumption is the intractability of the decomposition problem, which is more general than the conjugator search problem.

In 2006, Baumslag et al. [19] suggested potential cryptosystems using linear groups. In 2007, they [17] further suggested to use the classical modular group as a platform for cryptography. In fact, Yamamura [299] should be credit for introducing the extended modular group $SL_2(\mathbb{Z})$ into cryptography in 1998. But Yamamura's scheme was shown to have loopholes [135]. In [17] attacks based on these loopholes were closed [112].

---

[2]In group theory, a *normal series* is a series of normal subgroups of a group G,

$$G = H_1 \rhd H_2 \rhd \ldots \rhd H_{n+1} = \{1\}.$$

If each term of the series is normal not only in the whole group but also in the preceding term, then the series is called *subnormal*. A group G is called *polycyclic* if it has a subnormal series with cyclic factors, i.e., H_i/H_{i+1} is cyclic for $i = 1, \ldots, n$. A group G is called *nilpotent* if it has a normal series so that every quotient H_i/H_{i+1} lies in the centre of G/H_{i+1} (a so-called central series). The length of a shortest central series of a nilpotent group is called its class (or degree of nilpotency). Finitely-generated nilpotent groups are polycyclic groups and, moreover, have a central series with cyclic factors.

Also in 2006, Dehornoy [94] proposed an authentication scheme based on the left self-distributive (LD) systems. This idea was further developed by introducing the concept of the one-way LD system [287]. An algebraic system $(A, *)$ is called *left self-distributive* if for all $a, b, c \in A$,

$$a * (b * c) = (a * b) * (a * c). \tag{5.1}$$

An LD system is said one-way if it is intractable to extract a for giving $a * b$ and b. In general, an LD system is much different from (semi) groups. In fact, it is even not associative. However, one can easily define non-trivial LD systems over any noncommutative group G via the mapping

$$a * b \triangleq aba^{-1}. \tag{5.2}$$

Moreover, if the conjugator search problem (CSP) over G is intractable, then the derived LD system is one-way. Therefore, cryptosytems over one-way LD systems admit many kinds of implementations.

In 2006, we [73] proposed a method to use polynomials over noncommutative rings or (semi) groups to build cryptographic schemes. This method is now referred to as the $\mathbb{Z}$-modular method (See Section 5.3). In 2008, $\mathbb{Z}$-modular method was used to build signature schemes over noncommutative groups and division *semirings*[3] [231], respectively. Moreover, if we restrict the polynomials used in $\mathbb{Z}$-modular method to be monomials, then the conjugacy related assumptions can be viewed as special cases of the assumptions defined in $\mathbb{Z}$-modular method.

In 2009, Vats [277] proposed a cryptosystem, named as NNRU, and claimed that it is a noncommutative analogue of the well-known NTRU cryptosystem [143]. NNRU

---

[3]A semiring is a natural noncommutative generalization of a ring in the sense that in both the binary operations $+$ and $\cdot$ are not required to be commutative.

operates in the noncommutative ring $\mathbf{M} = M_k(\mathbb{Z})[X]/(X^n - I_{k \times k})$, where $\mathbf{M}$ is a matrix ring of $k \times k$ matrices of polynomials in $R = \mathbb{Z}[X]/(X^n - 1)$. The main enhancement is that the lattice-based attack which is the biggest threat to NTRU fail for NNRU.

In 2010, finite noncommutative groups of the four-dimension vectors over the ground field were constructed for implementing the cryptographic protocols [213].

In 2011, inspired by the recent success of applying the problems of learning parity with noise (LPN) [30] and learning with errors (LWE) [232] to a variety of cryptographic constructions, Baumslag et al. [18] proposed the generalized learning problem over noncommutative groups. Their work opens a new avenue for developing cryptography based on combinatorial group theory.

Of course, the above list is far from complete. Here, we just pick one or two typical examples in each year to illustrate that the research on noncommutative cryptography during the first decade of the new century is considerably attractive. We would like to point out that almost all initial attempts mentioned above have been shown to be insecure, but the experiences from these early attempts have a great significance. Readers who want to pay more attention to various attacks against these attempts and their security can refer to the recently published book *Group-based Cryptography (by Mayasnikov, Shpilrain, and Ushakov) [218].*

The rest of this chapter is organized as follows: In Section 5.2, we give a brief introduction on the well-known braid-based cryptography, including related fundamentals, cryptographic assumptions and related constructions, as well as our own contributions; In Section 5.3, we describe the $\mathbb{Z}$-modular method that is used to develop cryptographic schemes based on noncommutative rings or (semi) groups; In Section 5.4, by using monomials, instead of polynomials, we develop several cryptosystems based on intractability of the conjugator search problem and related assumptions; In Section 5.5, an improved key agreement protocol over Thompson's group is presented; And finally, further remarks are addressed in Section 5.6.

5.2 Braid-Based Cryptography

5.2.1 Basic Definitions

For index $n \geq 2$, the braid group B_n is defined by the generators $\sigma_1, \sigma_2, \ldots, \sigma_{n-1}$ and the relations $\sigma_i \sigma_j = \sigma_j \sigma_i$ for $|i - j| > 1$ and $\sigma_i \sigma_j \sigma_i = \sigma_j \sigma_i \sigma_j$ for $|i - j| = 1$ ($1 \leq i, j \leq n-1$). This definition is called the *Artin presentation* and the generators are called *Artin's generators*. Intuitively, a geometrical illustration of the identity, denoted by e, of the braid group B_4 and the Artin generators (e.g., $\sigma_2^{\pm 1}$) can be shown as Figure 5.2. Geometrically, the product of two braids is the braid obtained by merging the tail of the first braid with the head of the second braid. For example, Figure 5.3 shows the braid $\sigma_1 \sigma_2 \sigma_1^{-1} \sigma_3^{-1} \sigma_2 \sigma_3 \sigma_2 \sigma_3^{-1} \sigma_2^{-1} \sigma_1$. From the definition, we know that there is a natural automorphism from B_2 to the integer additive group $\mathbb{Z}$. For $n \geq 3$, the braid group B_n is infinite and *noncommutative*. For each $m(\leq n)$, the identity mapping on $\{\sigma_1, \ldots, \sigma_{m-1}\}$ naturally induces an embedding of B_m into B_n. By this approach, it is easy to define the limit group B_∞.

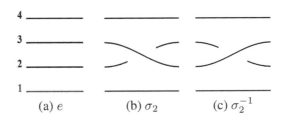

$$(a)\ e \qquad\qquad (b)\ \sigma_2 \qquad\qquad (c)\ \sigma_2^{-1}$$

Figure 5.2: Geometrical illustration on identity and Artin generators

The relations used in the above definition shine the lights of topology on braids, that is, continuously moving the curves with their ends fixed does not change the value of a braid. But algebraically, having a *normal form (i.e., a unique presentation)* for each braid is very useful, since it lets us compare two braids without resorting on topological

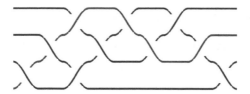

Figure 5.3: An example of geometric braids

images. For braid groups, there are some known normal forms, one of which, *the (left)*
canonical form, is commonly used in cryptography. With respect to the Artin presen-
tation, the complexity of transforming a braid into its canonical form is $\mathcal{O}(|w|^2 n \log n)$
where $|w|$ is the length of w, i.e., the number of the Artin generators in w [26].

5.2.2 Conjugacy and Related Problems

For arbitrary two braids $x, y \in B_n$, we say that they are *conjugate*, written as $x \sim$
y, if $y = a^{-1}xa$ for some $a \in B_n$. Here a or a^{-1} is called a *conjugator*. In the
braid group B_n, we can define the following cryptographic problems that are related to
conjugacy [173, 286]:

- Conjugacy deciding problem (CDP): Determine whether $x \sim y$ for a given instance $(x, y) \in B_n^2$.

- Conjugator searching problem (CSP): Find a braid $z \in B_n$ so that $y = z^{-1}xz$ for
  a given instance $(x, y) \in B_n^2$ with $x \sim y$.

- Matched conjugate searching problem (MCSP): Find a braid $y' \in B_n$ so that
  $y \sim y'$ and $xy \sim x'y'$ for a given instance $(x, x', y) \in B_n^3$ with $x \sim x'$.

- Matched triple searching problem (MTSP): Find a triple $(\alpha, \beta, \gamma) \in B_n^3$ so that
  $\alpha \sim x$, $\beta \sim y$, $\alpha\beta \sim xy$, $\gamma \sim y$ and $\alpha\gamma \sim x'y$ for a given instance $(x, x', y) \in$
  B_n^3 with $x \sim x'$.

- k-simultaneous conjugator searching problem (k-SCSP): Given k pairs (x_1, x_1'), $\cdots$, $(x_k, x_k') \in B_n^2$ with $x_i' = s^{-1} x_i s$ for some $s \in B_n$ and $1 \leq i \leq k$, find a braid $z \in B_n$ so that $x_i' = z^{-1} x_i z$ for $1 \leq i \leq k$.

- Decomposition problem (DP): Given $(x, y) \in B_n^2$ and $S_1, S_2 \subseteq B_n$, find $z_1 \in S_1$ and $z_2 \in S_2$ so that $y = z_1 x z_2$.

- Root problem (RP): Given a positive integer m and a braid $p \in B_n$ so that p is an mth power in B_n, find an mth root of p, i.e., find a braid r satisfying $r^m = p$.

By employing the terminologies used in [173], we say that a problem is

- *solvable*, if there is a deterministic finite algorithm that outputs an accurate solution.

- *unsolvable*, if it is not solvable.

- *tractable*, if there is a probabilistic polynomial-time algorithm that outputs an accurate solution with non-negligible probability (with respect to the length of description of the input instances).

- *intractable*, if it is not tractable.

At present, we know that all the above problems are solvable. However, we do not know whether they are tractable in general cases. It seems that all of them are intractable in the worst cases. According to [173, 286], we always take the following reduction relations into consideration:

$$CDP \prec MCSP \simeq MTSP \preceq CSP, \tag{5.3}$$

where $\prec$, $\preceq$, and $\simeq$ denote the relations of "easier than," "not harder than," and "as hard as," respectively.

In [173], Ko et al. also described an efficient algorithm to solve CDP with *over-whelming accuracy*[4]. This algorithm is the common basis for all existing braid-based signatures because the verification algorithms need to determine whether two given braids are conjugate.

Also, MCSP *seems* as hard as CSP according to the heuristic analysis in [173]. In our opinion, this is why MCSP is used to design signature schemes in [173].

At present, the relationship between k-SCSP and CSP is unclear. Although we have not proven rigorously which of them is easier, some approximate methods and heuristic analysis suggest that the former *seems* easier than the latter [286]. Therefore, if a cryptographic scheme lays its security on the assumption of the intractability of k-SCSP, we say that it is suffered from the weakness of k-simultaneous conjugacy.

The relation between CSP and DP is also unclear at present. On one hand, a solution of CSP is also a solution for DP *only if* $z^{-1} \in S_1$ and $z \in S$ hold simultaneously; On the other hand, a solution for DP becomes a solution for CSP *only if* $z_1 = z_2^{-1}$ holds.

According to [260], the only known algorithm for RP consists of explicitly enumerating several conjugacy classes related with the initial braid p. This enumerating process is exponential in essence and therefore becomes infeasible when the lengths of the braids are large enough. In practice, RP appears as even more difficult than CSP. Note that the braid groups are torsion-free, i.e., if b is a nontrivial braid, then b^m for every $m \geq 2$, is not trivial.

Although some algorithms for solving CSP were proposed [108,113,118,125], none of them has been proven implementable with a polynomial-time complexity (with respect to the braid index n). Until now, Gebhardt's algorithm [118], which was proposed in 2003 but formally published in 2005, is the most efficient method for solving CSP in braid groups. This algorithm has not yet been proven implementable within polynomial time. Subsequently, CSP in braid groups is classified for further study. According

---

[4]We say an algorithm has *overwhelming accuracy* if its outputs are accurate with the probabilities that are negligibly close to 1.

to Garber's report [117], within polynomial time, we can only solve CSP in periodic braids. There exist two kinds of challenges in obtaining ultimate solutions for CSP in braid groups: one is how to solve CSP in rigid braids within polynomial time, and the other is how to find polynomial boundaries of the complexity of Gebhardt's method.

Considering that the assumption of intractability of CSP plays a central role in braid-based cryptography, Ko et al. [173] introduced the concept of *CSP-hard pair*. Let S_1 and S_2 be two subgroups of B_n. A pair $(x, x') \in S_1 \times S_2$ is said *CSP-hard* if $x \sim x'$ and this CSP instance is intractable. It is more meaningful to define a special sampling algorithm, *CSP-hard pair generator*, rather than to define a particular CSP-hard pair. CSP-hard pair generator $\mathcal{K}_{csp}$ can be defined as follows:

- $\mathcal{K}_{csp}(n)$ is a probabilistic polynomial-time algorithm that takes the security parameter n as input, and outputs a triple $(p, q, w) \in B_n^3$ so that $q = w^{-1}pw$ holds and the CSP instance (p, q) is intractable.

For details regarding the methods to construct such a CSP-hard pair generator, please refer to [93], [173], [206] and [174]. In particular in [174], Ko et al. proposed several promising ways to generate hard instances of the conjugacy problems for braid cryptography.

5.2.3   Key Exchange, Encryption and Authentication

Now, let us have a quick review of the cryptographic protocols from braid groups. The first protocol was proposed theoretically by Anshel et al. in arbitrary Garside groups in 1999 [6], and implemented in braid groups in 2001 [5]. This protocol assumed that the CSP problem is difficult enough.

Suppose $\Lambda = \{l_1, \cdots, l_k\}$ be an alphabet. For a given word u on Λ, let $u(p_1, \ldots, p_k)$ denote the substitution of each l_i in u by p_i (for all $1 \leq i \leq k$). Suppose that Alice's secret key is a word u on an alphabet of size k, while Bob's secret key is a word v on a different alphabet of size m. The key exchange protocol works as follows:

Protocol 5.2.1. Anshel et al.'s Key Exchange:

- Public settings: $p_1, \ldots, p_k$ and $q_1, \ldots, q_m$ in B_n.

  Private keys: Alice: u; Bob: v.

- Exchange: Alice computes $s = u(p_1, \ldots, p_k)$, and sends Bob the conjugates $q'_1 = sq_1 s^{-1}, \ldots, q'_m = sq_m s^{-1}$; meanwhile, Bob computes $r = v(q_1, \ldots, q_m)$, and sends Alice the conjugates $p'_1 = rp_1 r^{-1}, \ldots, p'_k = rp_k r^{-1}$.

- Key deriving: Alice computes $K_A = H(su(p'_1, \ldots, p'_k)^{-1})$, and Bob computes $K_B = H(v(q'_1, \ldots, q'_m)r^{-1})$, where H is the colored Burau representation defined by Morton [215]. (Essentially, H is a key deriving function and can be instantiated by proper hashes.)

After the execution of this protocol, both Alice and Bob share the same key $K_A = K_B$, since the following equalities hold:

$$
\begin{aligned}
su(p'_1, \ldots, p'_k)^{-1} &= sru(p_1, \ldots, p_k)^{-1}r^{-1} \\
&= srs^{-1}r^{-1} \\
&= sv(q_1, \ldots, q_m)s^{-1}r^{-1} \\
&= v(q'_1, \ldots, q'_m)r^{-1}.
\end{aligned}
$$

The security of this protocol is based on the difficulty of a variant of CSP in B_n, namely k-SCSP (see the above subsection). In [5], it is suggested to work in B_{80} with $k = m = 20$ and short initial braids p_i, q_j with length 5 or 10.

The second protocol is proposed by Ko et al. [175]. Let LB_n (resp. UB_n), a subgroup of B_n, be generated by $\sigma_1, \ldots, \sigma_{m-1}$ (resp. $\sigma_{m+1}, \ldots, \sigma_{n-1}$) with $m = \lfloor \frac{n}{2} \rfloor$. Then, every braid in LB_n commutes with every braid in UB_n.

Protocol 5.2.2. Ko et al.'s Key Exchange:

- Public settings: a braid p in B_n.

  Private keys: Alice: $s \in LB_n$; Bob: $r \in UB_n$.

- Exchange: Alice sends Bob $p' = sps^{-1}$; meanwhile, Bob sends Alice $p'' = rpr^{-1}$.

- Key deriving: Alice computes $K_A = sp''s^{-1}$; Bob computes $K_B = rp'r^{-1}$.

After the execution of this protocol, both Alice and Bob share the same key $K_A = K_B = srpr^{-1}s^{-1}$, since s and r commute. The security of this protocol is based on the difficulty of the so-called Diffie–Hellman like conjugacy problem (DHCP). DHCP is to find the braid $rp'r^{-1}$, or equivalently, $sp''s^{-1}$ for given braids $p, p', p'' \in B_n$, where $p' = sps^{-1}$ and $p'' = rpr^{-1}$ for some $s \in LB_n$ and $r \in UB_n$. The suggested parameters are $n = 80$, and the working braids are specified by normal sequences of length 12.

Furthermore, in the same paper [175], Ko et al. gave a fully fledged public-key encryption scheme based on the above Diffie–Hellman like key exchange protocol. Suppose a hash function h, which maps a braid into a bit-string with proper length, is at hand, and Alice's public key is the pair $(p, p') \in B_n^2$ with $p' = sps^{-1}$, where $s \in LB_n$ is Alice's private key. For sending a message m to Alice, Bob chooses a random braid $r \in UB_n$ and sends the ciphertext $c = m \oplus h(rp'r^{-1})$, together with the additional datum $p'' = rpr^{-1}$. Now, Alice computes the plaintext $m = c \oplus h(sp''s^{-1})$. Apparently, this scheme is consistent and its security is also based on the difficulty of the DHCP problem. (Unfortunately, a deterministic polynomial time algorithm for solving the DHCP problem over braid groups was found by Cha et al. [82] in 2003. Thus, Ko et al.'s key exchange protocol, as well as the derived encryption scheme, is no longer secure.)

In 2006, Sibert et al. [260] designed three authentication schemes by using braid word reduction. The first scheme is based on DHCP, the second is based on CSP, and the third is based on CSP and RP. Considering that the first becomes insecure[5] according to the aforementioned Cha et al.'s polynomial-time algorithm, we merely introduce the second and the third herein.

Protocol 5.2.3. Sibert et al.'s Authentication Scheme II:

- Key generation:

  - A(lice) chooses a public braid $b \in B_n$, so that the CSP problem for b is hard;

  - A chooses a secret braid $s \in B_n$ as her private key; she computes $b' = sbs^{-1}$ and publishes the pair (b, b') as her public key.

- Authentication: Repeat the following exchanges k times, where k is a polynomial function of the size of the braid specified:

  - A chooses a random braid r, and sends $x = rbr^{-1}$ to B(ob);

  - B sends a random bit ϵ to A;

  - For $\epsilon = 0$, A sends $y = r$ to B, and B checks $x = yby^{-1}$;

  - For $\epsilon = 1$, A sends $y = rs^{-1}$ to B, and B checks $x = yb'y^{-1}$.

Protocol 5.2.4. Sibert et al.'s Authentication Scheme III:

- Key generation:

  - A(lice) chooses a secret braid $s \in B_n$ and computes $b = s^2$, so that both the CSP problem for s and b, and the RP problem for b are difficult; public key is b and private key is s.

- Authentication: Repeat the following exchanges k times:

---

[5] In fact, Sibert et al.'s work was originally proposed in 2002 — before Cha et al.'s work published in 2003.

- A chooses a random braid r, and sends $x = rbr^{-1}$ to B(ob);

- B sends a random bit ϵ to A;

- For $\epsilon = 0$, A sends $y = r$ to B, and B checks $x = yby^{-1}$;

- For $\epsilon = 1$, A sends $y = rsr^{-1}$ to B, and B checks $x = y^2$.

Sibert et al. proved that when the probability distribution of r picked by Alice at the authentication step is right-invariant, the above authentication schemes are zero-knowledge interactive proof of knowledge of s, under the corresponding intractability assumptions.

5.2.4   Braid-Based Signatures

In their pioneering paper on braid-based cryptography [175], Ko et al. pointed out that there are several challenges in finding a new digital signature scheme by using hard problems in braid groups. Two years later, Ko's team [173] reported two of the earliest braid-based signature schemes: The first scheme, denoted by SCSS, is based on the MCSP problem, and the second one, denoted by TCSS, is based on the MTSP problem. However, Ko et al. have not presented security proof on SCSS. As for TCSS, it is merely proved to be, in the random oracle model, secure against *no-message attacks*. The situation of lacking security proof for braid-based signature lasts about several years. In 2007, we [284], enlightened by the idea of "One-More-RSA-Inversion Problems" [22], defined the so-called one-more matching conjugate problem (OM-MCP) and proved that in the random oracle models both SCSS and TCSS are *existentially unforgeable against adaptively chosen message attacks (EUF-CMA)* assuming that the OM-MCP is intractable (See the next subsection).

Protocol 5.2.5. Simple Conjugate Signature Scheme (SCSS):

- System parameters: A noncommutative group G where CSP is intractable but CDP is tractable; A hash function h that maps a message m into an element in G.

- Key generation: A public key is a CSP-hard pair $(x, x') \in G^2$, and a secret key is $a \in G$ so that $x' = a^{-1}xa$.

- Signing: Given a message m, a signature s is given by $s = a^{-1}ya$ so that $y = h(m)$.

- Verifying: A signature s is valid if and only if $s \sim y$ and $x's \sim xy$.

Protocol 5.2.6. Triple Conjugate Signature Scheme (TCSS):

- System parameters: A noncommutative group G where CSP is intractable but CDP is tractable; A hash function h that maps a message m into an element in G.

- Key generation: A public key is a CSP-hard pair $(x, x') \in G^2$, and a secret key is $a \in G$ so that $x' = a^{-1}xa$.

- Signing: Given a message m, choose $b \in G$ at random and let $\alpha = b^{-1}xb$ and $y = h(m||\alpha)$, then a signature s is given by a triple $s = (\alpha, \beta, \gamma)$ where $\beta = b^{-1}yb$ and $\gamma = b^{-1}aya^{-1}b$.

- Verifying: A signature $s = (\alpha, \beta, \gamma)$ is valid if and only if $\alpha \sim x$, $\beta \sim \gamma \sim y$, $\alpha\beta \sim xy$ and $\alpha\gamma \sim x'y$.

5.2.5   One-More Like Assumptions and Provable Security

Suppose that N is the security parameter. A one-more matching conjugate problem attacker (om-mcp attacker for short) is a probabilistic polynomial-time algorithm $\mathcal{A}$ that gets input p, q and has access to two oracles: the matching conjugate oracle $\mathcal{O}_{mc}(\cdot)$ and the challenge oracle $\mathcal{O}_{ch}()$. We say that the attacker $\mathcal{A}$ *wins* the game if it succeeds in matching conjugate with all $n(N)$ braids output by the challenge oracle, but submits at most $m(N)$ queries to the matching conjugate oracle, where $m, n : \mathbb{N} \to \mathbb{N}$ are two polynomial functions defined over $\mathbb{N}$ so that $m(N) < n(N)$ holds. More formally, $\mathcal{A}$ is invoked in the following experiment.

Experiment $\mathbf{Exp}_{K_{csp},\mathcal{A}}^{om-mcp}(N)$

$(p, q, w) \xleftarrow{\$} K_{csp}(N); n \leftarrow 0; m \leftarrow 0$

$(r_1, \cdots, r_{n'}) \xleftarrow{\$} \mathcal{A}^{\mathcal{O}_{mc}, \mathcal{O}_{ch}}(p, q, N)$

If $n' = n$ and $m < n$ and $\forall\, i = 1, \cdots, n : (r_i \sim c_i) \wedge (qr_i \sim pc_i)$

Then return 1 else return 0

where the oracles are defined as

Oracle $\mathcal{O}_{mc}(b)$	Oracle $\mathcal{O}_{ch}()$
$m \leftarrow m + 1$	$n \leftarrow n + 1; c_n \xleftarrow{\$} B_N$
Return wbw^{-1}	Return c_n

The om-mcp advantage of $\mathcal{A}$, denoted by $\mathbf{Adv}_{K_{csp},\mathcal{A}}^{om-mcp}(N)$, is the probability that the above experiment returns 1, taking over the coins of K_{csp}, the coins of $\mathcal{A}$, and the coins used by the challenge oracle across its invocations. The *one-more matching conjugate assumption* says that the one-more matching conjugate problem associated with K_{csp} is hard, i.e., the function $\mathbf{Adv}_{K_{csp},\mathcal{A}}^{om-mcp}(N)$ is negligible with respect to the security parameter N for all probabilistic polynomial-time adversaries $\mathcal{A}$.

Now, let us consider the modes for sampling each c_i. If the adversary $\mathcal{A}$ can find some $k \in \mathbb{Z}$ and $j \in \{1, \cdots, m(N) + 1\}$ so that

$$c_j = p^k. \tag{5.4}$$

Then, $\mathcal{A}$ can set $r_j = q^k$. That is, the adversary can compute r_j correctly without querying the oracle $\mathcal{O}_{mc}$ and can win the game. Similarly, if the adversary $\mathcal{A}$ can find some $j \in \{1, \cdots, m(N) + 1\}$ and $\alpha_i, i = 1, \cdots, j - 1, j + 1, \cdots, m(N) + 1$, so that

$$c_j = \prod_{i=1, i \neq j}^{m(N)+1} c_i^{\alpha_i}, \tag{5.5}$$

then

$$r_j = \prod_{i=1, i \neq j}^{m(N)+1} r_i^{\alpha_i}. \tag{5.6}$$

Thus, the adversary can query $c_i, i = 1, \cdots, j-1, j+1, \cdots, m(N)+1$, one by one and then obtain the corresponding r_i. Now, he can also compute r_j correctly without querying the oracle again. But he can still win the game.

However, if each c_i is sampled randomly from B_N, the probability that the adversary can find k, j and $\alpha_i, i = 1, \cdots, j-1, j+1, \cdots, m(N)+1$, so that the equality (5.4) or the equality (5.5) holds is negligible with respect to the system parameter N. If $\mathcal{A}$ tries to deduce the equation (5.5) from right to left, he has to try and try, with little advantage over guessing at random. If he tries to deduce the equation (5.5) from left to right, he has to solve a root problem, which is intractable according to [93], for each $\alpha_i(> 0)$. If the adversary $\mathcal{A}$ tries to find k and j so that the equality (5.4) holds, he has to test each $p^k, k \in \mathbb{Z}$, by checking whether $p^k = c_j$ holds. However, since $\langle p \rangle$ is an infinite subgroup of B_N, the probability that the adversary can find such a pair (k, j) is also negligible with respect to N.

If the adversary is permitted to choose these c_i, he can select c_i so that the equality (5.4) or the equality (5.5) holds and this situation cannot be detected easily. Therefore, in our proposal the users, including the adversaries, are not allowed to choose these c_i by themselves. The users are allowed to perform queries on adaptively chosen messages and we employ a one-way hash function by which each message is mapped to a braid randomly sampled from B_N. The one-wayness of the hash function excludes the possibility for choosing c_i at the users' will.

Theorem 5.2.7. *In the random oracle model, the braid-based digital signature scheme SCSS is existentially unforgeable against adaptively chosen-message attack assuming that the one-more matching conjugate problem (OM-MCP) is intractable. More specifically, suppose that there is a forger $\mathcal{F}$ that can (t, q_h, q_s, ϵ)-break SCSS, then there* *exists an om-mcp attacker $\mathcal{A}$ that can win the one-more matching conjugate experiment with the probability at least ϵ' within the time t', where*

$$\epsilon' = \epsilon, \tag{5.7}$$

$$t' = t + t_s \cdot q_s + t_h \cdot q_h + t_{mc} \cdot (n(N) - q_s), \tag{5.8}$$

where $n : \mathbb{N} \to \mathbb{N}$ is a polynomial function defined over $\mathbb{N}$, while t_s, t_h, and t_{mc} are time for answering a signing oracle query, a hash oracle query and a matching conjugate oracle query, respectively.

Proof. That the forger $\mathcal{F}$ can (t, q_h, q_s, ϵ)-break SCSS means that $\mathcal{F}$ can output a forged signature (m^*, r^*) successfully with probability at least ϵ after he has made q_h hash queries and q_s signing queries, and then obtained the corresponding signatures $\sigma_i = (m_i, r_i), i = 1, \cdots, q_s$. The successful forgery (m^*, r^*) means that $r^* \sim H(m^*)$ and $qr^* \sim pH(m^*)$ hold while $\mathcal{F}$ has never made signing query on the message m^*. Now, let us construct another algorithm $\mathcal{A}$ so that $\mathcal{A}$ can win the experiment $\mathbf{Exp}_{K_{csp}, \mathcal{A}}^{om-mcp}(N)$ with the probability at least $\epsilon' = \epsilon$.

Without loss of generality, we assume that q_h and q_s are bounded by $n(N)$. The challenge the adversary $\mathcal{A}$ faced is, for given $c_i, i = 1, \cdots, n(N) + 1$, to output $r_i, i = 1, \cdots, n(N) + 1$, so that $r_i \sim c_i$ and $qr_i \sim pc_i$ hold, with the access to the matching conjugate oracle $\mathcal{O}_{mc}$ at most $n(N)$ times.

The algorithm $\mathcal{A}$ is constructed as follows:

- Initialize. We define a hash list H-List as follows: H-List includes three fields *m-field*, *r-field*, and *c-field*. At the beginning, let H-List contain $n(N) + 1$ items. For each item, the m-field and r-field are set to empty while the c-field is set to c_i accordingly. Then, set $i = 0$.

- Let $\mathcal{A}$ play the following interactive game with the forger $\mathcal{F}$.

(1) Answering $\mathcal{F}$'s hash queries and signing queries as follows:

- Hash query on m: Locate m in m-field of H-List. If found, i.e., $H(m)$ has been asked before, return the corresponding c-field as the reply; otherwise, set $i = i + 1$ and then fill m into the m-field of the ith item of the H-List, and then return the corresponding c-field as the reply.

- Signing query on m: Assume that $\mathcal{F}$ has asked $H(m)$ before (Otherwise, the algorithm $\mathcal{A}$ can make such a query on behalf of the forger $\mathcal{F}$). Then, there exists some $i \in \{1, \cdots, n(N)+1\}$ so that $c_i = H(m)$ holds. Now, let the algorithm $\mathcal{A}$ make a query on c_i toward the matching conjugate oracle $\mathcal{O}_{mc}$ and obtain the response r_i. Finally, let the algorithm $\mathcal{A}$ fill r_i into the r-field of the ith item of the H-List and forward r_i as the signing reply to the forger $\mathcal{F}$.

Clearly, the algorithm $\mathcal{A}$ provides perfect simulations on hash queries and signing queries for $\mathcal{F}$.

(2) Suppose that after q_s times signing queries, the forger $\mathcal{F}$ outputs a forged signature r^* on message m^*. Without loss of generality, suppose that $\mathcal{F}$ has made a hash query on m^* (if not, let $\mathcal{A}$ execute the query on behalf of $\mathcal{F}$). Then, there exists some $j \in \{1, \cdots, n(N)+1\}$ so that $c_j = H(m^*)$. Now, we set $r_j = r^*$.

(3) If $\mathcal{V}(m^*, r^*, pk) = 0$, then let $\mathcal{A}$ abort. Otherwise, let $\mathcal{A}$ continue.

(4) Locate the pair (m^*, r^*) in (m, r)-fields of H-List. If found, i.e., $\mathcal{F}$ has made signing query on m^* before, then let $\mathcal{A}$ abort; otherwise, let $\mathcal{A}$ fill $r_j = r^*$ into the r-field of the jth item of H-List and continue.

- For each $c_i, i = 1, \cdots, j-1, j+1, \cdots, n(N)+1$, if the r-field of the ith item of H-List is empty, let the algorithm $\mathcal{A}$ invoke matching conjugate query $\mathcal{O}_{mc}$ and fill the corresponding response r_i into the r-field of the ith item of H-List.

- Finally, let the algorithm $\mathcal{A}$ read each r_i, $i = 1, \cdots, n(N)+1$, from the r-field of H-List and output it as the ith response for the challenge c_i. Apparently, each r_i satisfies $(r_i \sim c_i)$ and $(qr_i \sim pc_i)$ while $\mathcal{A}$ has never made queries on c_j toward the matching conjugate oracle $\mathcal{O}_{mc}$. That is, the algorithm $\mathcal{A}$ needs to query the matching conjugate oracle $\mathcal{O}_{mc}$ exactly $n(N)$ times, which is strictly less than the times that $\mathcal{A}$ succeeds in matching conjugates with all $n(N)+1$ braids output by the challenge oracle. Therefore, $\mathcal{A}$ *wins* the experiment $\mathbf{Exp}_{K_{csp},\mathcal{A}}^{om-mcp}(N)$.

The above reduction shows that as soon as the forger $\mathcal{F}$ outputs a successful forgery (i.e., the algorithm $\mathcal{A}$ has not aborted before the end of its running), the algorithm $\mathcal{A}$ wins the experiment $\mathbf{Exp}_{K_{csp},\mathcal{A}}^{om-mcp}(N)$ with probability 1. That is, the probability for $\mathcal{A}$ to win the experiment is just the probability for $\mathcal{F}$ to output a successful forgery, so $\epsilon' = \epsilon$. According to the construction of $\mathcal{A}$, the total running time of $\mathcal{A}$ is $\mathcal{F}$'s running time plus the time for answering $\mathcal{F}$'s signing and hash queries and the time for $\mathcal{A}$ to make the remained matching conjugate queries. Thus, we have $t' = t + t_s \cdot q_s + t_h \cdot q_h + t_{mc} \cdot (n(N) - q_s)$. $\qquad\square$

Similarly, a provable EUF-CMA security reduction for TCSS can also be constructed by using a one-more like assumption and a random oracle model.

5.2.6 New Cryptographic Problems in Braid Groups

Finding new hard problems, as well as the variants, is always an interesting practice for (public-key) cryptography. In 2006, Dehornoy [94] suggested a new braid-based authentication scheme based on the shifted conjugate search problem. Let $x, y \in B_\infty$. The shifted conjugate operation is defined by

$$x * y = x \cdot \mathrm{d}y \cdot \sigma_1 \cdot \mathrm{d}x^{-1}, \tag{5.9}$$

where dx is the *shift* of x in B_∞. That is, the operator d can be viewed as an injective function on B_∞ which sends the generator σ_i to the generator σ_{i+1} for each $i \geq 1$.

Problem 5.2.8 (Shifted Conjugate Search Problem, Shifted-CSP). Let $s, p \in B_\infty$ and $p' = s * p$. Find a braid $\tilde{s}$ satisfying $p' = \tilde{s} * p$.

One cannot use the summit sets theory to attack Shifted-CSP, but one can apply the length-based attack [117] to try to solve Shifted-CSP. So it is interesting to study the following issues.

(1) How to evaluate the length-based attacks on Shifted-CSP?

(2) How to develop a theory for Shifted-CSP that will be parallel to the summit sets theory for CSP?

(3) How to design new cryptosystems based on shifted-CSP that will be secure against all currently known attacks?

A different type of problem consists in finding the shortest words representing a given braid. We consider this problem in B_∞ which is the group generated by an infinite sequences of generators $\{\sigma_1, \sigma_2, \dots\}$ subject to the usual braid relations.

Problem 5.2.9 (Shortest Word Problem, SWP). Given a word w that is represented by a sequence of $\sigma_i^{\pm 1}$ ($i = 1, 2, \cdots$), find the shortest word w' so that $w' \equiv w$.

Paterson and Razborov [226] proved that the SWP problem is co-NP-complete. This suggests us designing new cryptographic schemes in which the secret key is a short braid word, and the public key is another longer equivalent braid word. It must be noted that the co-NP-hardness of SWP holds in B_∞ only, but it is not known in B_n for fixed n. Polynomial-time algorithms for the SWP problem (with Artin presentation or Band presentation) in B_n for small fixed n were already reported [24, 25, 167, 297]. Also, an unpublished work indicates that a heuristic algorithm based on a random walk on the Cayley graph of the braid group might give good results in solving the SWP problem [117]. Thus, the following problems are interesting.

(4) How to design a cryptosystem based on the SWP in B_∞ by using the results from Paterson and Razborov?

(5) How hard is the SWP problem in B_n for a fixed and moderate scale n, say $n \geq 50$?

In short, to resist these attacks toward those schemes, one can try to change the distribution of the generators [112]. In this aspect, Maffre [206] proposed a new random algorithm for generating keys which are secure against the length-based attack from Hofheinz and Steinwandt [144].

5.3 $\mathbb{Z}$-Modular Method

In 2006, we [73] proposed a method for building public key cryptosystems by using noncommutative rings. Given a noncommutative ring R, our proposal can work over a derived $\mathbb{Z}$-modular structure $\mathbb{Z}[r]$, or a similar $\mathbb{Z}$-modular structure $\mathbb{Z}^+[r]$ (where $r \in R$ is undetermined), provided that the corresponding derived cryptographic assumptions are reasonable. Moreover, the $\mathbb{Z}$-modular method is extended to noncommutative groups and noncommutative semigroups. The method developed in our schemes is called $\mathbb{Z}$-modular method.

Afterwards, we use $\mathbb{Z}^+$ (resp. $\mathbb{Z}^-$) to denote the set of all positive (resp. negative) integers.

5.3.1 $\mathbb{Z}$-Modular Method over Noncommutative Rings

In this subsection, we give a very simple review on basic concepts about $\mathbb{Z}$-modular over noncommutative rings. Readers who know these concepts can go to the next subsection directly.

Suppose that R is a ring with $(R, +, \mathbf{0})$ and $(R, \cdot, \mathbf{1})$ as its additive abelian group and multiple non-abelian semigroup, respectively. Let us consider integral coefficient polynomials with ring assignment. The notion of scalar multiplication over R is already well-defined. For $k \in \mathbb{Z}^+$ and $r \in R$,

$$(k)r \triangleq \underbrace{r + \cdots + r}_{k \ times}. \tag{5.10}$$

When $k \in \mathbb{Z}^-$, we define

$$(k)r \triangleq (-k)(-r) = \underbrace{(-r) + \cdots + (-r)}_{-k \ times}. \tag{5.11}$$

For $k = 0$, it is natural to define $(k)r = \mathbf{0}$.

Proposition 5.3.1. $(a)r^m \cdot (b)r^n = (ab)r^{m+n} = (b)r^n \cdot (a)r^m, \forall a, b, m, n \in \mathbb{Z}$ and $\forall r \in R$.

Proof. According to the definition of scalar multiplication, the distributivity of multiplication with respect to addition, and commutativity of addition, this statement is concluded immediately. □

Remark 5.3.2. In general, $(a)r \cdot (b)s \neq (b)s \cdot (a)r$ when $r \neq s$, since multiplication in R is noncommutative.

Recall a polynomial with positive integral coefficient $f(x) = a_0 + a_1 x + \cdots + a_n x^n \in \mathbb{Z}^+[x]$. If we assign the undetermined element x as an element $r \in R$, then we obtain a well-defined element in R as shown below:

$$f(r) = \sum_{i=0}^{n} (a_i)r^i = (a_0)\mathbf{1} + (a_1)r + \cdots + (a_n)r^n. \tag{5.12}$$

Further, let r be undetermined, then $f(r)$ is univariable polynomial over R. The set of all (positive integral coefficient) univariable polynomials over R is denoted by $\mathbb{Z}^+[r]$.

Suppose that $f(r) = \sum_{i=0}^{n}(a_i)r^i \in \mathbb{Z}^+[r]$, $h(r) = \sum_{j=0}^{m}(b_j)r^j \in \mathbb{Z}^+[r]$ and $n \geq m$, then

$$\left(\sum_{i=0}^{n}(a_i)r^i\right) + \left(\sum_{j=0}^{m}(b_j)r^j\right) = \left(\sum_{i=0}^{m}(a_i + b_i)r^i\right) + \left(\sum_{i=m+1}^{n}(a_i)r^i\right), \quad (5.13)$$

and according to Property 5.3.1 as well as the distributivity, we have

$$\left(\sum_{i=0}^{n}(a_i)r^i\right) \cdot \left(\sum_{j=0}^{m}(b_j)r^j\right) = \sum_{i=0}^{n+m}(p_i)r^i, \quad (5.14)$$

where $p_i = \sum_{j=0}^{i} a_j b_{i-j} = \sum_{j+k=i} a_j b_k$. And then, we have the following theorem.

Theorem 5.3.3. $f(r) \cdot h(r) = h(r) \cdot f(r), \forall f(r), h(r) \in \mathbb{Z}^+[r]$.

5.3.2   New Problems over Noncommutative Rings

We would like to introduce the following cryptographic problems over a noncommutative group G:

- **Symmetrical Decomposition Problem (SDP):** Given $(x, y) \in G \times G$ and $m, n \in \mathbb{Z}$, find $z \in G$ so that $y = z^m x z^n$.

- **Generalized Symmetrical Decomposition Problem (GSDP):** Given $(x, y) \in G \times G$, $S \subseteq G$ and $m, n \in \mathbb{Z}$, find $z \in S$ so that $y = z^m x z^n$.

Clearly, GSDP can be viewed as a constrained variation of SDP. In general, if the size of S is large enough and its membership information *does not* help one to extract z from $z^m x z^n$, we believe that GSDP is at least as hard as SDP.

In the above definition of GSDP, if the parameters m and n are fixed, we define a new function $\hat{e}_{m,n}$ by

$$\hat{e}_{m,n} : G \times S, \quad (x, z) \mapsto z^m x z^n. \tag{5.15}$$

Further, if we denote $z^m x z^n$ as a new form x^z, then the above function can be viewed as a newly introduced exponential operation on G with respect to its subset S [6]. Similarly, if $y = z^m x z^n$, then z can be viewed as the discrete logarithm of y with respect to the base x, i.e., z can be denoted by $\log_x y$.

Now, we can regard GSDP as the discrete logarithm (DL) problem over G. Based on this observation, we introduce a new generalized comutational Diffie–Hellman (GCDH) problem over G:

- **Generalized Computational Diffie–Hellman Problem (GCDH)**: Compute $x^{z_1 z_2}$ (or $x^{z_2 z_1}$) given x, x^{z_1}, $x^{z_2} \in G$, where G is a noncommutative group and $z_1, z_2 \in S \subseteq G$.

Note that if z_1 lies in the center of z_2, i.e., z_1 commutes z_2, $x^{z_1 z_2} = x^{z_2 z_1}$ holds. It is clear that if GSDP, i.e., DL problem over G is tractable, so is GCDH problem over G. But the converse is not true. At present, there is no clue to solve this kind of GCDH problem without extracting z_1 (or z_2) from x and x^{z_1} (or x^{z_2}).

The GCDH assumption over G says that GCDH problem over G is intractable, i.e., there is no probabilistic polynomial time algorithm which can solve GCDH problem over G with non-negligible advantage with respect to the parameters of the problem scale.

Likewise, we arrive at the concept of the GSD (also DL) and GCDH assumptions over a noncommutative semigroup G and $m, n \in \mathbb{Z}^-$.

---

[6] In sequel we omit the clause of "with respect to its subset S" for visual comfort, unless the set S is explicitly specified.

Now, suppose that $(R, +, \cdot)$ is a noncommutative ring. For any $a \in R$, we define a set $P_a \subseteq R$ by

$$P_a \triangleq \{f(a) : f(r) \in \mathbb{Z}^+[r]\}.$$

Then, let us consider the new versions of GSD and GCDH problems over $(R, \cdot)$ with respect to its subset P_a, and name them as polynomial symmetric decomposition (PSD) problem and polynomial Diffie–Hellman (PDG) problem respectively:

- **Polynomial Symmetrical Decomposition (PSD) Problem**: Given $(a, x, y) \in R^3$ and $m, n \in \mathbb{Z}$, find $z \in P_a$ so that $y = z^m x z^n$, where R is a noncommutative ring.

- **Polynomial Diffie–Hellman (PDH) Problem**: Compute $x^{z_1 z_2}$ (or $x^{z_2 z_1}$) for given a, x, x^{z_1}, and x^{z_2}, where $a, x \in R, z_1, z_2 \in P_a$, and R is a noncommutative ring.

Accordingly, the PSD (PDH, respectively) cryptographic assumption says that PSD (PDH, respectively) problem over the ring $(R, \cdot)$ is intractable, i.e., there does not exist probabilistic polynomial time algorithm which can solve PSD (PDH, respectively) problem over $(R, \cdot)$ with non-negligible advantage with respect to the parameters of the problem scale.

5.3.3 Diffie–Hellman-Like Key Agreement Protocol

Now, let us consider Diffie–Hellman-like key agreement protocol over a noncommutative ring R.

Protocol 5.3.4. Diffie–Hellman-Like Key Agreement Protocol over Noncommutative Ring:

(0) Alice sends two random small, positive integers (say, less than 10) $m, n \in \mathbb{Z}^+$ and two random elements $a, b \in R$ to Bob.

(1) Alice chooses a random polynomial $f(x) \in \mathbb{Z}^+[x]$ so that $f(a) \neq \mathbf{0}$ and then takes $f(a)$ as her private key.

(2) Bob chooses a random polynomial $h(x) \in \mathbb{Z}^+[x]$ so that $h(a) \neq \mathbf{0}$ and then takes $h(a)$ as his private key.

(3) Alice computes $r_A - f(a)^m \cdot b \cdot f(a)^n$ and sends r_A to Bob.

(4) Bob computes $r_B = h(a)^m \cdot b \cdot h(a)^n$ and sends r_B to Alice.

(5) Alice computes $K_A = f(a)^m \cdot r_B \cdot f(a)^n$ as the shared session key.

(6) Bob computes $K_B = h(a)^m \cdot r_A \cdot h(a)^n$ as the shared session key.

In practice, the steps (0), (1), and (3) can be finished simultaneously and require only one pass communication from Alice to Bob. After that, the steps (2) and (4) can be finished in one pass communication from Bob to Alice. Finally, Alice and Bob can execute the steps (5) and (6), respectively. See Figure 5.4 for more details.

Pass	Alice	Bob
	$m, n \xleftarrow{\$} \mathbb{Z}^+$	
	$a, b \xleftarrow{\$} R$	
	$f(x) \xleftarrow{\$} \mathbb{Z}^+[x]$	
1	$\xrightarrow{\quad m,n,a,b,f(a)^m b f(a)^n \quad}$	
		$h(x) \xleftarrow{\$} \mathbb{Z}^+[x]$
2	$\xleftarrow{\quad h(a)^m b h(a)^n \quad}$	
	$K_A = f(a)^m h(a)^m b h(a)^n f(a)^n \quad = \quad K_B = h(a)^m f(a)^m b f(a)^n h(a)^n$	

Figure 5.4: Diffie–Hellman-Like Key Agreement over Noncommutative Rings

Clearly, the above key agreement protocol can resist passive adversary under the PDH assumption over the noncommutative monoid $(R, \cdot)$. It is similar to the standard Diffie–Hellman protocol that the protocol depicted in Figure 5.4 cannot resist the man-in-the-middle (MIM) attack. But it is easy to improve it so that the improvement can resist the MIM attack. This is left for interested readers.

5.3.4 ElGamal-Like Encryption Scheme

By the above key agreement protocol, we have the following ElGamal-like encryption scheme.

Protocol 5.3.5. ElGamal-Like Encryption (Basic) Scheme over Noncommutative Ring:

- **Setup:** Suppose that SDP is intractable on the monoid $(R, \cdot)$, where $(R, +, \cdot)$ is a noncommutative ring. Pick two small positive integers $m, n \in \mathbb{Z}^+$. Let $H : R \to \mathcal{M}$ be a cryptographic hash function which maps R to the message space $\mathcal{M}$. Set the public parameters of the system as the tuple $\langle R, m, n, \mathcal{M}, H \rangle$.

- **Key generation:** Each user chooses two random elements $p, q \in R$ and a random polynomial $f(x) \in \mathbb{Z}^+[x]$ so that $f(p) \neq \mathbf{0}$ and then takes $f(p)$ as his private key, computes $y = f(p)^m \cdot q \cdot f(p)^n$ and publishes his public key $(p, q, y) \in R^3$.

- **Encryption:** Given a message $M \in \mathcal{M}$ and the receiver's key $(p, q, y) \in R^3$, the sender chooses a random polynomial $h(x) \in \mathbb{Z}^+[x]$ so that $h(p) \neq \mathbf{0}$ and computes

$$c = h(p)^m \cdot q \cdot h(p)^n, \ d = H(h(p)^m \cdot y \cdot h(p)^n) \oplus M.$$

  Output the ciphertext $(c, d) \in R \times \mathcal{M}$.

- **Decryption:** Upon receiving a ciphertext $(c, d) \in R \times \mathcal{M}$, the receiver, by using his private key $f(p)$, computes the plaintext

$$M = H(f(p)^m \cdot c \cdot f(p)^n) \oplus d$$

Now, we prove that the above basic encryption scheme is "all-or-nothing" secure. The proof is very similar to that of Theorem 8.3 [209].

Theorem 5.3.6. *For a plaintext message uniformly distributed in the plaintext message space, the above encryption is "all-or-nothing" secure against CPA under the PDH assumption over the noncommutative ring $(R, +, \cdot)$ provided that H is a random oracle.*

Proof. On one hand, if PDH problem is tractable for any given ciphertext pair (c, d) and the corresponding public key (p, q, y), it is easy to extract $k = q^{(\log_q c)(\log_q y)}$ from the triple (q, c, y) and then compute the plaintext $M = d \oplus H(k)$.

On the other hand, suppose that there exists an efficient adversary $\mathcal{A}$, with access to the random oracle H, who is able to break the above cryptosystem, i.e., given any public key $(p, q, y = f(p)^m q f(p)^n)$ and ciphertext (c, d), $\mathcal{A}$ outputs

$$M \leftarrow \mathcal{A}^H(p, q, y, c, d)$$

with a non-negligible advantage ϵ so that M satisfies

$$M = d \oplus H(y^{\log_q c}) = d \oplus H(q^{(\log_q y)(\log_q c)}),$$

i.e.,

$$M = d \oplus H(h^m y h^n) \text{ and } c = h^m q h^n$$

for some $h \in P_p$. Then, for an arbitrary PDH instance (a, x, x^{z_1}, x^{z_2}), we set (a, x, x^{z_1}) as public key and (x^{z_2}, d) as ciphertext pair for a random $d \in \mathcal{M}$. Then, with the advantage ϵ, $\mathcal{A}$ outputs

$$M \leftarrow \mathcal{A}^H(a, x, x^{z_1}, x^{z_2}, d)$$

so that

$$M = d \oplus H(x^{z_1 z_2}), \text{ i.e., } M = d \oplus H(z_2^m z_1^m x z_1^n z_2^n)$$

for some $z_2 \in P_a$. Recall that $z_1 \in P_a$, thus $z_2 z_1 = z_1 z_2$ by Theorem 5.3.3. Then,

$$M = d \oplus H(z_2^m z_1^m x z_1^n z_2^n) = d \oplus H(x^{z_1 z_2}) = d \oplus H(x^{z_2 z_1}).$$

Clearly, if the adversary $\mathcal{A}$'s advantage ϵ is non-negligible, then $\mathcal{A}$ must make corresponding H-query on $x^{z_1 z_2}$. Otherwise, without knowing the hash value $H(x^{z_1 z_2})$, $\mathcal{A}$'s advantage for computing correct M should be *negligible*, since H is modeled as a cryptographic hash.

With the random oracle assumption on H, we can setup a H-list which contains two fields (r_i, h_i) and is initialized with empty. Whenever the adversary $\mathcal{A}$ makes a H-query with input r, we examine whether there exists the pair (r, h) in H-list. If so, return h as the answer to $\mathcal{A}$; otherwise, randomly pick $h \in \mathcal{M}$, put the pair (r, h) into H-list and return h as the answer to $\mathcal{A}$. Clearly, the simulation on H is perfect. Finally, when $\mathcal{A}$ outputs M, we can retrieve the correct item $x^{z_1 z_2} = r_i$ by checking the equality $M = d \oplus h_i$. Thus, we can solve PDH problem with the non-negligible probability ϵ. This contradicts the PDH assumption. $\qquad \square$

The above theorem shows that the basic scheme is of the weakest security, i.e., the OW-CPA security. Although we can use a technique proposed by Fjisaki and Okamoto (at CRYPTO'99) [116] to convert the above basic scheme into a chosen ciphertext secure system in the random oracle, we would like to adopt another technique also proposed by Fjisaki and Okamoto (at PKC'99) [115] because the latter is more compact.

First, we need to prove that the above basic scheme is of IND-CPA security.

Theorem 5.3.7. *Let H be a random oracle from $\mathcal{R}$ to $\mathcal{M}$. Let $\mathcal{A}$ be an IND-CPA adversary that has advantage ϵ against the above basic scheme within t steps. Suppose $\mathcal{A}$ makes a total of $q_H > 0$ queries to H. Then there is an algorithm $\mathcal{B}$ that solves PDH problem over the noncommutative ring $\mathcal{R}$ with advantage at least ϵ' within t' steps,* *where*

$$\epsilon' = \frac{2\epsilon}{q_H}, \text{ and } t' = \mathcal{O}(t).$$

Proof. The algorithm $\mathcal{B}$ takes as input a 4-tuple (a, x, y_1, y_2) with $y_i = x^{z_i} = z_i^m x z_i^n$ for *unknown* $z_i \in P_a$, $i = 1, 2$, i.e., a PDH instance. Let $y = x^{z_1 z_2}$ denote the solution to the PDH instance.

- **Setup.** $\mathcal{B}$ sets the system parameters as $\langle R, m, n, \mathcal{M}, H \rangle$ and (a, x, y_1) as a public key.

- **H-queries.** $\mathcal{B}$ sets up a H-list which contains two fields (r_j, h_j) and is initialized with empty. Whenever the adversary $\mathcal{A}$ makes a H-query with input r, $\mathcal{B}$ examines whether there exists the pair (r, h) in H-list. If so, return h as the answer to $\mathcal{A}$; otherwise, randomly pick $h \in \mathcal{M}$, put the pair (r, h) into H-list and return h as the answer to $\mathcal{A}$. Clearly, the simulation on H is perfect.

- **Challenge.** When $\mathcal{A}$ outputs two messages M_0 and M_1, $\mathcal{B}$ picks randomly a string $d \in \mathcal{M}$ and sets C as the ciphertext pair (y_2, d). It then gives C to $\mathcal{A}$ as the challenge. Notice that, the plaintext corresponding to C is
  $$d \oplus H(x^{(\log_x y_1)(\log_x y_2)}) = d \oplus H(x^{z_1 z_2}) = d \oplus H(y).$$ (Recall that z_1, z_2, and y are all unknown to $\mathcal{A}$ and y is just the solution to the above PDH instance.)

- **Guess.** $\mathcal{A}$ outputs its guess $b' \in \{0, 1\}$. At this point, $\mathcal{B}$ picks a random tuple (r_j, h_j) from the H-list and outputs r_j as the solution to the given PDH instance.

It is easy to see that $\mathcal{A}$'s view in $\mathcal{B}$'s simulation is the same as that in a real attack, in other words, the simulation is perfect. So $\mathcal{A}$'s advantage in this simulation will be ϵ.

We let $\mathcal{H}$ be the event that y is queried to H oracle during $\mathcal{B}$'s simulation. Notice that $H(y)$ is independent of $\mathcal{A}$'s view. If $\mathcal{A}$ never queries y to the H oracle in the above simulation, the plaintext corresponding to C is also independent of $\mathcal{A}$'s view. Therefore, in the simulation we have $\Pr[b = b' | \neg \mathcal{H}] = 1/2$. We know that in the real attack (and also in the simulation) $| \Pr[b = b'] - 1/2 | \geq \epsilon$. We have

$$
\begin{aligned}
\Pr[b = b'] &= \Pr[b = b'|\neg\mathcal{H}]\Pr[\neg\mathcal{H}] + \Pr[b = b'|\mathcal{H}]\Pr[\mathcal{H}] \\
&\leq \Pr[b = b'|\neg\mathcal{H}]\Pr[\neg\mathcal{H}] + \Pr[\mathcal{H}] \\
&= \frac{1}{2}\Pr[\neg\mathcal{H}] + \Pr[\mathcal{H}] \\
&= \frac{1}{2} + \frac{1}{2}\Pr[\mathcal{H}],
\end{aligned}
$$

$$
\begin{aligned}
\Pr[b = b'] &\geq \Pr[b = b'|\neg\mathcal{H}]\Pr[\neg\mathcal{H}] \\
&= \frac{1}{2}\Pr[\neg\mathcal{H}] \\
&= \frac{1}{2}(1 - \Pr[\mathcal{H}]) \\
&= \frac{1}{2} - \frac{1}{2}\Pr[\mathcal{H}]).
\end{aligned}
$$

Hence, $| \Pr[b = b'] - 1/2 | \leq \frac{1}{2}\Pr[\mathcal{H}]$. Since $| \Pr[b = b'] - 1/2 | \geq \epsilon$, $\Pr[\mathcal{H}] \geq 2\epsilon$. Furthermore, by the definition of the event $\mathcal{H}$, we know that y is in some tuple on the H-list with probability at least 2ϵ. It follows that $\mathcal{B}$ outputs the correct answer to the above PDH instance with probability at least $2\epsilon/q_H$. $\square$

At PKC'99, Fujisaki and Okamoto [115] introduced a method to convert an IND-CPA encryption scheme into an IND-CCA2 scheme. For self-contained, we rephrase their main idea as follows.

Suppose that $\Pi := \{\mathcal{K}, \mathcal{E}, \mathcal{D}\}$ is an IND-CPA secure public-key encryption scheme with key generation algorithm $\mathcal{K}(1^k)$, encryption algorithm $\mathcal{E}_{pk}(m, s)$ and decryption algorithm $\mathcal{D}_{sk}(y)$, where pk and sk are a public key and the corresponding private key, m is a message with $k + k_0$ bits, s is a random string with l bits and y is a ciphertext.

The converted public-key encryption scheme $\bar{\Pi} := \{\bar{\mathcal{K}}, \bar{\mathcal{E}}, \bar{\mathcal{D}}\}$ is defined by

$$
\begin{aligned}
\bar{\mathcal{K}}(1^k) \ &:= \ \mathcal{K}(1^{k+k_0}), \\[2mm]
\bar{\mathcal{E}}_{pk}(x, r) \ &:= \ \mathcal{E}_{pk}(x \parallel r, H(x \parallel r)), \\[2mm]
\bar{\mathcal{D}}_{sk}(y) \ &:= \ \begin{cases} MSB_k(\mathcal{D}_{sk}(y)), & \text{if } y = \mathcal{E}_{pk}(\mathcal{D}_{sk}(y), H(\mathcal{D}_{sk}(y))) \\ \bot, & \text{otherwise} \end{cases}
\end{aligned}
$$

where $MSB_k(\cdot)$ returns the leading k bits of the input bit-string, H is a random function of $\{0,1\}^{k+k_0} \to \{0,1\}^l$, x is a message with k bits, r is a random string with k_0 bits and $\parallel$ denotes concatenation.

Theorem 5.3.8 (Fujisaki-Okamoto Theorem [115]). *Suppose that $\Pi(1^{k+k_0})$ is an IND-CPA secure scheme and $\bar{\Pi}$ is the converted scheme. If there exists a (t, q_H, q_D, ϵ)-breaker $\mathcal{A}$ for $\bar{\Pi}(1^k)$ in the sense of IND-CCA2 in the random oracle model, there exist constant c and a $(t', 0, 0, \epsilon')$-breaker $\mathcal{A}'$ for $\Pi(1^{k+k_0})$ where*

$$
\begin{aligned}
\epsilon' \ &= \ (\epsilon - q_H \cdot 2^{-(k_0-1)}) \cdot (1 - 2^{-l_0})^{q_D} \ \text{and} \\[2mm]
t' \ &= \ t + q_H \cdot (T_{\mathcal{E}}(k) + c \cdot k).
\end{aligned}
$$

Here, (t, q_H, q_D, ϵ)-breaker $\mathcal{A}$, means that $\mathcal{A}$ stops within t steps, succeeds with probability at least ϵ, makes at most q_H queries to random oracle H, and makes at most q_D queries to decryption oracle D_{sk}. $T_{\mathcal{E}}(k)$ denotes the computational time of the encryption algorithm $\mathcal{E}_{pk}(\cdot)$, and

$$
l_0 := \log_2 \left(\min_{x \in \{0,1\}^{k+k_0}} \{ \#\{E_{pk}(x, r) | r \in \{0,1\}^l \} \} \right).
$$

Proof. See Theorem 3 in [115]. $\qquad\qquad\qquad\qquad\qquad\qquad\qquad\qquad\square$

According to Fujisaki-Okamoto [115], we can convert our basic encryption into a new which is of IND-CCA2 security by padding each plaintext with k_0 random bits. The new scheme has the same system parameters $\langle R, m, n, \mathcal{M} \rangle$. The cryptographic hash function H in basic scheme is replaced with two new cryptographic hash functions $H_1 : \{0,1\}^{k+k_0} \to \mathbb{Z}^+[x]$ and $H_2 : R \to \{0,1\}^{k+k_0}$, where k is the length of an original message.

Protocol 5.3.9. Elgamal-Like Encryption (Improved) Scheme over Non-commutative Ring:

- **Setup:** System public parameters include $\langle R, m, n, \mathcal{M} \rangle$ and k_0, H_1, H_2.

- **Key generation:** See the basic scheme.

- **Encryption:** Given a message $M \in \mathcal{M}$ and receiver's key $(p, q, y = f(p)^m \cdot q \cdot f(p)^n) \in R^3$, the sender chooses a random $r \in \{0,1\}^{k_0}$ and *extracts*[7] a polynomial $h(x) = H_1(M \parallel r) \in \mathbb{Z}^+[x]$ so that $h(p) \neq \mathbf{0}$ and then computes

$$c = h(p)^m \cdot q \cdot h(p)^n, \ d = H_2(h(p)^m \cdot pk \cdot h(p)^n) \oplus (M \parallel r).$$

  Output the ciphertext $(c, d) \in R \times \{0,1\}^{k+k_0}$.

- **Decryption:** Upon receiving a ciphertext $(c, d) \in R \times \{0,1\}^{k+k_0}$, the receiver, by using his private key $f(p)$, computes

$$M' = H_2(f(p)^m \cdot c \cdot f(p)^n) \oplus d.$$

  Extract $g(x) = H_1(M') \in \mathbb{Z}^+[x]$ and check whether $c = g(p)^m \cdot q \cdot g(p)^n$ holds. If so, output the first k bits of M'; otherwise, output $\perp$.

By Theorem 5.3.7 and Theorem 5.3.8, we have

---

[7]See Remark 5.3.5 for further discussion.

Theorem 5.3.10. *Let H_1 and H_2 be random oracles. Then the new scheme is an adaptively chosen ciphertext secure (IND-CCA2) encryption under the PDH assumption. More specifically, suppose there is an IND-CCA2 adversary $\mathcal{A}$ that has advantage ϵ against the new scheme within t steps. Suppose $\mathcal{A}$ makes at most q_D decryption queries, and at most q_{H_1}, q_{H_2} queries to the hash functions H_1, H_2, respectively. Then there is an algorithm $\mathcal{B}$ which can solve PDH with the probability at least ϵ' within t' steps, where*

$$\epsilon' = \frac{2}{q_{H_1}} \left\{ \frac{\epsilon}{(1 - 2^{-l_0})q_D} + q_{H_2} \cdot 2^{-(k_0-1)} \right\}, \ and$$

$$t' = \mathcal{O}(t - q_{H_2} \cdot (T_{\mathcal{E}}(k) \mid c \cdot k))$$

where c is a constant and $T_{\mathcal{E}}(k)$ denotes the computational time of the encryption algorithm $\mathcal{E}_{pk}(\cdot)$ in our basic scheme, and

$$l_0 := \log_2(\min_{x \in \{0,1\}^{k+k_0}} [\#\{E_{pk}(x,r) | r \in \{0,1\}^l\}]).$$

Proof. By the theorems 5.3.7 and 5.3.8, it can be immediately concluded that the new encryption is of IND-CCA2 security in the random oracle model under the PDH assumption. $\square$

5.3.5 Instantiation and Illustration (I)

In this subsection, we describe the method to construct a cryptographic hash that maps a binary string to a polynomial, such as $H_1 : \{0,1\}^{k+k_0} \to \mathbb{Z}^+[x]$. In particular, the resulting polynomials should satisfy more conditions, such as the condition $h(p) \neq 0$ and so on. We employ the so-called *divide-and-conquer* strategy to solve this problem: At first, we extract a polynomial $h(x) \in \mathbb{Z}^+[x]$ from a binary string in $\{0,1\}^{k+k_0}$; Then, we adopt a deterministic way to transform $h(x)$ to $\tilde{h}(x)$ so that $\tilde{h}(x)$ satisfies the desired condition C. In other words, we need to consider the following issues:

- **Initialization.** Suppose that there is a cryptographic hash function H_1 which maps $m \in \{0,1\}^{k+k_0}$ to $(z_0, z_1, \cdots, z_{d_H}) \in \mathbb{Z}_{c_M}^{d_H+1}$, where $d_H \cdot c_M$ is large enough to resist brute force attack.

- **Transformation.** For a polynomial $h(x) = z_0 + z_1 x + \cdots + z_{d_H} x^{d_H}$, it can be transformed to $\tilde{h}(x) = h(x) + \Delta$ with

$$\Delta = \min\{\delta \in \mathbb{Z}_{\geq 0} : h(x) + \delta \cdot 1_R \in \mathbb{Z}^+[x] \cap C\},$$

where $\mathbb{Z}^+[x] \cap C$ is the set of polynomials in $\mathbb{Z}^+[x]$ satisfying the given condition C.

Now, let us illustrate the above method by using a special matrix ring: $M_2(\mathbb{Z}_N)$, where $N = p \cdot q$, p and q are two large secure primes. We have solid reasons to believe that SDP over $M_2(\mathbb{Z}_N)$ is intractable, since it is infeasible to extract

$$A = \begin{pmatrix} a & 0 \\ 0 & 0 \end{pmatrix} \in M_2(\mathbb{Z}_N), a \in \mathbb{Z}_N$$

from

$$A^2 = \begin{pmatrix} a^2 \bmod N & 0 \\ 0 & 0 \end{pmatrix} \in M_2(\mathbb{Z}_N)$$

without knowing the factoring of N.

Example 1. Diffie–Hellman-Like key agreement over matrix rings. Let $N = 7 \cdot 11$. Alice chooses

$$m = 3, n = 5, A = \begin{pmatrix} 2 & 5 \\ 7 & 4 \end{pmatrix}, B = \begin{pmatrix} 1 & 9 \\ 3 & 2 \end{pmatrix}, \text{ and } f(x) = 3x^3 + 4x^2 + 5x + 6.$$

She computes

$$f(A) = 3 \cdot \begin{pmatrix} 2 & 5 \\ 7 & 4 \end{pmatrix}^3 + 4 \cdot \begin{pmatrix} 2 & 5 \\ 7 & 4 \end{pmatrix}^2 + 5 \cdot \begin{pmatrix} 2 & 5 \\ 7 & 4 \end{pmatrix} + 6 \cdot I = \begin{pmatrix} 35 & 12 \\ 63 & 9 \end{pmatrix},$$

and

$$r_A = \begin{pmatrix} 35 & 12 \\ 63 & 9 \end{pmatrix}^3 \begin{pmatrix} 1 & 9 \\ 3 & 2 \end{pmatrix} \begin{pmatrix} 35 & 12 \\ 63 & 9 \end{pmatrix}^5 = \begin{pmatrix} 49 & 53 \\ 42 & 31 \end{pmatrix}.$$

Then, she sends m, n, A, B, and r_A to Bob. Upon receiving m, n, A, B, and r_A from Alice, Bob chooses another polynomial $h(x) = x^5 + 5x + 1$ and computes

$$h(A) = \begin{pmatrix} 2 & 5 \\ 7 & 4 \end{pmatrix}^5 + 5 \cdot \begin{pmatrix} 2 & 5 \\ 7 & 4 \end{pmatrix} + I = \begin{pmatrix} 64 & 13 \\ 49 & 23 \end{pmatrix},$$

and

$$r_B = \begin{pmatrix} 64 & 13 \\ 49 & 23 \end{pmatrix}^3 \begin{pmatrix} 1 & 9 \\ 3 & 2 \end{pmatrix} \begin{pmatrix} 64 & 13 \\ 49 & 23 \end{pmatrix}^5 = \begin{pmatrix} 29 & 40 \\ 52 & 6 \end{pmatrix}.$$

Then, he sends r_B to Alice. Finally, Alice extracts the session key as

$$K_A = \begin{pmatrix} 35 & 12 \\ 63 & 9 \end{pmatrix}^3 \begin{pmatrix} 29 & 40 \\ 52 & 6 \end{pmatrix} \begin{pmatrix} 35 & 12 \\ 63 & 9 \end{pmatrix}^5 = \begin{pmatrix} 28 & 37 \\ 14 & 40 \end{pmatrix},$$

and Bob extracts the session key as

$$K_B = \begin{pmatrix} 64 & 13 \\ 49 & 23 \end{pmatrix}^3 \begin{pmatrix} 49 & 53 \\ 42 & 31 \end{pmatrix} \begin{pmatrix} 64 & 13 \\ 49 & 23 \end{pmatrix}^5 = \begin{pmatrix} 28 & 37 \\ 14 & 40 \end{pmatrix}.$$

Example 2. Encryption/Decryption over matrix rings for the basic scheme. Suppose that in the basic scheme, the message space is $M_2(\mathbb{Z}_N)$ and

$$H : M_2(\mathbb{Z}_N) \to M_2(\mathbb{Z}_N), m_{ij} \mapsto 2^{m_{ij}} \bmod N,$$

is a cryptographic hash function. Suppose that the remained system parameters are

$$m = 3, \; n = 5, \; p = \begin{pmatrix} 2 & 5 \\ 7 & 4 \end{pmatrix}, \; q = \begin{pmatrix} 1 & 9 \\ 3 & 2 \end{pmatrix}.$$

Suppose that the polynomial $f(x)$ picked by Alice is just that of in example 1. That is, Alice's private key is $f(p) = \begin{pmatrix} 35 & 12 \\ 63 & 9 \end{pmatrix}$, and the corresponding public key is $pk \triangleq f(p)^3 q f(p)^5 = \begin{pmatrix} 49 & 53 \\ 42 & 31 \end{pmatrix}$. Bob picks a random message $M = \begin{pmatrix} 27 & 19 \\ 34 & 8 \end{pmatrix}$. Suppose the random polynomial he picked is $h(p) = \begin{pmatrix} 64 & 13 \\ 49 & 23 \end{pmatrix}$. Now, he computes the ciphertext (c, d) as follows:

$$c = h(p)^3 q h(p)^5 = \begin{pmatrix} 64 & 13 \\ 49 & 23 \end{pmatrix}^3 \begin{pmatrix} 1 & 9 \\ 3 & 2 \end{pmatrix} \begin{pmatrix} 64 & 13 \\ 49 & 23 \end{pmatrix}^5 = \begin{pmatrix} 29 & 40 \\ 52 & 6 \end{pmatrix},$$

$$\begin{aligned}
d &= H(h(p)^3 \cdot pk \cdot h(p)^5) \oplus M \\
&= H\left(\begin{pmatrix} 64 & 13 \\ 49 & 23 \end{pmatrix}^3 \begin{pmatrix} 49 & 53 \\ 42 & 31 \end{pmatrix} \begin{pmatrix} 64 & 13 \\ 49 & 23 \end{pmatrix}^5 \right) \oplus \begin{pmatrix} 27 & 19 \\ 34 & 8 \end{pmatrix} \\
&= \left(\begin{pmatrix} 2^{28} & 2^{37} \\ 2^{14} & 2^{40} \end{pmatrix} \bmod N \right) \oplus \begin{pmatrix} 27 & 19 \\ 34 & 8 \end{pmatrix} \\
&= \begin{pmatrix} 58 & 51 \\ 60 & 23 \end{pmatrix} \oplus \begin{pmatrix} 27 & 19 \\ 34 & 8 \end{pmatrix} = \begin{pmatrix} 33 & 32 \\ 30 & 31 \end{pmatrix}.
\end{aligned}$$

Upon receiving the above ciphertext, Alice decrypts it as follows:

$$
\begin{aligned}
M' &= H(f(p)^3 \cdot c \cdot f(p)^5) \oplus d \\
&= H\left(\begin{pmatrix} 35 & 12 \\ 63 & 9 \end{pmatrix}^3 \begin{pmatrix} 29 & 40 \\ 52 & 6 \end{pmatrix} \begin{pmatrix} 35 & 12 \\ 63 & 9 \end{pmatrix}^5 \right) \oplus \begin{pmatrix} 33 & 32 \\ 30 & 31 \end{pmatrix} \\
&= H\left(\begin{pmatrix} 28 & 37 \\ 14 & 40 \end{pmatrix} \right) \oplus \begin{pmatrix} 33 & 32 \\ 30 & 31 \end{pmatrix} \\
&= \left(\begin{pmatrix} 2^{28} & 2^{37} \\ 2^{14} & 2^{40} \end{pmatrix} \bmod N \right) \oplus \begin{pmatrix} 33 & 32 \\ 30 & 31 \end{pmatrix} \\
&= \begin{pmatrix} 58 & 51 \\ 60 & 23 \end{pmatrix} \oplus \begin{pmatrix} 33 & 32 \\ 30 & 31 \end{pmatrix} = \begin{pmatrix} 27 & 19 \\ 34 & 8 \end{pmatrix} = M.
\end{aligned}
$$

Example 3. Encryption/Decryption over matrix rings for the new scheme. Suppose that in the new scheme, the message space is $\mathcal{M} = \left\{ \begin{pmatrix} a & b \\ c & 0 \end{pmatrix} : a, b, c \in \mathbb{Z}_N \right\}$, and

$$
H_1 : \mathcal{M} \times \mathbb{Z}_N \to \mathbb{Z}_N^+[x], \quad \left(\begin{pmatrix} a & b \\ c & 0 \end{pmatrix}, r \right) \mapsto 2^r + 2^a x + 2^b x^2 + 2^c x^3,
$$

For more simplicity, we define

$$
(M \parallel r) \triangleq \begin{pmatrix} a & b \\ c & 0 \end{pmatrix} + \begin{pmatrix} 0 & 0 \\ 0 & r \end{pmatrix} = \begin{pmatrix} a & b \\ c & r \end{pmatrix}.
$$

In addition, $H_2 : M_2(\mathbb{Z}_N) \to \mathcal{M} \times \mathbb{Z}_N$ can be defined as

$$
\begin{pmatrix} a & b \\ c & d \end{pmatrix} \mapsto (M \parallel r),
$$

where $M = \begin{pmatrix} 2^a & 2^b \\ 2^c & 0 \end{pmatrix} \bmod N, r = 2^d \bmod N$. Suppose that Alice's private/public keys are as the same as that in the basic scheme. Bob picks a message $M = \begin{pmatrix} 27 & 19 \\ 34 & 0 \end{pmatrix}$. Suppose $r = 35$. Then, he extracts a polynomial $h(x)$ and compute $h(p)$ as follows:

$$
\begin{aligned}
h(x) &= 2^{35} + 2^{27}x + 2^{19}x^2 + 2^{34}x^3 \bmod N \\
&= 32 + 29x + 72x^2 + 16x^3,
\end{aligned}
$$

$$h(p) = 32 \cdot I + 29 \cdot \begin{pmatrix} 2 & 5 \\ 7 & 4 \end{pmatrix} + 72 \cdot \begin{pmatrix} 2 & 5 \\ 7 & 4 \end{pmatrix}^2 + 16 \cdot \begin{pmatrix} 2 & 5 \\ 7 & 4 \end{pmatrix}^3 = \begin{pmatrix} 37 & 30 \\ 42 & 49 \end{pmatrix} \neq \mathbf{0}.$$

Note that if $h(x)$ does not satisfy the condition $h(p) \neq \mathbf{0}$, he should transform $h(x)$ to
$\tilde{h}(x) = h(x) + \Delta$, where

$$\Delta = \min\left\{ \delta \in \mathbb{Z}_{\geq 0} : h(p) + \delta \cdot \begin{pmatrix} 1 & 0 \\ 0 & 1 \end{pmatrix} \neq \mathbf{0} \right\}.$$

Then, the ciphertext is (c, d), where

$$c = h(p)^3 q h(p)^5 = \begin{pmatrix} 37 & 30 \\ 42 & 49 \end{pmatrix}^3 \begin{pmatrix} 1 & 9 \\ 3 & 2 \end{pmatrix} \begin{pmatrix} 37 & 30 \\ 42 & 49 \end{pmatrix}^5 = \begin{pmatrix} 65 & 37 \\ 35 & 7 \end{pmatrix},$$

$$\begin{aligned}
d &= H(h(p)^3 \cdot pk \cdot h(p)^5) \oplus (M \parallel r) \\
&= H\left(\begin{pmatrix} 37 & 30 \\ 42 & 49 \end{pmatrix}^3 \begin{pmatrix} 49 & 53 \\ 42 & 31 \end{pmatrix} \begin{pmatrix} 37 & 30 \\ 42 & 49 \end{pmatrix}^5 \right) \oplus \begin{pmatrix} 27 & 19 \\ 34 & 35 \end{pmatrix} \\
&= H\left(\begin{pmatrix} 21 & 14 \\ 21 & 7 \end{pmatrix} \right) \oplus \begin{pmatrix} 27 & 19 \\ 34 & 35 \end{pmatrix} \\
&= \left(\begin{pmatrix} 2^{21} & 2^{14} \\ 2^{21} & 2^7 \end{pmatrix} \bmod N \right) \oplus \begin{pmatrix} 27 & 19 \\ 34 & 35 \end{pmatrix} \\
&= \begin{pmatrix} 57 & 60 \\ 57 & 51 \end{pmatrix} \oplus \begin{pmatrix} 27 & 19 \\ 34 & 35 \end{pmatrix} = \begin{pmatrix} 34 & 47 \\ 27 & 16 \end{pmatrix}.
\end{aligned}$$

Upon receiving the above ciphertext, Alice decrypts it as follows:

$$\begin{aligned}
M' &= H(f(p)^3 \cdot c \cdot f(p)^5) \oplus d \\
&= H\left(\begin{pmatrix} 35 & 12 \\ 63 & 9 \end{pmatrix}^3 \begin{pmatrix} 65 & 37 \\ 35 & 7 \end{pmatrix} \begin{pmatrix} 35 & 12 \\ 63 & 9 \end{pmatrix}^5 \right) \oplus \begin{pmatrix} 34 & 47 \\ 27 & 16 \end{pmatrix}
\end{aligned}$$

$$
\begin{aligned}
&= H\left(\begin{pmatrix} 21 & 14 \\ 21 & 7 \end{pmatrix}\right) \oplus \begin{pmatrix} 34 & 47 \\ 27 & 16 \end{pmatrix} \\
&= \left(\begin{pmatrix} 2^{21} & 2^{14} \\ 2^{21} & 2^7 \end{pmatrix} \bmod N\right) \oplus \begin{pmatrix} 34 & 47 \\ 27 & 16 \end{pmatrix} \\
&= \begin{pmatrix} 57 & 60 \\ 57 & 51 \end{pmatrix} \oplus \begin{pmatrix} 34 & 47 \\ 27 & 16 \end{pmatrix} = \begin{pmatrix} 27 & 19 \\ 34 & 35 \end{pmatrix} \\
&= \begin{pmatrix} 27 & 19 \\ 34 & 0 \end{pmatrix} + \begin{pmatrix} 0 & 0 \\ 0 & 35 \end{pmatrix} = M \parallel r.
\end{aligned}
$$

5.3.6   ℤ-Modular Method over Noncommutative Groups/ Semigroups

Suppose that $(G, \cdot, 1_G)$ is a noncommutative group, $(R, +, \cdot, 1_R)$ is a ring and there is monomorphism $\tau : (G, \cdot, 1_G) \to (R, \cdot, 1_R)$. Then, the inverse map $\tau^{-1} : \tau(G) \to G$ is also a well-defined monomorphism. For $a, b \in G$, if $\tau(a) + \tau(b) \in \tau(G)$, we define

$$
a \boxplus b \triangleq \tau^{-1}(\tau(a) + \tau(b)), \tag{5.16}
$$

and call it the *quasi-sum* of a and b. Similarly, for $k \in R$ and $a \in G$, if $k \cdot \tau(a) \in \tau(G)$, we define

$$
k \boxtimes a \triangleq \tau^{-1}(k \cdot \tau(a)), \tag{5.17}
$$

and call it the k *quasi-multiple* of a.

Given $a, b \in G$ and $k \cdot \tau(a) + \tau(b) \in \tau(G)$, it is easy to see that the monomorphism τ is *quasi-linear*, which satisfies that the following equality

$$
\begin{aligned}
\tau(k \boxtimes a \boxplus b) &= \tau((k \boxtimes a) \boxplus b) \overset{d \leftarrow k \boxtimes a}{=\!=\!=} \tau(d \boxplus b) \\
&= \tau(\tau^{-1}(\tau(d) + \tau(b))) \\
&= \tau(\tau^{-1}(\tau(\tau^{-1}(k \cdot \tau(a))) + \tau(b))) \\
&= \tau(\tau^{-1}(k \cdot \tau(a) + \tau(b))) = k \cdot \tau(a) + \tau(b).
\end{aligned}
$$

Furthermore, for $f(x) = z_0 + z_1 x + \cdots + z_n x^n \in \mathbb{Z}[x]$ and $a \in G$, if $f(\tau(a)) = z_0 \cdot 1_R + z_1 \cdot \tau(a) + \cdots + z_n \cdot \tau(a)^n \in \tau(G)$, we define

$$\hat{f}(a) \triangleq \tau^{-1}(f(\tau(a))) = \tau^{-1}(z_0 \cdot 1_R + z_1 \cdot \tau(a) + \cdots + z_n \cdot \tau(a)^n), \qquad (5.18)$$

and call it the *quasi-polynomial* of f with respect to a. (Notice that, for arbitrary $a, b \in G$, $k \in R$ and $f(x) \in \mathbb{Z}[x]$, $a \boxplus b$, $k \boxtimes a$ and $\hat{f}(a)$ are not always well-defined)

Theorem 5.3.11. *Suppose that $(G, \cdot, 1_G)$ is a noncommutative group, $(R, +, \cdot, 1_R)$ is a ring and $\tau : (G, \cdot, 1_G) \to (R, \cdot, 1_R)$ is a monomorphism. For some $a \in G$ and some $f(x), h(x) \in \mathbb{Z}[x]$, if both $\hat{f}(a)$ and $\hat{h}(a)$ are well-defined, we have*

$$\hat{f}(a) \cdot \hat{h}(a) = \hat{h}(a) \cdot \hat{f}(a).$$

Proof.

$$
\begin{aligned}
\hat{f}(a) \cdot \hat{h}(a) &= \tau^{-1}(f(\tau(a))) \cdot \tau^{-1}(h(\tau(a))) \\
&= \tau^{-1}(f(\tau(a)) \cdot h(\tau(a))) \\
&= \tau^{-1}(h(\tau(a)) \cdot f(\tau(a))) \\
&= \tau^{-1}(h(\tau(a))) \cdot \tau^{-1}(f(\tau(a))) \\
&= \hat{h}(a) \cdot \hat{f}(a).
\end{aligned}
$$

Now, similar to the discussion in the subsection 5.3.2, we can discuss new problems over the noncommutative group $(G, \cdot, 1_G)$. Suppose that $(R, +, \cdot, 1_R)$ is a ring and $\tau : (G, \cdot, 1_G) \to (R, \cdot, 1_R)$ is a monomorphism. For any randomly picked element $a \in G$, we define a set $P_a \subseteq G$ by

$$P_a \triangleq \{\hat{f}(a) : f(x) \in \mathbb{Z}[x], f(\tau(a)) \in \tau(G)\}.$$

Then, we can define the PSD and PDH problems over $(G, \cdot)$ by a similar way:

- **Polynomial Symmetrical Decomposition (PSD*) Problem**: Given $(a, x, y) \in G^3$ and $m, n \in \mathbb{Z}$, find $z \in P_a$ so that $y = z^m x z^n$, where G is a noncommutative group.

- **Polynomial Diffie–Hellman (PDH*) Problem**: Given a, x, x^{z_1}, x^{z_2}, $a, x \in G$ and $z_1, z_2 \in P_a$, compute $x^{z_1 z_2}$ or $x^{z_2 z_1}$, where G is a noncommutative group.

Accordingly, the PSD* (PDH*, respectively) cryptographic assumptions over $(G, \cdot)$ says that PSD* (PDH*, respectively) problems over $(G, \cdot)$ is intractable, i.e., there does not exist probabilistic polynomial time algorithm which can solve PSD* (PDH*, respectively) problems over $(G, \cdot)$ with non-negligible advantage with respect to the parameters of the problem scale.

Under these assumptions, we have

Protocol 5.3.12. Diffie–Hellman-Like Key Agreement Protocol over Noncommutative Group:

(0) Suppose that $(G, \cdot, 1_G)$ is a noncommutative group, $(R, +, \cdot, 1_R)$ is a ring and $\tau : (G, \cdot, 1_G) \to (R, \cdot, 1_R)$ is a monomorphism. Alice sends two random small, positive integers (say, less than 10) $m, n \in \mathbb{Z}$ and two random elements $a, b \in G$ to Bob.

(1) Alice chooses $f(x) \in \mathbb{Z}[x]$ at random so that $\hat{f}(a)$ is well-defined, i.e., $f(\tau(a)) \in \tau(G)$. Then, Alice takes $\hat{f}(a)$ as her private key.

(2) Bob chooses $h(x) \in \mathbb{Z}[x]$ at random so that $\hat{h}(a)$ is well-defined, i.e., $h(\tau(a)) \in \tau(G)$. Then, Bob takes $\hat{h}(a)$ as his private key.

(3) Alice computes $r_A = \hat{f}(a)^m \cdot b \cdot \hat{f}(a)^n$ and sends r_A to Bob.

(4) Bob computes $r_B = \hat{h}(a)^m \cdot b \cdot \hat{h}(a)^n$ and sends r_B to Alice.

(5) Alice computes $K_A = \hat{f}(a)^m \cdot r_B \cdot \hat{f}(a)^n$ as the shared session key.

(6) Bob computes $K_B = \hat{h}(a)^m \cdot r_A \cdot \hat{h}(a)^n$ as the shared session key.

In practice, the steps (0), (1), and (3) can be finished simultaneously and require only one pass communication from Alice to Bob. After that, the steps (2) and (4) can be finished simultaneously and require another pass communication from Bob to Alice. Finally, Alice and Bob can execute the steps (5) and (6), respectively. We can depict the protocol in Figure 5.5.

Pass	Alice	Bob
	$m, n \xleftarrow{\$} \mathbb{Z}$ $a, b \xleftarrow{\$} G$ $f(x) \xleftarrow{\$} \mathbb{Z}[x]$ s.t. $f(\tau(a)) \in \tau(G)$	
1	$\xrightarrow{\quad m,n,a,b,\hat{f}(a)^m b \hat{f}(a)^n \quad}$	
		$h(g) \xleftarrow{\$} \mathbb{Z}[x]$ s.t. $h(\tau(a)) \in \tau(G)$
2	$\xleftarrow{\quad \hat{h}(a)^m b \hat{h}(a)^n \quad}$	
	$K_A = \hat{f}(a)^m \hat{h}(a)^m b \hat{h}(a)^n \hat{f}(a)^n \quad = \quad K_B = \hat{h}(a)^m \hat{f}(a)^m b \hat{f}(a)^n \hat{h}(a)^n$	

Figure 5.5: Diffie–Hellman-Like Key Agreement over Noncommutative Groups

Similarly, it is easy to describe ElGamal-like encryption schemes, including the basic scheme and the improved scheme as well, by using noncommutative groups as the underlying algebraic structure.

Protocol 5.3.13. ElGamal-Like Encryption (Basic) Scheme over Noncommutative Group:

- **Setup:** Suppose that $(G, \cdot, 1_G)$ is a noncommutative group, $(R, +, \cdot, 1_R)$ is a ring and $\tau : (G, \cdot, 1_G) \to (R, \cdot, 1_R)$ is a monomorphism. We assume that SDP on G is intractable. Pick two small positive integers $m, n \in \mathbb{Z}$ and two elements $p, q \in G$ at random. Let $H : G \to \mathcal{M}$ be a cryptographic hash function which maps G

to the message space $\mathcal{M}$. Then, set the tuple $\langle G, R, \tau, m, n, p, q, \mathcal{M}, H \rangle$ as the public parameters of the system.

- **Key generation:** Each user chooses a random polynomial $f(x) \in \mathbb{Z}[x]$ so that $f(\tau(p)) \in \tau(G)$ and takes $sk \triangleq \hat{f}(p)$ as his private key, then computes and publishes his public key $pk \triangleq \hat{f}(p)^m \cdot q \cdot \hat{f}(p)^n \in G$.

- **Encryption:** Given a message $M \in \mathcal{M}$ and receiver's key $pk \in G$, the sender chooses a random polynomial $h(x) \in \mathbb{Z}[x]$ so that $h(\tau(p)) \in \tau(G)$. Compute

$$c = \hat{h}(p)^m \cdot q \cdot \hat{h}(p)^n, \ d = H(\hat{h}(p)^m \cdot pk \cdot \hat{h}(p)^n) \oplus M,$$

and output the ciphertext $(c, d) \in G \times \mathcal{M}$.

- **Decryption:** Upon receiving a ciphertext $(c, d) \in R \times \mathcal{M}$, the receiver, by using his private key $\hat{f}(p)$, computes the plaintext

$$M = H(\hat{f}(p)^m \cdot c \cdot \hat{f}(p)^n) \oplus d$$

Protocol 5.3.14. Improved ElGamal-Like Encryption Scheme over Noncommutative Group:

- **Setup:** The parameters $G, R, \tau, m, n, p, q, \mathcal{M}$ are set as that in Protocol 5.3.13. In addition, suppose that $H_1 : \{0,1\}^{k+k_0} \to \mathbb{Z}^+[x]$ and $H_2 : G \to \{0,1\}^{k+k_0}$ are two cryptographic hash functions, where k is the length of a message and k_0 is the length of random string for padding.

- **Key generation:** See Protocol 5.3.13.

- **Encryption:** Given a message $M \in \mathcal{M}$ and receiver's key $pk \in G$, the sender chooses a random $r \in \{0,1\}^{k_0}$ and *extracts* a polynomial $h(x) \in \mathbb{Z}[x]$ so that

$h(\tau(p)) \in \tau(G)$ and then computes

$$c = \hat{h}(p)^m \cdot q \cdot \hat{h}(p)^n, \ d = H_2(\hat{h}(p)^m \cdot pk \cdot \hat{h}(p)^n) \oplus (M \parallel r),$$

and outputs the ciphertext $(c, d) \in R \times \{0, 1\}^{k+k_0}$.

- **Decryption:** Upon receiving a ciphertext $(c, d) \in R \times \{0, 1\}^{k+k_0}$, the receiver, by using his private key $\hat{f}(p)$, computes

$$M' = H_2(\hat{f}(p)^m \cdot c \cdot \hat{f}(p)^n) \oplus d.$$

Extract $g(x) = H_1(M') \in \mathbb{Z}[x]$ and check whether $c = \hat{g}(p)^m \cdot q \cdot \hat{g}(p)^n$ holds. If so, output the first k bits of M'; otherwise, output $\perp$.

The security and related proofs of the above cryptosystems are very similar, except replacing the PDH assumption over noncommutative ring R with the PDH* assumption over noncommutative group G. The similar constructions can also be obtained for the case when G is a noncommutative semigroup, except replacing $\mathbb{Z}$ and $\mathbb{Z}[x]$ with $\mathbb{Z}^+$ and $\mathbb{Z}^+[x]$ respectively. The main differences between the protocols and schemes over noncommutative rings and semigroups and those over noncommutative groups are:

(1) $m, n \in \mathbb{Z}^+$ for the schemes over rings and semigroups, while $m, n \in \mathbb{Z}$ for the schemes over groups;

(2) $f, h \in \mathbb{Z}^+[x]$ for the schemes over rings and semigroups, while $f, h \in \{g \in \mathbb{Z}[x] : g(\tau(a)) \in \tau(G)\} \subseteq \mathbb{Z}[x]$ for the schemes over groups.

5.3.7   Instantiation and Illustration (II)

To illustrate our method, we take the minimal noncommutative group, say, the symmetric group S_3, as the underlying algebraic structure. We choose $M_2(\mathbb{Z}_2)$ as the required ring. Define the required monomorphism $\tau : S_3 \to M_2(\mathbb{Z}_2)$ as follows:

$$\begin{pmatrix} 1\,2\,3 \\ 1\,2\,3 \end{pmatrix} \mapsto \begin{pmatrix} 1 & 0 \\ 0 & 1 \end{pmatrix}, \quad \begin{pmatrix} 1\,2\,3 \\ 1\,3\,2 \end{pmatrix} \mapsto \begin{pmatrix} 1 & 1 \\ 0 & 1 \end{pmatrix}, \quad \begin{pmatrix} 1\,2\,3 \\ 2\,1\,3 \end{pmatrix} \mapsto \begin{pmatrix} 0 & 1 \\ 1 & 0 \end{pmatrix},$$

$$\begin{pmatrix} 1\,2\,3 \\ 2\,3\,1 \end{pmatrix} \mapsto \begin{pmatrix} 0 & 1 \\ 1 & 1 \end{pmatrix}, \quad \begin{pmatrix} 1\,2\,3 \\ 3\,1\,2 \end{pmatrix} \mapsto \begin{pmatrix} 1 & 1 \\ 1 & 0 \end{pmatrix}, \quad \begin{pmatrix} 1\,2\,3 \\ 3\,2\,1 \end{pmatrix} \mapsto \begin{pmatrix} 1 & 0 \\ 1 & 1 \end{pmatrix}.$$

Example 4. Diffie–Hellman-Like Key Agreement over S_3

Alice chooses

$$m = 3, n = 5, A = \begin{pmatrix} 1\,2\,3 \\ 2\,3\,1 \end{pmatrix}, B = \begin{pmatrix} 1\,2\,3 \\ 2\,1\,3 \end{pmatrix},$$

and picks a polynomial $f(x) = 4x^2 + x + 2$. Clearly, $f(A) \in \tau(S_3)$. Then, Alice computes

$$\begin{aligned} f(A) &= \tau^{-1}(f(\tau(A))) \\ &= \tau^{-1}\left(4 \cdot \begin{pmatrix} 0 & 1 \\ 1 & 1 \end{pmatrix}^2 + \begin{pmatrix} 0 & 1 \\ 1 & 1 \end{pmatrix} + 2 * I \right) \\ &= \tau^{-1}\left(\begin{pmatrix} 0 & 1 \\ 1 & 1 \end{pmatrix}\right) = \begin{pmatrix} 1\,2\,3 \\ 2\,3\,1 \end{pmatrix} \end{aligned}$$

$$r_A = \begin{pmatrix} 1\,2\,3 \\ 2\,3\,1 \end{pmatrix}^3 \circ \begin{pmatrix} 1\,2\,3 \\ 2\,1\,3 \end{pmatrix} \circ \begin{pmatrix} 1\,2\,3 \\ 2\,3\,1 \end{pmatrix}^5 = \begin{pmatrix} 1\,2\,3 \\ 3\,2\,1 \end{pmatrix}.$$

Send m, n, A, B, and r_A to Bob.

Upon receiving m, n, A, B, and $f(A)$ from Alice, Bob chooses another random polynomial $h(x) = 4x^4 + x^3 + 4x^2 + 3x + 4$. Clearly, $h(A) \in \tau(S_3)$. Then, Bob computes

$$
\begin{aligned}
h(A) \;&=\; \tau^{-1}(h(\tau(A))) \\
&=\; \tau^{-1}\left(4 \cdot \begin{pmatrix} 0 & 1 \\ 1 & 1 \end{pmatrix}^4 + \begin{pmatrix} 0 & 1 \\ 1 & 1 \end{pmatrix}^3 + 4 \cdot \begin{pmatrix} 0 & 1 \\ 1 & 1 \end{pmatrix}^2 + 3 \cdot \begin{pmatrix} 0 & 1 \\ 1 & 1 \end{pmatrix} + 4 * I \right) \\
&=\; \tau^{-1}\left(\begin{pmatrix} 1 & 1 \\ 1 & 0 \end{pmatrix} \right) = \begin{pmatrix} 1\,2\,3 \\ 3\,1\,2 \end{pmatrix}, \\
r_B \;&=\; \begin{pmatrix} 1\,2\,3 \\ 3\,1\,2 \end{pmatrix}^3 \circ \begin{pmatrix} 1\,2\,3 \\ 2\,1\,3 \end{pmatrix} \circ \begin{pmatrix} 1\,2\,3 \\ 3\,1\,2 \end{pmatrix}^5 = \begin{pmatrix} 1\,2\,3 \\ 1\,3\,2 \end{pmatrix}.
\end{aligned}
$$

Send r_B to Alice.

Finally, Alice extracts the session key

$$
K_A = \begin{pmatrix} 1\,2\,3 \\ 2\,3\,1 \end{pmatrix}^3 \circ \begin{pmatrix} 1\,2\,3 \\ 1\,3\,2 \end{pmatrix} \circ \begin{pmatrix} 1\,2\,3 \\ 2\,3\,1 \end{pmatrix}^5 = \begin{pmatrix} 1\,2\,3 \\ 2\,1\,3 \end{pmatrix},
$$

and Bob extracts the session key

$$
K_B = \begin{pmatrix} 1\,2\,3 \\ 3\,1\,2 \end{pmatrix}^3 \circ \begin{pmatrix} 1\,2\,3 \\ 3\,2\,1 \end{pmatrix} \circ \begin{pmatrix} 1\,2\,3 \\ 3\,1\,2 \end{pmatrix}^5 = \begin{pmatrix} 1\,2\,3 \\ 2\,1\,3 \end{pmatrix}.
$$

Example 5. Encryption/Decryption over S_3

We can implement encryption/decryption by ways which are very similar to those in Example 2 and Example 3.

At first, let us choose two prime p and q so that $q|p-1$ and set Z_p as the message space $\mathcal{M}$. Also, we assume that g is a generator of order q. Then, we define

$$
H : S_3 \to \mathcal{M}, \quad \begin{pmatrix} 1 & 2 & 3 \\ \sigma_1 & \sigma_2 & \sigma_3 \end{pmatrix} \mapsto g^{\sigma_1 + 2 \cdot \sigma_2 + 2^2 \cdot \sigma_3} \bmod p,
$$

$$
H_1 : \mathcal{M} \times \mathbb{Z}_N \to \mathbb{Z}^+[x], (M, r) \mapsto r_0 + r_1 x + r_2 x^2 + \cdots + r_n x^n,
$$

where all coefficients $r_i, 0 \le i \le k$, are determined by the following process[8]:

$$
\begin{aligned}
M &= rk_0 + r_0, \ 0 < r_0 < r, \\
r &= r_0 k_1 + r_1, \ 0 < r_1 < r_0, \\
r_0 &= r_1 k_2 + r_2, \ 0 < r_2 < r_1, \\
&\cdots \quad \cdots \cdots \cdots \cdots \cdots \cdots \\
r_{l-2} &= r_{l-1} k_l + r_l, \ 0 < r_l < r_{l-1}, \\
&\cdots \quad \cdots \cdots \cdots \cdots \cdots \cdots \\
r_{n-1} &= r_n k_{n+1} + r_{n+1}, \ r_{n+1} = 0.
\end{aligned}
$$

For simplicity, we define

$$
M \parallel r \triangleq r \cdot p + M \in \mathbb{Z}.
$$

It is easy to extract M from $M \parallel r$ by $M = (M \parallel r) \bmod p$. Another required hash $H_2 : S_3 \to \mathcal{M} \times \mathbb{Z}_N$ can be defined as

$$
\begin{pmatrix} 1 & 2 & 3 \\ \sigma_1 & \sigma_2 & \sigma_3 \end{pmatrix} \mapsto (g^{\sigma_1 + 2 \cdot \sigma_2 + 2^2 \cdot \sigma_3} \bmod p, \ g^{\sigma_3 + 2 \cdot \sigma_1 + 2^2 \cdot \sigma_2} \bmod p).
$$

Next, let $p = 23, q = 11$ and $g = 2$. Suppose that other system parameters are

$$
m = 3, n = 5, p = \begin{pmatrix} 1 & 2 & 3 \\ 2 & 3 & 1 \end{pmatrix}, q = \begin{pmatrix} 1 & 2 & 3 \\ 2 & 1 & 3 \end{pmatrix}.
$$

Suppose that the polynomial $f(x)$ picked by Alice is just that in example 4. Then, Alice's private key is $\hat{f}(p) = \begin{pmatrix} 1 & 2 & 3 \\ 2 & 3 & 1 \end{pmatrix}$, and the corresponding public key is

$$
pk \triangleq \hat{f}(p)^3 q \hat{f}(p)^5 = \begin{pmatrix} 1 & 2 & 3 \\ 2 & 3 & 1 \end{pmatrix}^3 \circ \begin{pmatrix} 1 & 2 & 3 \\ 2 & 1 & 3 \end{pmatrix} \circ \begin{pmatrix} 1 & 2 & 3 \\ 2 & 3 & 1 \end{pmatrix}^5 = \begin{pmatrix} 1 & 2 & 3 \\ 3 & 2 & 1 \end{pmatrix}.
$$

---

[8]Here, we assume that $\gcd(M, r) \ne r$; Otherwise, we set $r = g^r \bmod p$ and then resume the process. Also, we assume that $M > r$; Otherwise, we can swap them in advance.

Bob picks a message $M = 17$ and a polynomial $h(x) = 4x^4 + x^3 + 4x^2 + 3x + 4$.

Then, the permutation $\hat{h}(p) = \left(\begin{smallmatrix} 1\,2\,3 \\ 3\,1\,2 \end{smallmatrix}\right)$. Compute the ciphertext (c, d) as follows:

$$
\begin{aligned}
c &= \hat{h}(p)^3 q \hat{h}(p)^5 = \begin{pmatrix} 1\,2\,3 \\ 3\,1\,2 \end{pmatrix}^3 \circ \begin{pmatrix} 1\,2\,3 \\ 2\,1\,3 \end{pmatrix} \circ \begin{pmatrix} 1\,2\,3 \\ 3\,1\,2 \end{pmatrix}^5 = \begin{pmatrix} 1\,2\,3 \\ 1\,3\,2 \end{pmatrix}, \\
d &= H(\hat{h}(p)^3 \cdot pk \cdot \hat{h}(p)^5) \oplus M \\
&= H\left(\begin{pmatrix} 1\,2\,3 \\ 3\,1\,2 \end{pmatrix}^3 \circ \begin{pmatrix} 1\,2\,3 \\ 3\,2\,1 \end{pmatrix} \circ \begin{pmatrix} 1\,2\,3 \\ 3\,1\,2 \end{pmatrix}^5 \right) \oplus 17 \\
&= H\left(\begin{pmatrix} 1\,2\,3 \\ 2\,1\,3 \end{pmatrix} \right) \oplus 17 \\
&= \left(2^2 + 2^{2\cdot1} + 2^{2^2\cdot3} \bmod 23 \right) \oplus 17 \\
&= 18 \oplus 17 = 3.
\end{aligned}
$$

On receiving the ciphertext, Alice decrypts it as follows:

$$
\begin{aligned}
M' &= H(\hat{f}(p)^3 \cdot c \cdot \hat{f}(p)^5) \oplus d \\
&= H\left(\begin{pmatrix} 1\,2\,3 \\ 2\,3\,1 \end{pmatrix}^3 \circ \begin{pmatrix} 1\,2\,3 \\ 1\,3\,2 \end{pmatrix} \circ \begin{pmatrix} 1\,2\,3 \\ 2\,3\,1 \end{pmatrix}^5 \right) \oplus 3
\end{aligned}
$$

$$
= H\left(\begin{pmatrix} 1\,2\,3 \\ 2\,1\,3 \end{pmatrix} \right) \oplus 3 = \left(2^2 + 2^{2\cdot1} + 2^{2^2\cdot3} \bmod 23 \right) \oplus 3 = 18 \oplus 3 = 17 = M.
$$

For the improved encryption, Bob picks $M = 19$ and sets $r = 7$. Then, he extracts a polynomial as follows:

$$
h(x) \triangleq H_1(19, 7) = 5 + 2x + x^2 + x^3.
$$

Clearly, $h(\tau(p)) \in \tau(S_3)$. Note that if $h(x)$ does not satisfy the condition of $h(\tau(p)) \in \tau(S_3)$, he should transform $h(x)$ to $\tilde{h}(x) = h(x) + \Delta$, where

$$
\Delta = \min\left\{ \delta \in \mathbb{Z}_{\geq 0} : h(\tau(p)) + \delta \cdot \begin{pmatrix} 1 & 0 \\ 0 & 1 \end{pmatrix} \in \tau(S_3) \right\}.
$$

Then,

$$
\begin{aligned}
\hat{h}(p) &= \tau^{-1}(h(\tau(p))) \\
&= \tau^{-1}\left(5 \cdot \begin{pmatrix} 1 & 0 \\ 0 & 1 \end{pmatrix} + 2 \cdot \begin{pmatrix} 0 & 1 \\ 1 & 1 \end{pmatrix} + \begin{pmatrix} 0 & 1 \\ 1 & 1 \end{pmatrix}^2 + \begin{pmatrix} 0 & 1 \\ 1 & 1 \end{pmatrix}^3\right) \\
&= \tau^{-1}\left(\begin{pmatrix} 1 & 1 \\ 0 & 1 \end{pmatrix}\right) = \begin{pmatrix} 1 & 2 & 3 \\ 3 & 1 & 2 \end{pmatrix}.
\end{aligned}
$$

Consequently, the corresponding ciphertext (c, d) is

$$
\begin{aligned}
c &= \hat{h}(p)^3 q \hat{h}(p)^5 = \begin{pmatrix} 1 & 2 & 3 \\ 3 & 1 & 2 \end{pmatrix}^3 \circ \begin{pmatrix} 1 & 2 & 3 \\ 2 & 1 & 3 \end{pmatrix} \circ \begin{pmatrix} 1 & 2 & 3 \\ 3 & 1 & 2 \end{pmatrix}^5 = \begin{pmatrix} 1 & 2 & 3 \\ 1 & 3 & 2 \end{pmatrix}, \\
d &= H(\hat{h}(p)^3 \cdot pk \cdot \hat{h}(p)^5) \oplus (M \parallel r) \\
&= H\left(\begin{pmatrix} 1 & 2 & 3 \\ 3 & 1 & 2 \end{pmatrix}^3 \circ \begin{pmatrix} 1 & 2 & 3 \\ 3 & 2 & 1 \end{pmatrix} \circ \begin{pmatrix} 1 & 2 & 3 \\ 3 & 1 & 2 \end{pmatrix}^5\right) \oplus (19 + 5 \cdot 23) \\
&= H\left(\begin{pmatrix} 1 & 2 & 3 \\ 2 & 1 & 3 \end{pmatrix}\right) \oplus 134 \\
&= \left(2^2 + 2^{2 \cdot 1} + 2^{2^2 \cdot 3} \bmod 23\right) \oplus 134 = 18 \oplus 134 = 148.
\end{aligned}
$$

On receiving the ciphertext, Alice decrypts it as follows:

$$
\begin{aligned}
M' &= H(f(p)^3 \cdot c \cdot f(p)^5) \oplus d \\
&= H\left(\begin{pmatrix} 1 & 2 & 3 \\ 2 & 3 & 1 \end{pmatrix}^3 \circ \begin{pmatrix} 1 & 2 & 3 \\ 1 & 3 & 2 \end{pmatrix} \circ \begin{pmatrix} 1 & 2 & 3 \\ 2 & 3 & 1 \end{pmatrix}^5\right) \oplus 148 \\
&= H\left(\begin{pmatrix} 1 & 2 & 3 \\ 2 & 1 & 3 \end{pmatrix}\right) \oplus 3 \\
&= \left(2^2 + 2^{2 \cdot 1} + 2^{2^2 \cdot 3} \bmod 23\right) \oplus 148 = 18 \oplus 148 = 134.
\end{aligned}
$$

Then, $M = M' \bmod 23 = 19$.

5.4   Using Monomials in $\mathbb{Z}$-Modular Method

If the polynomials used in $\mathbb{Z}$-modular method are constrained to be monomials, then the conjugacy related assumptions can be viewed as special cases of the assumptions defined in $\mathbb{Z}$-modular method. Under these assumptions, we [287] recently proposed several noncommutative cryptosystems.

5.4.1   Conjugate Left Self-Distributed System (Conj-LD)

At first, let us recall the definition of the conjugacy search problem and the so-called left self-distributive system [94].

Let $(G, \circ, 1)$ be a noncommutative monoid with identity 1. For $a \in G$, if there exists an element $b \in G$ so that $a \circ b = 1 = b \circ a$, then we say that a is *invertible*, and call b *an inverse of* a. Note that not all elements in G are invertible. If the inverse of a, exists, it is unique, and denoted by a^{-1}. In a monoid, one can define positive integer powers of an element a: $a^1 = a$, and $a^n = \underbrace{a \circ \cdots \circ a}_{n \text{ times}}$ for $n > 1$. If b is the inverse of a, one can also define negative powers of a by setting $a^{-1} = b$ and $a^{-n} = \underbrace{b \circ \cdots \circ b}_{n \text{ times}}$ for $n > 1$. In addition, let us denote $a^0 = 1$ for each element $a \in G$. Throughout this section, let G^{-1} be the set of all invertible elements in G, i.e.,

$$G^{-1} \triangleq \{a \in G : \exists b \in G \text{ so that } a \circ b = 1 = b \circ a\}. \tag{5.19}$$

In fact, $(G^{-1}, \circ, 1)$ forms a group. For clarity, we omit "$\circ$" in the following presentation, i.e., writing $a \circ b$ as ab directly.

The conjugacy problem is extensively studied in *group theory*. But in this section, we would like to extend the conjugacy concept to monoids by a similar manner: Given a monoid G, for $\forall a \in G$ and $\forall x \in G^{-1}$, xax^{-1} is *conjugate* to a, and call x a conjugator of the pair (a, xax^{-1}).

Definition 5.4.1 (Conjugacy Search Problem, CSP). Let G be a noncommutative monoid. Given two elements $a, b \in G$ so that $b = xax^{-1}$ for some unknown element $x \in G^{-1}$, the objective of the conjugacy search problem in G is to find $x' \in G^{-1}$ so that $b = x'ax'^{-1}$ holds. Here, x' is not required to be x.

Definition 5.4.2 (Left self-distributive system, LD [94]). Suppose that S is a non-empty set, $F : S \times S \rightarrow S$ is a well-defined function and let us denote $F(a, b)$ by $F_a(b)$. If the following rewritten formula holds,

$$F_r(F_s(p)) = F_{F_r(s)}(F_r(p)), \quad (\forall p, r, s \in S) \tag{5.20}$$

then, we call $F.(\cdot)$ a *left self-distributive system*, abbreviated as LD system.

The terminology "left self-distributive" arises from the following analogical observation: If we consider $F_r(s)$ as a binary operation $r * s$, then the formula (5.20) becomes

$$r * (s * p) = (r * s) * (r * p), \tag{5.21}$$

i.e., the operation "$*$" is left distributive with respect to itself [94].

Combining the above two concepts, one can define the following LD system, named as Conj-LD system.

Theorem 5.4.3 (Conj-LD System). *Let G be a noncommutative monoid. The binary function F, given by the following conjugate operation*

$$F : G^{-1} \times G \rightarrow G, \quad (a, b) \mapsto aba^{-1}, \tag{5.22}$$

is an LD system, abbreviated as Conj-LD.

Proof. It is easy to see that F satisfies the rewritten formula (5.20). Thus, the above definitions $F_a(b)$ is an LD system. $\qquad\square$

In what follows, we prove some properties of the Conj-LD system. These properties are useful from the cryptographic viewpoint, providing that the related hardness assumptions hold.

Proposition 5.4.4. *Let F be a Conj-LD system defined over a noncommutative monoid G. Given $a \in G^{-1}$ and $b, c \in G$, we have*

(i) $F_a(a) = a$;

(ii) $F_a(b) = c \Leftrightarrow F_{a^{-1}}(c) = b$;

(iii) $F_a(bc) = F_a(b)F_a(c)$.

Proof. Firstly, since $aaa^{-1} = a$, we have $F_a(a) = a$. Next, $F_a(b) = c \Leftrightarrow c = aba^{-1} \Leftrightarrow a^{-1}ca = b \Leftrightarrow F_{a^{-1}}(c) = b$. Finally, $F_a(bc) = a(bc)a^{-1} = (aba^{-1})(aca^{-1}) = F_a(b)F_a(c)$. $\square$

Proposition 5.4.5 (Power Law). *Let F be a Conj-LD system defined over a noncommutative monoid G. Suppose that $a \in G^{-1}$ and $b \in G$. Then, for arbitrary three positive integers m, s, t so that $m = s + t$, we have*

$$F_a(b^m) = F_a(b^s)F_a(b^t) = F_a^m(b) \quad and \quad F_{a^m}(b) = F_{a^s}(F_{a^t}(b)). \qquad (5.23)$$

Proof. See the property (iii) in Proposition 1 and the definition of the Conj-LD system.

Remark 5.4.6. The term of left self-distributive system is due to the following reasons:

- The binary function defined by the formula (5.22) *does* satisfy the left self-distributiveness defined by the formula (5.20). In addition, it is more convenient to use $F_a(b)$ than to use aba^{-1} in the sequel presentation, especially when some additional operations are exerted on a and a^{-1} simultaneously from both sides.

- In 2006, Dehorney [94] proposed an authentication scheme based on left self-distributive systems in braid groups. Although some cryptanalysis on Dehorney's authentication scheme were reported [202], we find that Dehorney's work is still meaningful at least in the following two aspects: (1) self-distributive systems can be defined over arbitrary noncommutative monoids, rather than braid groups; (2) self-distributive systems have the potential for building variety of cryptographic schemes, rather than authentication schemes. Therefore, we would like to use the terminology of (left) self-distributive system to give Dehornoy the credit for coining this concept.

Now, using the notation of $F.(\cdot)$, the intractability assumption of the CSP problem in G can be re-formulated as follows: It is hard to retrieve a' from the given pair $(a, F_a(b))$ so that $F_a(b) = F_{a'}(b)$. We must take care of the relationship between the intractability assumption of CSP and the hardness CSP instances. The CSP problem is defined here as a *worst-case* problem, whereas from the cryptographic viewpoint, one needs *average-case* hardness for CSP instances. Therefore, we need a practical sampling algorithm that can produce hard CSP instances over G. For a generic noncommutative monoid G without explicit definition or presentation, it is difficult to determine whether we can sample hard CSP instances over G. Afterwards, the CSP instances used in our proposal are always assumed to be hard.

5.4.2 New Assumptions in Conj-LD Systems

Let G be a noncommutative monoid. Given a Conj-LD system F over a noncommutative monoid G (cf. Definition 5.4.3). Let $a \in G^{-1}$ and G_a denote the subgroup generated by $\{a, a^{-1}\}$, i.e., $G_a \triangleq \langle a, a^{-1} \rangle$. Now, given $b \in G$, we define the following notations:

- $\mathbb{T} \triangleq \{1, \cdots, n\}$ is a *finite* subset of $\mathbb{Z}$, where n is the order of $G_a{}^9$.

---

[9]In case G_a is *infinite*, we can set n as a fixed integer that is large enough, say 160 bits, to resist exhaustive

- The symbols "$\in \mathbb{T}$" and "$\xleftarrow{\$} \mathbb{T}$" always indicate sampling procedures that pick random integers uniformly from $\mathbb{T}$.

- $\mathcal{K}_{[a,b]} \triangleq \{F_{a^i}(b) : i \in \mathbb{T}\}$ is a *finite* subset of G.

Note that in cryptographic applications, a, b should be chosen appropriately so that n and $\mathcal{K}_{[a,b]}$ are large enough to resist exhaustive attacks.

Definition 5.4.7 (CSP-based Computational Diffie–Hellman Problem: CSP-CDH). Let F be a Conj-LD system over a noncommutative monoid G and let $\mathcal{A}$ be an adversary. For given $a \in G^{-1}$ and $b \in G$, consider the following experiment:

$$\text{Experiment } \mathbf{Exp}_{F,\mathcal{A}}^{csp-cdh}$$
$$i \xleftarrow{\$} \mathbb{T}; X \leftarrow F_{a^i}(b);$$
$$j \xleftarrow{\$} \mathbb{T}; Y \leftarrow F_{a^j}(b);$$
$$Z \leftarrow \mathcal{A}(X, Y);$$
$$\text{if } Z = F_{a^{i+j}}(b) \text{ then } b \leftarrow 1 \text{ else } b \leftarrow 0;$$
$$\text{return } b.$$

Now define the *advantage* of $\mathcal{A}$ violating the CSP-based computational Diffie–Hellman assumption as

$$\mathbf{Adv}_{F,\mathcal{A}}^{csp-cdh} = \Pr[\mathbf{Exp}_{F,\mathcal{A}}^{csp-cdh} = 1]. \tag{5.24}$$

The CSP-CDH assumption states, roughly, that given $F_{a^i}(b)$ and $F_{a^j}(b)$, where i, j are drawn at random from $\mathbb{T}$, it is hard to compute $F_{a^{i+j}}(b)$. Suppose the CSP-CDH assumption is moderate, the adversary is possible to compute something interesting, such as the most significant bit of $F_{a^{i+j}}(b)$ given $F_{a^i}(b)$ and $F_{a^j}(b)$. This means the CSP-CDH problem is too weak to be directly used in cryptographic applications. Thus, we need the following stronger variants.

---

attacks. Furthermore, to ensure the randomness of sampling from $\mathcal{K}_{[a,b]}$, n should be the order of a factor-group of G_a modulo the centralizer of b in G_a.

Definition 5.4.8 (CSP-based Decisional Diffie–Hellman Problem: CSP-DDH). Let F be a Conj-LD system over a noncommutative monoid G and let $\mathcal{A}$ be an adversary. Given $a \in G^{-1}$ and $b \in G$, consider the following two experiments in a parallel manner

Experiment $\mathbf{Exp}_{F,\mathcal{A}}^{csp-ddh-real}$	Experiment $\mathbf{Exp}_{F,\mathcal{A}}^{csp-ddh-rand}$
$i \xleftarrow{\$} \mathbb{T}; X \leftarrow F_{a^i}(b);$	$i \xleftarrow{\$} \mathbb{T}; X \leftarrow F_{a^i}(b);$
$j \xleftarrow{\$} \mathbb{T}; Y \leftarrow F_{a^j}(b);$	$j \xleftarrow{\$} \mathbb{T}; Y \leftarrow F_{a^j}(b);$
$Z \leftarrow F_{a^{i+j}}(b);$	$\ell \xleftarrow{\$} \mathbb{T}; Z \leftarrow F_{a^\ell}(b);$
$b \leftarrow \mathcal{A}(X, Y, Z);$	$b \leftarrow \mathcal{A}(X, Y, Z);$
return b.	return b.

Now define the *advantage* of $\mathcal{A}$ violating the CSP-based decisional Diffie–Hellman assumption as

$$\mathbf{Adv}_{F,\mathcal{A}}^{csp-ddh} = |\Pr[\mathbf{Exp}_{F,\mathcal{A}}^{csp-ddh-real} = 1] - \Pr[\mathbf{Exp}_{F,\mathcal{A}}^{csp-ddh-rand} = 1]|. \quad (5.25)$$

That is, the CSP-DDH assumption states that the distributions $(F_{a^i}(b), F_{a^j}(b), F_{a^{i+j}}(b))$ and $(F_{a^i}(b), F_{a^j}(b), F_{a^\ell}(b))$ are computationally indistinguishable when i, j, ℓ are drawn at random from $\mathbb{T}$.

We now move on to another assumption that will be used in our proposal. Here, the adversary is allowed to access a restricted CSP-DDH oracle $\mathcal{O}_v(\cdot, \cdot)$, which behaves as follows:

$$\mathcal{O}_v(X, U) = \begin{cases} 1, \text{if } U = F_v(X); \\ 0, \text{otherwise.} \end{cases} \quad (5.26)$$

That is, the oracle tells whether the second argument is conjugate to the first argument by v. This oracle can be treated as a restricted CSP-DDH oracle with one of its arguments being fixed as $F_v(b)$.

Definition 5.4.9 (CSP-based Strong Diffie–Hellman Problem: CSP-SDH). Let F be a Conj-LD system over a noncommutative monoid G and let $\mathcal{A}$ be an adversary. Given

$a \in G^{-1}$ and $b \in G$, consider the following experiment

> Experiment $\mathbf{Exp}_{F,\mathcal{A}}^{csp-sdh}$
>
> $i \xleftarrow{\$} \mathbb{T}; X \leftarrow F_{a^i}(b)$;
>
> $j \xleftarrow{\$} \mathbb{T}; Y \leftarrow F_{a^j}(b)$;
>
> $\mathcal{O}_v(X, U) \overset{def}{=} (U = F_v(X))$;
>
> $Z \leftarrow \mathcal{A}^{\mathcal{O}_v(\cdot,\cdot)}(X, Y)$;
>
> if $Z = F_{a^{i+j}}(b)$ then $b \leftarrow 1$ else $b \leftarrow 0$;
>
> return b.

Now define the *advantage* of $\mathcal{A}$ violating the CSP-based strong Diffie–Hellman assumption as

$$\mathbf{Adv}_{F,\mathcal{A}}^{csp-sdh} = \Pr[\mathbf{Exp}_{F,\mathcal{A}}^{csp-sdh} = 1]. \tag{5.27}$$

The intuition behind the CSP-SDH assumption is that the restricted CSP-DDH oracle is useless because the adversary already "knows" the answer to almost any query it will ask. Similar to [77], it is also worth mentioning that the CSP-SDH assumption is different from (and weaker than) the so-called *gap CSP-CDH assumption* where an adversary gets access to a *full* CSP-DDH decision oracle.

Remark 5.4.10. At present, it is unclear whether the CSP-DDH problem is indeed hard or not. Intuitively, we cannot solve the CSP-DDH problem without solving the CSP problem if G is a generic monoid. In fact, the CSP problem and the CSP-DDH problem of Conj-LD systems over a noncommutative monoids are counterparts of the DLP problem and the DDH problem over finite fields, respectively. According to [212], we know that the DLP problem and the DDH problem are polynomially equivalent in a generic cyclic group. Therefore, we speculate that the CSP problem and the CSP-DDH problem in a generic noncommutative monoid are polynomially equivalent.

5.4.3 Cryptosystems from Conj-LD Systems

In this section, we always assume that F is a Conj-LD system defined over a non-commutative monoid G, and $a \in G^{-1}$ and $b \in G$.

As a warmup step, let us at first describe a Diffie–Hellman-like key agreement protocol by using the CSP-DDH assumption over a noncommutative monoid G. Assume that Alice and Bob want to negotiate a common session key. Then, Alice (resp. Bob) picks $s \in \mathbb{T}$ (resp. $t \in \mathbb{T}$). Then, send $F_{a^s}(b)$ (resp. $F_{a^t}(b)$) to Bob (resp. Alice). Finally, both of them can extract $F_{a^{s+t}}(b)$, by which a session key can be computed as

$$K_{session} = Kdf(F_{a^{s+t}}(b)), \tag{5.28}$$

where $Kdf(\cdot)$ is a key derivation function, such as KDF1 defined in IEEE Std 1363-2000.

The above protocol[10] immediately implies the following ElGamal-like construction, denoted by CSP-ElG.

Protocol 5.4.11. CSP-Based ElGamal-Like Encryption Scheme (CSP-ElG):

- **Key-generation.** Suppose that k is the security parameter, $\mathcal{M} = \{0,1\}^k$ is the message space, and $\mathcal{C} = \mathcal{K}_{[a,b]} \times \mathcal{M}$ is the ciphertext space. In addition, we need a cryptographic hash function $H : \mathcal{K}_{[a,b]} \to \mathcal{M}$. A user at first picks $s \in \mathbb{T}$, and then publishes $pk = F_{a^s}(b)$ as his public key, and keeps his secret key $sk = s$.

- **Encryption.** Given the public-key $pk \in \mathcal{K}_{[a,b]}$ and a message $m \in \mathcal{M}$, one picks $t \in \mathbb{T}$, and then constructs a ciphertext as follows:

$$c = (F_{a^t}(b), \ m \oplus H(F_{a^t}(pk))) \tag{5.29}$$

---

[10]This protocol cannot resist the so-called man-in-middle attack. But in this section we don't pay attention to how to remedy it.

- **Decryption.** Given the secret key $s \in \mathbb{T}$ and a ciphertext $c = (c_1, c_2) \in \mathcal{K}_{[a,b]} \times \mathcal{M}$, one can extract a plaintext as follows:

$$m = c_2 \oplus H(F_{a^s}(c_1)) \tag{5.30}$$

The consistency of the CSP-ElG scheme directly comes from the power law of the Conj-LD system. The security of the CSP-ElG scheme is captured by the following theorem.

Theorem 5.4.12 (IND-CPA of CSP-ElG). *Based on the CSP-DDH assumption, we claim that the ciphertexts of the encryption scheme CSP-ElG are indistinguishable under chosen plaintext attacks in the standard model.*

Proof. Assume that the CSP-DDH assumption holds for the underlying noncommutative monoid G. We will prove by contradiction that CSP-ElG is IND-CPA. Suppose that CSP-ElG is not IND-CPA, and let $\mathcal{A}$ be an algorithm which, on the system parameters $a \in G^{-1}$, $b \in G$ and a random public key $pk \in \mathcal{K}_{[a,b]}$, has probability non-negligibly greater than 1/2 of distinguishing random encryptions $Enc(m_0)$ and $Enc(m_1)$ of two messages m_0, m_1 of its choice. Let $Z = (F_{a^s}(b), F_{a^t}(b), F_{a^u}(b)) \in \mathcal{K}^3_{[a,b]}$ be either a random CSP-DDH triple or a random triple, with equal probability. We will produce an algorithm $\mathcal{B}$ which can distinguish between the two cases, using $\mathcal{A}$ as an oracle, with high probability.

The algorithm $\mathcal{B}$ who interacts with the algorithm $\mathcal{A}$ can be defined as follows:

- $\mathcal{B}$ at first picks a random integer $v \in \mathbb{T}$ and sets the public-key as $pk = F_{a^{s+v}}(b) \in \mathcal{K}_{[a,b]}$, which is then sent to $\mathcal{A}$.

- Upon receiving the public-key pk, $\mathcal{A}$ selects two messages $m_0, m_1 \in \mathcal{M}$ with equal length as the challenge messages, which are then sent to $\mathcal{B}$.

- Upon receiving the challenge messages pair (m_0, m_1), $\mathcal{B}$ randomly picks another integer $w \in \mathbb{T}$, flips a coin $\beta \in \{0, 1\}$ and then replies $\mathcal{A}$ with the challenge ciphertext as follows:

$$c_\beta^* = (F_{a^{t+w}}(b), m_\beta \oplus H(F_{a^{u+v+w}}(b))).$$

- Upon receiving the challenge ciphertext c_β^*, $\mathcal{A}$ replies $\mathcal{B}$ with $\hat{\beta} \in \{0, 1\}$, i.e., $\mathcal{A}$'s guess on β.

- Upon receiving $\hat{\beta} \in \{0, 1\}$, i.e., $\mathcal{A}$'s guess on β, $\mathcal{B}$ checks whether $\mathcal{A}$'s guess is correct, i.e., whether $\hat{\beta} = \beta$ holds.

- By repeating the above process with several different choices of random integers v, w, the algorithm $\mathcal{B}$ can determine with high probability whether or not $\mathcal{A}$ can determine the value of β. Based on the detailed analysis given below $\mathcal{B}$ can, in this way, determine whether or not Z is a CSP-DDH triple, thus violating the CSP-DDH assumption over G.

In detail, there are two cases to be taken into consideration during a single execution of the above interactive process between $\mathcal{B}$ and $\mathcal{A}$.

- Suppose that $u = s + t$. Then $u + v + w = (s + v) + (t + w)$, so Z is a CSP-DDH triple in $\mathcal{K}_{[a,b]}^3$. Moreover, all possible CSP-DDH triples w.r.t. (a, b) are equally likely to occur as Z, since v and w are random. Hence, c_β^* is a valid random encryption of m_β because $F_{a^{t+w}}(b)$ is random in $\mathcal{K}_{[a,b]}$. Under these conditions, the algorithm $\mathcal{A}$, by hypothesis, will succeed in outputting β with probability exceeding 1/2 by a non-negligible quantity.

- Suppose that u is random. Then Z is a random triple in $\mathcal{K}_{[a,b]}^3$, and all possible triples belonging to $\mathcal{K}_{[a,b]}^3$ occur with equal probability. In this situation, the probability distribution of c_0^* is identical to that of c_1^*, w.r.t. all possible random choices of v and w. It follows that the algorithm $\mathcal{A}$ cannot exhibit different be-havior for $\beta = 0$ and $\beta = 1$. Note that we can arrive at this conclusion even though the expression c_β^* is an invalid encryption of m_β — that is, even though there is no information about how $\mathcal{A}$ behaves on invalid inputs, it is certain that $\mathcal{A}$ cannot behave differently depending on the value of β.

The above analysis shows that if Z is a CSP-DDH triple then $\mathcal{A}$ with non-negligible probability exhibits different behavior depending on whether $\beta = 0$ or $\beta = 1$, whereas $\mathcal{A}$ must behave identically regardless of the value of β if Z is not a CSP-DDH triple. $\quad\square$

Based on the above scheme, it is not difficult to derive a CCA secure encryption scheme by employing the Fujisaki-Okamoto transformation. Here, we would like to give two different extensions that are enlightened by [1, 77, 91, 177]. The first extension is called the hashed ElGamal variant that is IND-CCA secure in the random oracle model, and the second is called the Cramer-Shoup-like variant that is IND-CCA secure in the standard model.

Protocol 5.4.13. CSP-hElG.

- **Key-generation.** Suppose that k is the security parameter, $\mathcal{M} = \{0, 1\}^k$ is the message space, and $\mathcal{C} = \mathcal{K}_{[a,b]} \times \mathcal{M}$ is the ciphertext space. In addition, we need a symmetric cipher $\Pi = (E, D)$ with the key space $\mathcal{K}$ and a hash function $H : \mathcal{K}_{[a,b]}^2 \to \mathcal{K}$. A user at first picks an integer $s \in \mathbb{T}$, and then publishes $pk = F_{a^s}(b)$ as his public key, and keeps his secret key $sk = s$.

- **Encryption.** Given the public-key $pk \in \mathcal{K}_{[a,b]}$ and a message $m \in \mathcal{M}$, one picks an integer $t \in \mathbb{T}$, and then constructs a ciphertext $c = (c_1, c_2)$ as follows:

$$c_1 := F_{a^t}(b), \ T := F_{a^t}(pk), \ K := H(c_1, T), \ c_2 := E_K(m) \qquad (5.31)$$

- **Decryption.** Given the secret key $s \in \mathbb{T}$ and a ciphertext $c = (c_1, c_2) \in \mathcal{K}_{[a,b]} \times$

$\mathcal{M}$, one can extract a plaintext as follows:

$$Z := F_{a^s}(c_1), \ K := H(c_1, Z), \ m := D_K(c_2) \tag{5.32}$$

The consistency of the CSP-hElG scheme can also be easily verified according to the power law of the Conj-LD system. The security of the CSP-hElG scheme is formulated by the following theorem.

Theorem 5.4.14 (IND-CCA of CSP-hElG). *If H is modeled as a random oracle, and the underlying symmetric cipher Π is secure against chosen ciphertext attacks, then the hashed ElGamal encryption scheme CSP-hElG is secure against chosen ciphertext attacks under the strong CSP-CDH assumption.*

Proof. See Theorem 2 in [1] and Theorem 1 in [177], as well as the improvement of DHIES given in Section 5.1 of [177].

Protocol 5.4.15. CSP-CS.

- **Key-generation.** Suppose that k is the security parameter, $\mathcal{M} = \{0, 1\}^k$ is the message space, and $\mathcal{C} = \mathcal{K}_{[a,b]}^2 \times \widetilde{\mathcal{K}_{[a,b]}} \times \mathcal{M}$ is the ciphertext space, where $\widetilde{\mathcal{K}_{[a,b]}}$ is defined as

$$\widetilde{\mathcal{K}_{[a,b]}} \triangleq \{F_{a^i}(F_{a^{j_1}}(b) F_{a^{j_2}}(b)) : i, j_1, j_2 \in \mathbb{T}\}.$$

In addition, we need a symmetric cipher $\Pi = (E, D)$ with the key space $\mathcal{K}$ and two hash functions $H : \mathcal{K}_{[a,b]}^2 \to \mathbb{T}$ and $H_1 : \mathcal{K}_{[a,b]} \to \mathcal{K}$. A user at first picks $x_1, x_2, x_3, x_4 \in \mathbb{T}$, and then publishes $pk = (X_1, X_2, X_3, X_4)$ as his public key, and keeps his secret key $sk = (x_1, x_2, x_3, x_4)$, where $X_i = F_{a^{x_i}}(b)$ for $i = 1, 2, 3, 4$.

- **Encryption.** Given the public-key $pk = (X_1, X_2, X_3, X_4) \in \mathcal{K}_{[a,b]}^4$ and a message $m \in \mathcal{M}$, one picks $t \in \mathbb{T}$, and then constructs a ciphertext $c = (c_1, c_2, c_3, c_4)$ as follows:

$$c_1 := F_{a^t}(b), \ c_2 := F_{a^t}(X_1), \ c_3 := F_{a^t}(F_{a^h}(X_2)X_3), \ c_4 := E_K(m),$$

where $h := H(c_1, c_2)$ and $K := H_1(F_{a^t}(X_4))$.

- **Decryption.** Given the secret key $(x_1, x_2, x_3, x_4) \in \mathbb{T}^4$ and a ciphertext $c = (c_1, c_2, c_3, c_4) \in \mathcal{K}_{[a,b]}^2 \times \widetilde{\mathcal{K}_{[a,b]}} \times \mathcal{M}$, one at first computes $h := H(c_1, c_2)$ and then tests if the following two equalities hold:

$$F_{a^{x_1}}(c_1) = c_2 \text{ and } F_{a^{h+x_2}}(c_1)F_{a^{x_3}}(c_1) = c_3.$$

If not, output $\perp$. Otherwise, compute $K := H_1(F_{a^{x_4}}(c_1))$ and output $m := D_K(c_4)$.

The consistency and the security of the CSP-CS scheme are formulated by the following theorems, respectively.

Theorem 5.4.16 (Consistency of CSP-CS). *The CSP-CS scheme is consistent.*

Proof. Suppose that $c = (c_1, c_2, c_3, c_4)$ is a well-formed ciphertext on the message m. Then, there exists some integer $t \in \mathbb{T}$ so that

$$c_1 := F_{a^t}(b), \ c_2 := F_{a^t}(X_1), \ c_3 := F_{a^t}(F_{a^h}(X_2)X_3), \ c_4 := E_K(m),$$

where $h := H(c_1, c_2)$ and $K := H_1(F_{a^t}(X_4))$. Since $X_i = F_{a^{x_i}}(b)$ for $i =$

$1, 2, 3, 4$, we have

$$
\begin{aligned}
K &= H_1(F_{a^t}(F_{a^{x_4}}(b))) = H_1(F_{a^{t+x_4}}(b)) \\
&= H_1(F_{a^{x_4}}(F_{a^t}(b)) = H_1(F_{a^{x_4}}(c_1)), \\
c_2 &= F_{a^t}(F_{a^{x_1}}(b)) = F_{a^{x_1+t}}(b) \\
&= F_{a^{x_1}}(F_{a^t}(b)) = F_{a^{x_1}}(c_1),
\end{aligned}
$$

$$
\begin{aligned}
c_3 &= F_{a^t}(F_{a^h}(F_{a^{x_2}}(b))F_{a^{x_3}}(b)) = F_{a^t}(F_{a^{h+x_2}}(b)F_{a^{x_3}}(b)) \\
&= F_{a^t}(F_{a^{h+x_2}}(b))F_{a^t}(F_{a^{x_3}}(b)) = F_{a^{t+h+x_2}}(b)F_{a^{t+x_3}}(b) \\
&= F_{a^{h+x_2}}(F_{a^t}(b))F_{a^{x_3}}(F_{a^t}(b)) = F_{a^{h+x_2}}(c_1)F_{a^{x_3}}(c_1).
\end{aligned}
$$

That is, the ciphertext c can stand the validation and the value of K used in encryption $c_4 = E_K(m)$ is exactly the value of K used in decryption $m = D_K(c_4)$. Since $\Pi = (E, D)$ is symmetric, the output m of the decryption is exactly the input m of the encryption. Therefore, the CSP-CS scheme is consistent.

Theorem 5.4.17 (IND-CCA of CSP-CS). *Suppose H is a target collision resistant hash function. Further, suppose the CSP-DDH assumption holds, and the symmetric cipher $\Pi = (E, D)$ is secure against chosen ciphertext attack. Then CSP-CS is secure against chosen ciphertext attack.*

Proof. See Theorem 13 in [77].

5.4.4   Security and Efficiency Issues on $F_{a^t}(b)$

For computing $F_{a^t}(b)$, we should at first compute a^t, and then plus one inversion and two multiplications in the underlying noncommutative monoid G. When t is large, say several hundreds of digits, rather than to multiply a for t times, a similar "successive doubling" method should be employed, and thus a factor of $\log t$ would be taken into consideration in performance evaluations. At present, it is enough to set t as an integer with 160 bits to resist exhaustive attacks.

It is necessary to assume that the basic monoid operations (i.e., multiplication and inversion) can be finished efficiently. This assumption implies that the lengths of the representations of all elements in G, including a, b, a^t, and $F_{a^t}(b)$, should be polynomial in the system security parameters, since the results have to be provided in bits by using classical computers.

Moreover, we require that there is a *secure efficient canonical form* for representing elements in G. This means that

(C-1) By using this form, the representation of an element in G is unique. Otherwise, the proposed schemes cannot work.

(C-2) The transformation from an element in G to its canonical form can be finished efficiently. Otherwise, the proposed schemes are impractical.

(C-3) By using this form, the length of an element $F_{a^t}(b)$ does not reveal any information about a^t. Otherwise, the developed assumptions could suffer from the so-called length-based attacks.

5.5 Improved Key Exchange over Thompson's Group

In 2005, Shpilrain and Ushakov proposed a key exchange protocol, referred to as SU05, based on the intractability of *decomposition problem* on Thompson's group [259]. They claimed that the decomposition problem has the same difficulty level with *conjugacy search problem* [6, 175], and the advantage of the proposed protocol is that it is invulnerable to to *length-based attack* according to their simulation results. In 2007, Ruinskiy et al. [235] proposed an improved length-based attack, referred to

as RST07, which does indeed threaten the security of SU05. They used five methods,
"using memory," "avoiding repetitions," "looking ahead," "automorphism attacks," and
"alternative solutions," to improve the basic length-based attack and tested their success rates of breaking SU05. From their simulation results, the best success rate is over
80%. By investigating the properties of SU05 and the basic length-based attack, as well
as the computational complexity of RST07, we [80] propose an improved key exchange
protocol over Thompson's Group. It can resist known length-based attacks, including
RST07 and other similar ones.

5.5.1 Thompson's Group and Decomposition Problem

Thompson's group F is well known in many areas of mathematics, including algebra, geometry, and analysis. This group is infinite and noncommutative. It is important
that Thompson's group has the following nice presentation in terms of generators and
relations:

$$F = \langle x_0, x_1, x_2, \cdots | x_i^{-1} x_k x_i = x_{k+1}(k > i)\rangle \tag{5.33}$$

This presentation is infinite. There are also finite presentations of this group, for example,

$$F = \langle x_0, x_1, x_2, x_3, x_4 | x_i^{-1} x_k x_i = x_{k+1}(k > i, k < 4)\rangle$$

For the infinite presentation, any $w \in F$ can be represented as a unique normal form

$$x_{i_1} \cdots x_{i_s} x_{j_t}^{-1} \cdots x_{j_1}^{-1} \tag{5.34}$$

so that

- (NF1) $i_1 \leq \cdots \leq i_s$ and $j_1 \leq \cdots \leq j_t$

- (NF2) if both x_i and x_i^{-1} occur, then either x_{i+1} or x_{i+1}^{-1} occurs, too.

The time complexity of reducing a word of length n to the normal form in Thompson's group is $O(|n| \log |n|)$, which is very efficient. Since each word has a unique normal form, we use "word length" to represent "length of the word's normal form" in the rest of this section.

Recall the decomposition problem over braid group (See Section 5.2.2), we define the decomposition problem over Thompson's group F as follows: Given a subset $A \subseteq F$ and a pair of element $(w, w') \in F^2$, find two elements $x, y \in A$ so that $w' = x \cdot w \cdot y$. In fact, the conjugacy search problem is a special case of the decomposition problem in the sense of taking $x = y^{-1}$.

5.5.2   Analysis of SU05 Protocol

Based on the intractability assumption of the decomposition problem over arbitrary (semi) group G, achieving key agreement is quite straightforward: given two subsets $A, B \subseteq G$ so that $ab = ba$ for any $a \in A, b \in B$, and given a public element $w \in G$. Alice selects private $a_1 \in A, b_1 \in B$ and sends the element $a_1 w b_1$ to Bob. Similarly, Bob selects private $b_2 \in B, a_2 \in A$ and sends the element $b_2 w a_2$ to Alice. Then Alice computes $K_{\text{Alice}} = a_1 b_2 w a_2 b_1$ and Bob computes $K_{\text{Bob}} = b_2 a_1 w b_1 a_2$. Since $a_i b_j = b_j a_i$ $(i, j = 1, 2)$ in G, one has $K_{\text{Alice}} = K_{\text{Bob}} = K$, which can be used as Alice's and Bob's common secret key.

The non-triviality of the SU05 protocol lies in instantiating with the Thompson's group F given by its standard infinite presentation (5.33) and the method for defining the suite subset $A, B \subseteq F$, as well as the method for sampling related elements in F.

Suppose that $s \in N^+$ is the first system parameter, which is relevant to the size of following *generating groups*. Let $S_A = \{x_0 x_1^{-1}, x_0 x_2^{-1}, \cdots, x_0 x_s^{-1}\}$, $S_B = \{x_{s+1}, x_{s+2}, \cdots, x_{2s}\}$, $S_W = \{x_1, x_2, \cdots, x_{s+2}\}$ be the generating groups, and define the groups generated by $S_A^{\pm 1}, S_B^{\pm 1}, S_W^{\pm 1}$ as A_s, B_s and W, respectively. Obviously, A_s and B_s are subgroups of F and according to [259], for $\forall \in A_s$ and $\forall b \in B_s$,

$ab = ba$.

Another system parameter of the protocol, $L \in N^+$, is chosen to control the word length. SU05 uses the same method to randomly generate $a_1, a_2 \in A_s, b_1, b_2, \in B_s$ and $w \in W$. We take a_1 as an example to analyze the generating process.

The generating process begins with an empty word e. Let $u_0 = e$. Firstly, randomly choose an element r_1 from $S_A^{\pm 1}$, and multiply u_0 on the right by r_1 and get u_1. For example, if $r_1 = x_0 x_1^{-1}$, then $u_1 = u_0 r_1 = x_0 x_1^{-1}$. Repeat the previous step until the word length reaches L. Then we have generated a random word $a_1 \in A_s$ of length L.

According to analysis of the generating process of a_1, we find that a_1 can always be generated within finite steps from an empty word to a word of length L, due to one property of Thompson's group: with very high probability, product of two words has larger length than any one of the two words. Thus after each multiplication, the word length increases with high probability. As long as the generators are randomly chosen, the word length will finally reach L within finite steps. Suppose the expectation of word length increment after each step is $E(\Delta l)$, then on average after $L/E(\Delta l)$ steps, a word of length L can be generated. Here $E(\Delta l)$ is a constant which only relates to subset $S_A^{\pm 1}$, so the generating time of a_1 is linear in L.

Although the generating process is efficient due to the property of Thompson's group we mentioned above, the protocol is also insecure under the length-based attacks due to the same reason.

5.5.3 Analysis of RST07 Attack

At first, let us review the process of the basic length-based attack and analyze the capability of SU05 for resisting the basic length-based attack. Let G be a subgroup of Thompson's group F, generated by the generators in $S_G = \{g_1^{\pm 1}, \cdots, g_k^{\pm 1}\}$. For any unknown $y, w \in G$, given the product $z = yw$ and a black box **BOX**, where for any $w' \in G$, if $w' = w$ holds, **BOX**$(w') = 1$, otherwise **BOX**$(w') = 0$. The purpose of length-based attack is to output the exact w.

Represent y as the product of generators, i.e., $y = y_1 y_2 \cdots y_n$, where $y_i \in S_G$. Then, we have $z = y_1 y_2 \cdots y_n w$. Let $\ell(z)$ be the word length of z. Since in Thompson's group, "with very high probability, product of two words has larger length than any one of the two words," the following inequality also holds with very high probability

$$\ell(y_1^{-1} z) < \ell(z) < \ell(y_j z), \tag{5.35}$$

where $y_j \in S_G$ and $y_j \neq y_1^{-1}$. It indicates that if we multiply each of the $2k$ generators on the left of z, except y_1^{-1}, other $2k - 1$ generators will make the length of product increase with very high probability.

Based on this observation, a basic length-based attack algorithm can be formalized as follows:

Algorithm 1 Basic Length-based Attack

Procedure Basic Length-based Attack

Input: Product $z = yw$, black box **BOX**

Output: w or $\perp$

1:    $z \leftarrow yw$

2:    **while** running time does not exceed limitation **do**

3:        choose $h \in S_G$ so that $\ell(h^{-1} z) \leq \ell(g^{-1} z)$ for all $g \in S_G$

4:        $z \leftarrow h^{-1} z$

5:        **if BOX**(z) = 1 **then**

6:            **return** z

7:        **end if**

8:    **end while**

9:    **return** $\perp$

Note that the algorithm needs a limitation of running time in order to stop looping if it keeps failing within a long time. As we can see, the algorithm is based on the assumption that the generator h which minimizes $\ell(h^{-1}z)$ is the right decomposition.

We take the following example to analyze the attack algorithm. Assume $k = 3$, i.e., $S_G = \{g_1^{\pm 1}, g_2^{\pm 1}, g_3^{\pm 1}\}$. Let $n = 3$ and $y = g_1 g_3 g_2^{-1}$. The process of generating y can be seen as operation on a $2k$-ary tree, as shown in Fig 5.6(a). It begins with an empty word e and one of the $2k(k - 3)$ child nodes, y_1, which is randomly chosen. Then, repeat the previous step until some y_n is chosen so that the length of $y = y_1 y_2 \cdots y_n$ is satisfied. Since the $2k$-ary tree has $(2k)^n$ leaf-nodes at the nth level, where each leaf node represents a possible value of y, it is impossible to find y with brute force search.

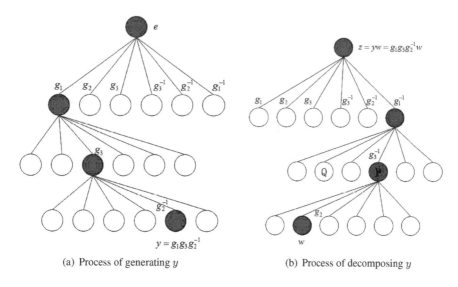

(a) Process of generating y (b) Process of decomposing y

Figure 5.6: An example for analyzing basic length-based attack

The process of attack is reverse for instance searching on a $2k$-ary tree. Dark nodes are the targets of attack which forms the optimal search path (as shown in Figure 5.6(b)). The best result of an attack algorithm is to find this path which indicates decomposition of y. However, mostly there are more than one node to choose in step 5 of algorithm ??.

For example, in Fig 5.6(b), during the process of searching in the 2nd level, if there are two child-nodes P, Q with equal length and the algorithm chooses node Q wrongly, then the algorithm has to search all the child-nodes of Q which has smaller length than P before turning back to P. It makes the algorithm fall into an exponential search in the worst case. If there are on average two candidate nodes in each level, then the time complexity of attack algorithm is $\mathcal{O}(2^n)$. This should be the main reason why basic length-based attack cannot defeat SU05 [259].

Next, let us analyze the performance of RST07, i.e., the improved length-based attack proposed by Ruinskiy et al. [235]. RST07 includes the following five improvements. We will analyze them one by one with the purpose of finding out methods for resisting this improved attack.

- **Using memory**. Using memory increases width of search. By using memory, the algorithm computes $2km$ child-nodes from M subtrees and chooses from them the shortest M ones as the roots of subtrees for the next loop. As a result, the search width is enhanced from 1 to M. By this, the algorithm becomes efficient, but it is still exponential. For example, if in any loop, the right node is not included in the M candidates, then the algorithm may degenerate to exponential. Suppose there exits a y_j so that $\ell(y_j w_{j+1}) < \ell(w_{j+1})$, where $w_{j+1} = y_{j+1} \cdots y_n$, say y_j left multiplies w_{j+1} makes its length decrease. In this situation, the right node may not be included in the M candidates, because the length of child-node, referred to as y_j, increases, but it happens to be correct. In fact, this is the most common situation in which the basic length-based attack fails, and the success of the attack is based on the core assumption that direction of length descent is correct. However, the above situation is just opposite. If we introduce more like situations, the success rate of the attack algorithm will obviously decrease.

- **Avoiding repetitions**. Avoiding repetitions is an improvement usually seen in search. Since two nodes in the search tree may have the same value, the algorithm maintains a hash table to record values of visited nodes. If the same value appears again, then the node will be canceled from the candidates list. Therefore, avoiding repetitions not only improves efficiency of the algorithm, but also prevents the algorithm from trapping into a closed loop. The extra time and space costs involved in the method are negligible.

- **Look-ahead**. Look-ahead increases depth of search. The original algorithm searches one level each time, while the method searches p level each time. The problem of look-ahead is obvious. On one hand, the time complexity it involves increases exponentially with p. Because the time cost of traversing a p-level subtree is $(2k)^p$, p should not be large. On the other hand, since p is not large, if multiple right nodes are increased in length, even if the algorithm looks ahead for p steps, it still cannot find the right path. For example, in a p-level subtree, if there are $p/2$ right nodes being increased in length, then the right node at level p is also increased in length, thus the algorithm will again fall into an exponential search. Simulation results of RST07 prove this claim, for the algorithm success rate increases slightly by using look-ahead.

- **Automorphism attacks**. From simulation results of RST07, the improvement of automorphism attack is not satisfied. It indicates that in Thompson's group, at least under their parameters selection, random automorphism functions are highly correlative, thus effect of this method is negligible.

- **Alternative solutions**. Alternative solution is not a general method, it is a specific solution to SU05. Although this improvement is very efficient, it does not change the search complexity. In fact, this method helps the algorithm to find more potential right nodes which vastly decreases search time, but it is based on the condition that the search direction is right. If the algorithm enters a wrong subtree, it still has large possibility to fall into an exponential search.

In brief, the basic length-based attack frequently falls into exponential search, so its success rate is very low, while the most significant improvements of RST07 are using memory and look-ahead, because to some extent they are able to avoid exponential search. However, the cost of look-ahead is too large to make it effective. Avoiding repetition is usually seen in search algorithm, although it is effective, the algorithm complexity is not changed. Automorphism attacks are restricted by the highly related automorphism functions, thus it is also not effective. Alternative solution not only enhances verification efficiency but also decreases search volume, thus it obviously increases success rate. But it does not change search complexity from exponential to polynomial, and it is not a general method.

Therefore, if we want to improve the security of the SU05, we need to make the attack algorithm fall into exponential search. In order to test efficiency of RST07, we realized SU05 and RST07 and acquired our own experiment data. Through the data analysis we propose our improved key exchange protocol on Thompson's group.

5.5.4   Tests and Improvements

We choose to realize the combination of "using memory + avoid repetitions + alternative solutions," which has the highest success rate in the original paper. The simulation results are shown in Table 5.2, where η_1 is the success rate of RST07 recovering a_1, $M = 1024$. We change s and L to analyze success rates under different parameters, and use them as reference of latter experiments. Each success rate is the average value of over 1000 test cases, and the failure condition is searching over 100,000 nodes.

Because of the possible differences between experiment equipments and between definitions of failure (how many nodes visited), our simulation results are different from those in the original paper. But we will continue to use our results as reference to

Table 5.2: Success rate of recovering a_1

L	$\eta_1/\%$					
	$s=3$	$s=4$	$s=5$	$s=6$	$s=7$	$s=8$
64	93.5	82.6	67.3	50.1	32.5	11.6
128	34.5	23.1	17.2	8.9	3.1	1.2
256	11.3	6.2	3.7	1.2	0.3	0
320	8.8	4.2	1.5	0.6	0.1	0

analyze our improvements of SU05.

As shown in Table 5.2, the success rate of the attack algorithm decreases dramatically with the increment of s and L. Irrespective of the increase in base number or exponent, efficiency of the algorithm decreases rapidly. The phenomenon itself shows the algorithm is exponential. Obviously, increasing protocol parameters is one of the direct methods to resist exponential attacks.

Based on the analysis aforementioned, we find out that randomly generating a_1 leads to nondecreasing growth of its length. As a result, when left multiplying a_1 to $a_1 w b_1$ repeatedly, the length of product decreases gradually, and this situation just satisfies the success condition of length-based attacks. Thus we need to control the process of generating a_1 to improve its robustness against the length-based attacks. Therefore, we propose an improved key exchange protocol over Thompson's group as follows:

1. Pick $s \in [6, 10]$ and $L \in \{512, 514, \cdots, 640\}$. Set the security parameters as N, T, R. Choose $w \in W$ so that $\ell(w) = L$. Publish s, L, w, N, T, R.

2. Alice randomly chooses secret $a_1 \in A_s, b_1 \in B_s$ so that $\ell(a_1) = L, \ell(b_1) = L$, and computes $u_1 = a_1 w b_1$.

3. Suppose $a_1 = r_1 r_2 \cdots r_n, r_i \in S_A^{\pm}, i = 1, \cdots, n, b_1 = t_1 t_2 \cdots t_m, t_j \in S_B^{\pm}, j = 1, \cdots, m$. Alice verifies the following conditions.

   (a) $n \geq N$ (and $m \geq N$)

   (b) $\ell(r_T^{-1} r_{T-1}^{-1} \cdots r_1^{-1} u_1) \geq \ell(u_1)$ (and $\ell(u_1 t_m^{-1} t_{m-1}^{-1} \cdots t_{m-T+1}^{-1}) \geq \ell(u_1)$)

(c) After a_1's first T factors are left multiplied to u_1 (b_1's first T factors are right multiplied to u_1), the length of the resulting word $r \geq R$.

If above conditions are all satisfied, Alice sends u_1 to Bob. Otherwise, go to step 2 to regenerate a_1, b_1 and compute u_1.

4. Bob generates a_2, b_2 and computes u_2 similarly as Alice does in step 2 and 3.

5. Alice computes $K_{\text{Alice}} = a_1 u_2 b_1$ after getting u_2 from Bob.

6. Bob computes $K_{\text{Bob}} = b_2 u_1 a_2$ after getting u_1 from Alice.

There are two differences between the new and the original. One is that the system parameters s and L are increased. The other is that a_1, b_1, a_2, b_2 should satisfy the following requirements.

We introduced three new parameters N, T, R in our protocol. Alice (and Bob) can decide them by herself (himself) to achieve different security levels. We take a_1 as an example to explain the three parameters. (1) Let the number of a_1's factors be greater than N. The more factors a_1 has, the larger the depth of the search tree is, the less likely the attack algorithm succeeds. (2) Suppose that the length of u_1 does not decrease when it is left multiplied by the inverses of the first T factors of a_1. It indicates that some of the first T factors increase the length of u_1. (3) Compute the maximum and minimum lengths in the process of left multiplying inverses of a_1's first T factors to u_1, and define r as the difference of maximum and minimum lengths. Let $r \geq R$. A protocol with larger r is harder to be defeated by the attack algorithm.

We repeat the simulation with our protocol to show its advantages. We still choose the combination of "using memory + avoid repetitions + alternative solutions" as attack algorithm, in which $M = 1024$.

The purpose of our simulation is to show how the success rates of the attack algorithm are affected by the three new parameters. In order to reduce the affection of other parameters, we choose $s \in (3, 4)$ and $L \in \{64, 128, 256\}$.

- **Given lower bound of factors' number (parameter N).** If N is too large, the efficiency of generating a_1 will be low. Thus we should choose a proper N. To test success rate of RST07 recovering a_1, we fix $T = 0, R = 0$ and change s, L, N respectively, as shown in Table 5.3.

Table 5.3: Success rate of recovering a_1 by RST07 with a restricted N

	N	$\eta_2/\%$	
		$s = 3$	$s = 4$
$L = 64$	120	55.6	45.9
	140	52.8	41.6
	160	49.0	40.2
	180	45.7	37.8
$L = 128$	240	24.3	17.4
	280	22.0	15.5
	320	20.4	14.9
	360	18.1	13.6
$L = 256$	480	7.6	4.4
	560	6.9	3.8
	640	6.7	3.5
	720	6.1	3.2

Obviously, for different L we need to choose different range of N. Comparing Table 5.3 with Table 5.2, we find that the success rate decreases rapidly. For the same pair of s and L, increasing N decreases the success rate slightly. Thus, we just need to choose a proper N.

- **Ensure the number of factors increasing (parameter T).** T is used to make sure that the length of the resulting word does not decrease when the inverses of the first T factors of a_1 are left multiplied to u_1. Similarly, we should choose a proper T to guarantee the efficiency of generating a_1. We fix $N = 0, R = 0$ and change s, L, T respectively, as shown in Table 5.4, to test success rate of RST07 recovering a_1.

From Table 5.4 we can see that this method greatly decreases success rates with an increment of T. Note that with the increment of T, the time of generating a_1 is increasing. When $T = 100$, it spends about 30 minutes to acquire a legal a_1. This time is not acceptable. We suggest that T be less than 85.

Table 5.4: Success rate of recovering a_1 by RST07 with a restricted T

T	$\eta_3/\%$					
	$s = 3$			$s = 4$		
	$L = 64$	$L = 128$	$L = 256$	$L = 64$	$L = 128$	$L = 256$
40	33.1	12.6	6.3	31.6	10.7	4.7
50	25.3	9.5	4.2	22.9	7.5	2.2
60	12.5	6.7	2.3	10.2	4.2	1.0

- **Restriction for parameter** R. R is the difference between the maximum length and the minimum length. To test success rate of RST07 recovering a_1, we fix $N = 0, T = 0$ and change s, L, R respectively.

Table 5.5: Success rate of recovering a_1 by RST07 with a restricted R

R	$\eta_4/\%$					
	$s = 3$			$s = 4$		
	$L = 64$	$L = 128$	$L = 256$	$L = 64$	$L = 128$	$L = 256$
6	36.4	18.6	8.7	32.5	13.7	4.9
8	32.1	16.9	8.0	29.8	11.5	4.5
10	27.5	13.9	6.8	26.6	9.5	3.9

As shown in Table 5.5, this method also effectively decreases success rates. The larger R is, the longer it takes for the attack algorithm to arrive at correct nodes. Similarly, the time of generating R increases when it becomes larger. So R should not be too large.

- **Combination of above improvements**. We take the three parameters into consideration simultaneously to design our last experiment. The choice of the parameters are shown in Table 5.6, where η_5 is the success rate of RST07 recovering a_1.

Table 5.6: Success rate of recovering a_1 by RST07 with restricted N, T, R

(N, T, R)	$\eta_5/\%$					
	$s = 3$			$s = 4$		
	$L = 64$	$L = 128$	$L = 256$	$L = 64$	$L = 128$	$L = 256$
$(120, 40, 6)$	19.8	6.5	1.4	14.4	4.8	0.4
$(140, 50, 8)$	9.5	3.9	0.5	7.9	2.2	0.1
$(160, 60, 10)$	4.1	1.2	0.2	3.2	0.6	0

Comparing this result with Table 5.2, we see the improvements are effective.

Although the improved length-based attack RST07 is effective, it does still not change its complexity from exponential to polynomial, and its success is based on the randomly generating process. If we modify the process of generating words, the attack algorithm becomes inefficient. In addition, the idea of our improved protocol can be extended to resist general length-based attacks against the protocols over other algebraic structures. The core idea is that length-based attacks focus on the change of word lengths, so it can be used in all kinds of protocols over noncommutative groups.

5.6 Notes

In this chapter, we open the windows to noncommutative cryptography for those who are not familiar with this subject. Our introduction is far from complete. In particular, we did not mention other attacks against the aforementioned noncommutative cryptosystems. We think that a cryptographic scheme or protocol that opens a new door for cryptographic applications has definitely great significance, even if the original version is insecure in some ways. We also introduce some of our contributions on this subject, which are published at AsiaCCS 2007, Inscrypt 2010, and SCIENCE CHINA Information Sciences, etc. We refer to [80, 284–287] for more technical details. In addition, the readers are also suggested to read the survey papers [112] and the books [218].

As for braid-based cryptography, one of the most challenging problem is how to sample hard CSP instances. It turns out that for a braid-based cryptographic scheme to be secure, it is essential that keys are selected from a "very small" subset rather than from the whole group. Constructing these subsets for a particular cryptographic scheme is usually a very challenging problem [218].

As for $\mathbb{Z}$-modular method, there are at least three interesting and challenging directions that are to be developed. The first is to find more promising instantiations for various algebraic structures. The second is to design new cryptographic schemes based on relevant cryptographic assumptions derived by $\mathbb{Z}$-modular method. The third is to extend the $\mathbb{Z}$-modular method from unitary polynomial to multivariate polynomial, and then bridge the $\mathbb{Z}$-modular method and the well-known multivariate cryptographic method.

Another interesting problem is to consider the braided Thompson group BV which was developed independently by Dehornoy and Brin. In geometry, an element in BV group can be viewed as the coupling of two elements in Thompson group F by using a braid (See Figure 5.7). According to Bux and Sonkin's suggestion, BV group does not admit non-trivial linear representation. Thus, from a cryptographic perspective, it possesses the advantages of braid group and Thompson's group simultaneously.

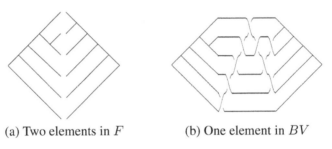

(a) Two elements in F (b) One element in BV

Figure 5.7: Geometric Illustration of Braided Thompson Group BV

Chapter 6

Perspectives

In this chapter, we give some perspectives including some applications and some open problems of the new directions of modern cryptography. The applications cover all the wired and wireless communication networks, satellite communication networks, multicast/broadcast and TV networks, computer networks, all newly emerging networks, such as Internet of Things, Cloud Computing, Social Networks, Named Data Networks, and the fields of the digital right management protection, radio-frequency identification, proof of taken path, conditional access, biometric authentication, and so on.

6.1 On Proxy Re-Cryptography

As we see, more and more researchers are working on proxy re-cryptography, especially on its applications [245]. It is safe to say that applications drive forward the research on proxy re-cryptography.

Proxy re-cryptography is being applied to data sharing [104, 151, 168], group key management [81], cloud storage [197], vehicular ad hoc network (VANET) [269], multicast [149], digital rights management (DRM) [163, 271], proof of taken path [11], anonymous routing [204], radio-frequency identification (RFID) [140, 170], certified mailing lists [171], and so on. These applications can be roughly classified into the following three kinds:

Application 1: Inter-domain scenarios. Proxy re-cryptography is originally designed for these scenarios, where the signature (ciphertext) is desired to be verifiable (decryptable) in many domains [11, 140, 163, 170, 197, 204, 269, 271]. When a signature (ciphertext) travels in a new domain, it can be transformed to the corresponding one in the new domain by using the transformation functionality of proxy re-cryptography. Once a signature (ciphertext) can travel across many domains as in its own domain, the problems of trust transfer, verifiability transfer, and decryption ability transfer can be solved easily, and the storage cost of different encryptions and signatures can be saved. Hence, the resulting system could become quite flexible and efficient.

Application 2: Multi-receiver scenarios. In these scenarios, one sender usually sends an encryption to many receivers via a third party or several third parties [81, 149, 168,216]. In an ideal situation, no matter how many receivers exist and how many new receivers appear after the encryption, the computational cost of the encryption on the sender side is constant, and only the intended receivers can recover the plaintext. By using proxy re-encryption, the above demand can be easily met. In particular, the sender, the third party, and the receivers act as the delegator, proxy, and delegatee(s) in proxy re-encryption, respectively. When a new receiver appears, the sender only needs to send the corresponding re-encryption key to the proxy, and the proxy transforms the encryption to another encryption that can be decrypted by the new receiver.

Application 3: Key revocation scenarios. In these scenarios, the data is usually encrypted under some keys and shared among a group of users via a server [81, 104, 151, 168]. Such scenarios require the backward security and the forward security, which essentially require a key revocation process. The backward security guarantees that the new user cannot access the old data, while the forward security prevents the user left from accessing new data. By using proxy re-encryption, once the data is re-encrypted by the server, it can be decrypted by the new decryption key (corresponding to the delegatee in proxy re-encryption), not the old decryption key (corresponding to the delegator in proxy re-encryption). Hence, the old decryption key is revoked.

Generally speaking, the normal proxy re-cryptographic schemes defined in Chapter 2 cannot be used in the above scenarios directly. For example, the cloud storage [197] requires the threshold proxy re-encryption where the role of the proxy is played by a group of sub-proxies. On the other hand, the current solutions based on proxy re-cryptography are not yet perfect. For instance, the solution to the proof of taken path [11] cannot efficiently prove the exact path the user took. It is because that we do not have any multiuse, unidirectional proxy re-cryptographic scheme with constant size.

Due to the above three situations in proxy re-cryptography, we can see a picture that there will be more new proxy re-cryptographic schemes designed for new applications, and more new applications based on proxy re-cryptography.

6.2 On Attribute-Based Cryptography

We present several application environments based on the research results in the attribute-based cryptography. As we know, attribute-based encryption could result in a perfect mergence of the encrypted storage and the access control [31, 55, 66, 188, 273, 280, 282, 306]. These applications can be roughly classified into the following three kinds:

Application 1: Multi and fuzzy targeted encryption. Multi and fuzzy targeted encryption [227], is described as a new attribute-based encryption scenario. It
means only one encrypted copy can be decrypted by different users with different private keys. Consider a scenario in which a large school periodically sends
updated schedules to all the professors. Those files are encrypted under some
access policies concerning a professor's gender $(Male, \neg Male)$, experience
$((> 5\ years), \neg(> 5\ years))$, department $(Computer, \neg(Computer))$, and
specialty $(Network, \neg(Network))$. The school manager probably constructs an
encrypted email with an access structure $(Male) \wedge (> 5\ years) \wedge (Computer)$
to inform these professors the location of recent conferences. Suppose there is
a male doctor, Jack, with seven years of work experience in a computer department. He is qualified to open the encrypted email since the attribute set of his
secret key satisfies the access structure. However, a male professor in the Mathematics department is not qualified to open the email. As a direct result, this kind
of encryption can be easily transformed to anonymous encryption schemes to preserve the receivers' privacy [32, 166, 189, 236]. We also want more applications
on preserving privacy through this multi and fuzzy targeted encryption.

Application 2: Trust authority application. In the ABE schemes, the central authority can decrypt all the ciphertexts. We first remove the central authority by the
threshold multi-authority [292]. Although no individual authority can decrypt all
ciphertexts in the system, each authority can individually decrypt a part of ciphertexts whose access structures are satisfied by the attributes managed by the
authority. We further solve this problem to achieve the trustfulness of the authority [201]. In addition, more scenarios can be applied to multi-authority ABE to
decentralize the threats from the authority.

Application 3: Conditional access system. Ciphertext policy attribute-based encryption can be used in the conditional access system [31] of digital television [114, 305]. Different users can be labeled with different attributes according to their reservation or purchase. TV programs are encrypted by an access structure and sent to all users. Only the users whose attributes satisfy the access structure can watch the TV programs.

Application 4: Location-based service. Interval encryption can be used not only in the broadcast encryption schemes but also in the vehicle networks [15, 147, 150, 238] and other location-based service networks [157,237]. The vehicle (equipped with the private key) can decrypt the ciphertext only if its private key belongs to the interval that is used in the encryption as a public key. No one else can decrypt without the related private keys. This kind of ciphertexts can be encrypted and stored at the service spots to construct networks for providing service. Interval encryption not only keeps the confidentiality of the message or service, but also provides a new solution to preserve the privacy in the vehicle networks [105].

Application 5: Named data network application. The emerging network, called data network (NDN) [310], focuses on the new Internet architecture that can capitalize on strengths and address weaknesses of the Internet's current host-based, point-to-point communication architecture in order to naturally accommodate patterns of communication. NDN has some innate privacy friendly features, such as lack of source and destination addresses in packets. In practice, this means that the adversary who eavesdrops on a link close to a content producer cannot immediately identify the consumer who expressed interest in that particular content. Moreover, two features of standard NDN routers, content caching and collapsing of redundant interests, reduce the utility of eavesdropping near a content producer since not all interests for the same content reach its producer [98]. However,

NDN provides no protection against an adversary that monitors local activities of a specific consumer. As most content names are expected to be semantically relevant to content itself, interests can leak a lot of information about the content they aim to retrieve. To mitigate this issue, "encrypted names" may be a solution. The producer encrypts the name. However, this simple approach does not provide names for routing or share the content in the requests. Attribute-based encryption could provide a perfect mergence of the encrypted storage and the access control.

Application 6: Selective attribute delegation. Combined with the proxy re-encryption, attribute-based encryption could also be used in an application scenario of personal information system in a university [100]. In this system, there are some confidential records such as the grades of every student. This information is encrypted under the access policy "$((AGE > 40) \wedge (Tenure))$". The professors who are older than 40 and have a tenure position would receive the secret keys corresponding to "$AGE > 40$", "$Tenure$", and thus they are qualified to retrieve the confidential records. Nevertheless, when these professors are on vacation, it is necessary to find some trustworthy substitutes who can decrypt these records in time. Therefore, ABPRE allows a qualified professor to freely designate a proxy who can translate these encrypted records to those encrypted with a different access policy (such as administrators with at least 10 years of working experience "$(Admin) \wedge (EXP > 10)$"). Hence, even if no qualified professor is available, some highly experienced administrator can open the confidential records with the help of the professor's proxy [14, 154, 190, 239, 256].

Application 7: Biometric authentication. An important application of FIBS deals with biometric authentication [185, 220, 301]. A biometric authentication system is essentially a pattern-recognition system that recognizes a person based on a feature vector derived from a specific physiological or behavioral characteristic that the person possesses. Since biometrics cannot be lost or forgotten like computer passwords, biometrics have the potential to offer higher security and more convenience for the users. However, two biometric feature scan results are not identical with high probability. As a result, it is not a good solution to directly choose biometric features as the password or authentication for comparison. FIBS system provides an attractive solution to biometric authentication using cryptographic techniques since two scan results have a certain (possibly high) proportion of overlaps for the same person while the proportion is low for the different persons. Using FIBS system [301], it is not necessary to store the biometric feature for comparison, which can save a lot of storage and achieve economic benefits.

6.3 On Batch Cryptography

Although the researches on batch cryptography started in 1989, there are still some open problems in the field. At the same time, applications based on the batch cryptography can be widely used. Let us conceive the perspective of batch cryptography. We present several application environments based on these research results on the batch cryptography.

Application 1: Batch request process. In the cloud computing environment [281, 292], the customer wants a handshake with the cloud server at the beginning of the communication. When the requests exceed the server's handling ability, the response will be delayed and the service would be influenced in the way similar to the deny of service attacks. It becomes worse when the customer uses encryption to send requests. Batch cryptography has resolved this problem. When the sender sends requests using the cloud server's public key, the cloud server can achieve batch key agreement to improve the efficiency and maintain the security as well.

It does not require the customers and the cloud server to do it in a point-to-point way. Here, we stress that it is an open problem how to apply the batch cryptography technique in every many-to-one communication environment to achieve efficiency and security.

Application 2: Batch packet authentication. In the wireless networks such as ad-hoc networks [307] and delay-tolerant networks [311, 312], packet authentication is a critical security service that ensures authenticity and integrity of packets during multi-hop transmissions. Public key signatures, which have been suggested in existing packet security protocol specification, achieve packet authentication at the cost of increased computational overhead and transmission overhead, and a higher energy consumption, which is not desirable for energy-constrained ad hoc networks and DTNs. Batch authentication schemes, through aggregate signature and batch verification, could achieve efficient packet authentication and cut down the communication cost in the resource-constrained networks.

Application 3: Batch network coding. In the network coding scenarios, each packet is transmitted in the encrypted way by our batch RSA schemes. Each node in the networks can be equipped with the software or hardware, which can be designed as universal chips to parallely decrypt the ciphertext in a fixed block. We can use the subset of solutions of Plus-Type Equations or Minus-Type Equations and use one modular operation to solve such a problem. These kinds of chips can also be used in the key exchange schemes for the multi-request scenarios in cloud computing.

6.4   On Noncommutative Cryptography

Although the beginning of the research on noncommutative cryptography can be traced back to 1984, this field is considerably young. As far as we know, the earliest formal appearance of the term "noncommutative cryptography" is at the second international conference on symbolic computation and cryptography (SCC 2010) (See `http://scc2010.rhul.ac.uk/`). Informally, Myasinikov et al. used this term even earlier and the new upsurge of study on noncommutative cryptography owe a great deal of credits to their efforts. Now, let us conceive the perspective of noncommutative cryptography.

Recently, Fine et al. [112] gave a list of open problems related to group-based cryptography. Most of these open problems are relevant to noncommutative cryptography, too. Instead of enumerating each problem based on concrete noncommutative algebraic structures, we classify the open problems of noncommutative cryptography into the following categories:

Open problem 1: What is the most appropriate noncommutative platform (group, semigroup, ring, or other algebraic structures) for cryptography?

> *In fact, this is one of the most common problems and it is very difficult to answer. At first, it seems that braid group might not be a good candidate, although it is probably the most "famous" platform for noncommutative cryptography. At present, we are even confronting the challenge of generating secure keys for braid-based cryptography, needless to say the recently published successful cryptanalysis on braid-based cryptographic schemes. Second, we believe that using special matrices in noncommutative cryptography has potential advantages. In particular, matrices have a ring structure, which means two binary operations rather than one can be used to "diffuse" transmissions. However, whenever matrices are used, there is always a danger of a linear algebra attack. A possible way to avoid this kind of attack might be using matrices over a rather peculiar semiring that does not admit*

an embedding into a field. Third, as the well-known noncommutative group, the symmetric group S_n should be explored carefully. On one hand, the operations over S_n can be implemented efficiently and certain hidden subgroup problem over S_n can resist known quantum algorithm attacks; on the other hand, the CSP problem over S_n seems tractable on average — thus we have to pay attention to other hard problems over S_n.

Open problem 2: How to estimate the hardness of the related cryptographic problems, especially when the problems are unsolvable or undecidable?

The classical complexity theory always stops at the line when we conclude that the given computational/decisional problem is unsolvable/undecidable. However, this kind of conclusions are not sufficient for cryptographic applications. For example, Diophantine equations are unsolvable in general, but almost all published cryptographic schemes based on Diophantine equations are broken. This situation also happens for the knapsack problem which in the worst case is proved to be NP-hard. Therefore, to reduce the security of a cryptographic scheme to some underlying problem P, we should be confident on at least one of the following: (1) P is intractable in average case; (2) P is intractable in the worst case and we have an efficient method to sample many of the worst instances of P; (3) P is intractable in the worst case and the reduction also worst-case, i.e., broking of the scheme can lead to solving the arbitrary instance of P.

Open problem 3: How to extend the theory of provable security so that it is suitable for noncommutative cryptography?

The concept of indistinguishability plays a key role in the theory of provable security. However, noncommutative cryptography always deals with finitely generated groups. Without proper sampling methods, it is hard to define a desired distribution, needless to say indistinguishability between distributions. In addition, random objects are very important in cryptography. Thus we need not only efficient methods to sample elements from a (probably infinite) space, but also well-established theory to estimate the randomness of the distributions obtained by given sampling methods. Fortunately, the probability theory on groups has been developed even before the birth of the public-key cryptography and now is available for analysis of probability measures and relevant characteristics of noncommutative cryptography.

Open problem 4: Is it possible to bridge noncommutative cryptography with quantum cryptography?

The starting point of noncommutative cryptography is to resist quantum attacks and we learn that quantum cryptography is unconditionally secure. Thus, it is meaningful to bridge between the two sides. On one hand, the development of quantum theory does indeed benefit from employing matrices theory and noncommutative multiplication [12]. In particular, each quantum gate can be expressed as a unitary matrix. The non-commutativity of matrix multiplication implies that quantum operations are essentially noncommutative too. Thus we can try to research quantum cryptography by employing the methods from noncommutative cryptography, such as the ideas behind braid-based cryptography. On the other hand, we have already developed some quantum algebraical objects and topological objects, such as quantum integers, quantum torus, quantum braids, and quan-

tum knots, etc. If the quantum versions of the corresponding classical
cryptographic problems, say conjugator search problem over quantum
braids, are intractable in the quantum Turing model, the quantum ver-
sions of the corresponding classical cryptographic schemes and proto-
cols, say AAG protocol over quantum braids, might be secure against
quantum attacks.

Open problem 5: Is it possible to bridge noncommutative cryptography with biologic
cryptography?

The field of DNA computing, or, more generally, biomolecular com-
puting, was established in Leonard M. Adleman's seminal paper [2].
DNA sequences are known to appear in the form of double helices
in living cells, in which one DNA strand is hybridized to its comple-
mentary strand through a series of hydrogen bonds. DNA computing
requires that the self-assembly of the oligonucleotide strands happen
in such a way that hybridization should occur in a manner compatible
with the goals of computation. There are multiple methods for building
a computing device based on DNA, each with its own advantages and
disadvantages. Among them, the concept of toehold exchange mani-
fests a noncommutative algebraic system. In this system, an input DNA
strand binds to a sticky end, or toehold, on another DNA molecule,
which allows it to displace another strand segment from the molecule.
This allows the creation of modular logic components such as AND,
OR, and NOT gates and signal amplifiers, which can be linked into
arbitrarily large computers. Clearly, this kind of binding operation
is noncommutative. Therefore, the remaining problem is to probe intractable[1] *problems based on toehold exchange system. Later, we can*

---

[1] Here, the term *intractability* should be considered in DNA computational environment. In fact, from the

design biologically cryptographic schemes.

To conclude the exploration on the subject of noncommutative cryptography, we would like to quote paragraphs from the Fields Lecture that was given by Atiyah [12] at the World Mathematical Year 2000 Symposium:

> *"A third theme is the shift from commutative to noncommutative. This is perhaps one of the most characteristic features of mathematics, particularly algebra, in the 20th century. The noncommutative aspect of algebra has been extremely prominent"* ⋯ *"All these are different ways or strands that form the basis of the introduction of noncommutative multiplication into algebra, which is the 'bread and butter' of 20th century algebraic machinery."*

Is it possible that the introduction of noncommutative operations into cryptography would become the "bread and butter" of cryptographic machinery in the 21st century? We hope so.

---

standpoint of computability theory, DNA computing does not provide any new capabilities. For example, if the space required for the solution of a problem grows exponentially with the size of the problem (i.e., EXPSPACE problems) on von Neumann machines, it still grows exponentially with the size of the problem on DNA machines. For very large EXPSPACE problems, the amount of DNA required is too large to be practical. (From Wikipedia, the free encyclopedia. http://en.wikipedia.org/wiki/DNA_computing)

Appendixes

Appendix A: All 122 solutions of $\sum_{1 \le i \le 8} \frac{1}{x_i} + \prod_{1 \le i \le 8} \frac{1}{x_i} = 1$

1:   (2,3,7,43,1807,3263443,10650056950807,113423713055421844361000443)
2:   (2,3,7,43,1807,3263443,10650057792155,13481173926138375 3719)
3:   (2,3,7,43,1807,3263443,10652778201539,41691378583707695)
4:   (2,3,7,43,1807,3263443,10699597306267,2300171639909623)
5:   (2,3,7,43,1807,3263447,2130014000915,226847982204674980901852 11)
6:   (2,3,7,43,1807,3263447,2130014387399,11739058070963394487)
7:   (2,3,7,43,1807,3263479,288182779055,243811701792623)
8:   (2,3,7,43,1807,3263483,260604226747,80249212735823)
9:   (2,3,7,43,1807,3263495,200947673239,67137380077902268343)
10:   (2,3,7,43,1807,3263495,200949404503,23316080984691959)
11:   (2,3,7,43,1807,3263531,119666789791,8081907028348841339)
12:   (2,3,7,43,1807,3263591,71480133827,76130202025687 7140089595)
13:   (2,3,7,43,1807,3263779,31834629787,4396910340967)
14:   (2,3,7,43,1807,3264187,14298637519,1523160211000302785506427)
15:   (2,3,7,43,1807,3316627,203509259,109643149191047)
16:   (2,3,7,43,1807,3586039,36800447,2550097247)
17:   (2,3,7,43,1811,655519,389313431,1507818475)
18:   (2,3,7,43,1811,713899,7813583,2409102303622951)
19:   (2,3,7,43,1811,793595,3722287,233296531681207)
20:   (2,3,7,43,1817,298637,279594269,3859101523354821017)
21:   (2,3,7,43,1819,252731,2134319143,6047845668256680791)
22:   (2,3,7,43,1823,193667,637617223447,4065557236356239093 38363)
23:   (2,3,7,43,1823,193667,637617223459,3127351720332887046 3055)
24:   (2,3,7,43,1823,193675,4683210919,754794584867)
25:   (2,3,7,43,1831,132347,231679879,1197240789041771)
26:   (2,3,7,43,1891,40379,9444811,55866875)
27:   (2,3,7,43,1943,25615,456729463,450222796871)
28:   (2,3,7,43,1951,30571,118463,14484098803019)
29:   (2,3,7,43,2105,12773,2775277,168100338289)
30:   (2,3,7,43,2137,16921,37501,49708999789)
31:   (2,3,7,43,2755,5407,172771,357538828973647)
32:   (2,3,7,43,2813,5045,692705317,188433744928309)

33: (2,3,7,43,3263,4051,2558951,61088439723561979)
34: (2,3,7,43,3559,3667,33816127,797040720326433787)
35: (2,3,7,47,395,779731,607979652631,369639258012703445569531)
36: (2,3,7,47,395,779731,607979652647,2174348576602536000683)
37: (2,3,7,47,395,779731,607979652683,6974325623477705424647)
38: (2,3,7,47,395,779731,607979653531,410254449012081168631)
39: (2,3,7,47,395,779731,607979655287,139119028839856004123)
40: (2,3,7,47,395,779731,607979697799,8183472856913555659)
41: (2,3,7,47,395,779731,607979793451,2624887933109395111)
42: (2,3,7,47,395,779731,607982046587,154405744751990423)
43: (2,3,7,47,395,779743,46768385339,1672627310178141725483)
44: (2,3,7,47,395,779747,35764242947,12154487527525118239)
45: (2,3,7,47,395,779827,6286857907,2158880732959)
46: (2,3,7,47,395,779831,6020372531,3660733426607933569531)
47: (2,3,7,47,395,781727,305967719,125881309327)
48: (2,3,7,47,395,782111,257276179,57278664659)
49: (2,3,7,47,395,782287,277442411,1701723083)
50: (2,3,7,47,395,782611,211810259,1592773460578079)
51: (2,3,7,47,395,816247,17428931,652510750371360683)
52: (2,3,7,47,395,1108727,2627707,140495574531059)
53: (2,3,7,47,403,19403,15435513367,238255072887400163323)
54: (2,3,7,47,403,19403,15435513395,8215692183434294399)
55: (2,3,7,47,403,19403,15435513463,2456237880094942747)
56: (2,3,7,47,403,19403,15435516179,84697872837562655)
57: (2,3,7,47,415,8111,6644612311,44150872756848148411)
58: (2,3,7,47,415,8111,6644612339,1522443894582665279)
59: (2,3,7,47,415,8111,6644613463,38292177286592827)
60: (2,3,7,47,415,8111,6644645747,1320426321921983)
61: (2,3,7,47,449,4477,12137,34035763385)
62: (2,3,7,47,583,1223,1407479767,1980999293106894523)
63: (2,3,7,47,583,1223,1407479807,48317057302587443)
64: (2,3,7,47,583,1223,1468268915,33995520959)
65: (2,3,7,47,583,1223,2202310039,3899834875)
66: (2,3,7,53,209,10589,19651,86321)
67: (2,3,7,53,269,817,7301713,48932949591475)
68: (2,3,7,53,401,409,351691,397617853)
69: (2,3,7,55,179,24323,10057317271,101149630679497570171)
70: (2,3,7,55,179,24323,10057317287,5949978284730273323)
71: (2,3,7,55,179,24323,10057317311,2467064172726591731)
72: (2,3,7,55,179,24323,10057317467,513449911932648503)
73: (2,3,7,55,179,24323,10057317967,145121431390804003)
74: (2,3,7,55,179,24323,10057320619,30202945461748519)
75: (2,3,7,55,179,24323,10057325347,12523178395739983)
76: (2,3,7,55,179,24323,10057454579,736667018400959)
77: (2,3,7,61,187,485,150809,971259409)
78: (2,3,7,61,293,457,551,21709309)
79: (2,3,7,65,121,6271,1579937,2869621)
80: (2,3,7,67,113,28925,48220169,4074021053)
81: (2,3,7,67,113,29153,3712777,45401353)

  82:     (2,3,7,67,113,34477,178945,178344158228021)
  83:     (2,3,7,67,187,283,334651,49836124516795)
  84:     (2,3,7,71,103,65059,1101031,4400294969594807)
  85:     (2,3,11,17,79,301,1049,3696653)
  86:     (2,3,11,17,97,151,444161,317361415625)
  87:     (2,3,11,17,101,149,3109,52495396603)
  88:     (2,3,11,23,31,47059,2214502423,4904020979258368507)
  89:     (2,3,11,23,31,47059,2214502427,980804197623275639)
  90:     (2,3,11,23,31,47059,2214502475,92528699894575367)
  91:     (2,3,11,23,31,47059,2214502687,18505741750517011)
  92:     (2,3,11,23,31,47059,2214502831,11990273552017987)
  93:     (2,3,11,23,31,47059,2214504467,2398056482005535)
  94:     (2,3,11,23,31,47059,2214524099,226233749172527)
  95:     (2,3,11,23,31,47059,2214610807,45248521436443)
  96:     (2,3,11,23,31,47059,2215070383,8636647107907)
  97:     (2,3,11,23,31,47059,2217342227,1729101023519)
  98:     (2,3,11,23,31,47059,2244604355,165128325167)
  99:     (2,3,11,23,31,47059,2294166883,63772955407)
100:     (2,3,11,23,31,47059,2365012087,34797266971)
101:     (2,3,11,23,31,47059,2446798471,23325584587)
102:     (2,3,11,23,31,47059,2612824727,14526193019)
103:     (2,3,11,23,31,47059,3375982667,6436718855)
104:     (2,3,11,23,31,47063,442938131,980970939025927675)
105:     (2,3,11,23,31,47063,447473399,43702604167)
106:     (2,3,11,23,31,47095,59897203,132743972247361531)
107:     (2,3,11,23,31,47119,36349891,4619150372467)
108:     (2,3,11,23,31,47131,30382063,67384091875543675)
109:     (2,3,11,23,31,47147,24928579,11061526082145911)
110:     (2,3,11,23,31,47243,12017087,26715920281613179)
111:     (2,3,11,23,31,47423,6114059,13644326865136507)
112:     (2,3,11,23,31,47479,5307047,2371471764522551)
113:     (2,3,11,23,31,47491,5161279,4952592862147)
114:     (2,3,11,23,31,49759,866923,2029951372029307)
115:     (2,3,11,23,31,60563,211031,6014327790177275)
116:     (2,3,11,23,31,74963,126415,259118345891)
117:     (2,3,11,23,31,84527,106159,84453127154999)
118:     (2,3,11,25,29,787,264841,2542873)
119:     (2,3,11,25,29,1097,2753,144508961851)
120:     (2,3,11,31,35,67,369067,1770735487291)
121:     (2,3,13,25,29,67,2981,11294561851)
122:     (2,5,7,11,17,157,961,4398619)

Appendix B: All solutions of $\sum\limits_{1 \leq i \leq N} \frac{1}{x_i} - \prod\limits_{1 \leq i \leq N} \frac{1}{x_i} = 1$ for $N \leq 7$

1:    (2,3,5)
2:    (2,3,11,13)
3:    (2,3,7,41)
4:    (2,3,11,17,59)
5:    (2,3,7,83,85)
6:    (2,3,7,43,1805)
7:    (2,3,11,23,31,47057)
8:    (2,3,7,71,103,61429)
9:    (2,3,7,53,271,799)
10:   (2,3,7,47,481,2203)
11:   (2,3,7,47,395,779729)
12:   (2,3,7,43,3611,3613)
13:   (2,3,7,43,3041,4447)
14:   (2,3,7,43,2501,6499)
15:   (2,3,7,43,2167,10841)
16:   (2,3,7,43,2053,15011)
17:   (2,3,7,43,1945,25271)
18:   (2,3,7,43,1901,36139)
19:   (2,3,7,43,1871,51985)
20:   (2,3,7,43,1825,173471)
21:   (2,3,7,43,1819,252701)
22:   (2,3,7,43,1811,654133)
23:   (2,3,7,43,1807,3263441)
24:   (2,3,11,23,31,94115,94117)
25:   (2,3,11,23,31,47059,2214502421)
26:   (2,3,7,71,103,67213,713863)
27:   (2,3,7,71,103,62857,2704339)
28:   (2,3,7,71,103,61955,7238201)
29:   (2,3,7,71,103,61559,29133437)
30:   (2,3,7,71,103,61477,79005919)
31:   (2,3,7,71,103,61441,319853515)
32:   (2,3,7,59,163,1381,775807)
33:   (2,3,7,55,179,24323,10057317269)
34:   (2,3,7,47,583,1223,1407479765)
35:   (2,3,7,47,449,3299,379591)
36:   (2,3,7,47,415,8111,6644612309)
37:   (2,3,7,47,403,19403,15435513365)
38:   (2,3,7,47,401,25535,1837531099)
39:   (2,3,7,47,395,779731,607979652629)
40:   (2,3,7,47,395,788491,70175789)
41:   (2,3,7,47,395,779819,6832003021)
42:   (2,3,7,47,395,1559459,1559461)
43:   (2,3,7,43,3307,3979,642279641)
44:   (2,3,7,43,2533,7807,32435)

45:    (2,3,7,43,2159,11047,98567401)
46:    (2,3,7,43,1907,43115,163073)
47:    (2,3,7,43,1907,34165,17766223)
48:    (2,3,7,43,1823,193667,637617223445)
49:    (2,3,7,43,1807,3263443,10650056950805)
50:    (2,3,7,43,1807,6526883,6526885)

Appendix C: All 550 solutions of $\sum\limits_{1 \le i \le 8} \frac{1}{x_i} - \prod\limits_{1 \le i \le 8} \frac{1}{x_i} = 1$

1: (2,3,7,43,1807,3263443,10650056950807,11342371305542184436100044l)
2: (2,3,7,43,1807,3263443,10650056950811,22684742611092888917760733)
3: (2,3,7,43,1807,3263443,10650056950847,2766432025752386503391041)
4: (2,3,7,43,1807,3263443,10650056950895,1274423742206281453625521)
5: (2,3,7,43,1807,3263443,10650056951011,553286405158997346238853)
6: (2,3,7,43,1807,3263443,10650056951251,254884748449776336285749)
7: (2,3,7,43,1807,3263443,10650056951353,207355965377032496263511)
8: (2,3,7,43,1807,3263443,10650056951413,186859494335874429774611)
9: (2,3,7,43,1807,3263443,10650056951839,109800303065259151146401)
10: (2,3,7,43,1807,3263443,10650056953541,41471193083926544813347)
11: (2,3,7,43,1807,3263443,10650056953841,37371898875694931515567)
12: (2,3,7,43,1807,3263443,10650056954455,31083505917860578820921)
13: (2,3,7,43,1807,3263443,10650056955925,22157396583303152145971)
14: (2,3,7,43,1807,3263443,10650056955971,21960060621571875789925)
15: (2,3,7,43,1807,3263443,10650056969051,6216701192092161324829)
16: (2,3,7,43,1807,3263443,10650056973233,5057462580561823763311)
17: (2,3,7,43,1807,3263443,10650056975693,4557548652728700190411)
18: (2,3,7,43,1807,3263443,10650056976401,4431479325180675989839)
19: (2,3,7,43,1807,3263443,10650056981857,3652819986026411796191)
20: (2,3,7,43,1807,3263443,10650056993159,2678056182713693394601)
21: (2,3,7,43,1807,3263443,10650056999489,2329842318137500089151)
22: (2,3,7,43,1807,3263443,10650057004829,2099544890708757769051)
23: (2,3,7,43,1807,3263443,10650057042743,1233711280926563627161)
24: (2,3,7,43,1807,3263443,10650057062941,1011492524632410313307)
25: (2,3,7,43,1807,3263443,10650057064837,994674380386454897891)
26: (2,3,7,43,1807,3263443,10650057075241,911509739065785598727)
27: (2,3,7,43,1807,3263443,10650057106061,730564005725327919883)
28: (2,3,7,43,1807,3263443,10650057160685,540424317300132443371)
29: (2,3,7,43,1807,3263443,10650057162571,535611245062784239565)
30: (2,3,7,43,1807,3263443,10650057194221,465968472147545578475)
31: (2,3,7,43,1807,3263443,10650057220921,419908986661797114455)
32: (2,3,7,43,1807,3263443,10650057282835,341607861336024725621)
33: (2,3,7,43,1807,3263443,10650057406397,248959522702338919291)
34: (2,3,7,43,1807,3263443,10650057410491,246742264705358286077)
35: (2,3,7,43,1807,3263443,10650057515857,200731826106380706191)
36: (2,3,7,43,1807,3263443,10650057520961,198934884597336540223)
37: (2,3,7,43,1807,3263443,10650057577837,180890131003613942891)
38: (2,3,7,43,1807,3263443,10650058000201,108084871980072049319)
39: (2,3,7,43,1807,3263443,10650058223897,89093180781187556791)
40: (2,3,7,43,1807,3263443,10650058610951,68321580787250505769)
41: (2,3,7,43,1807,3263443,10650058946809,56825432783897027351)
42: (2,3,7,43,1807,3263443,10650059165749,51208422358805751251)
43: (2,3,7,43,1807,3263443,10650059228761,49791913060513344503)
44: (2,3,7,43,1807,3263443,10650059714345,41042931721701387271)
45: (2,3,7,43,1807,3263443,10650059750899,40507134183243596501)

46:   (2,3,7,43,1807,3263443,10650059776061,40146373741321701883)
47:   (2,3,7,43,1807,3263443,10650060058039,36503135151956613401)
48:   (2,3,7,43,1807,3263443,10650060085961,36178034720768349223)
49:   (2,3,7,43,1807,3263443,10650060720223,30090529437288820961)
50:   (2,3,7,43,1807,3263443,10650061626077,24260361131432510491)
51:   (2,3,7,43,1807,3263443,10650062238733,21449571707804380811)
52:   (2,3,7,43,1807,3263443,10650063316261,17818644676283072003)
53:   (2,3,7,43,1807,3263443,10650066930821,11365095076824966115)
54:   (2,3,7,43,1807,3263443,10650067099565,11176127163949062571)
55:   (2,3,7,43,1807,3263443,10650068025521,10241692991806710895)
56:   (2,3,7,43,1807,3263443,10650070563995,8331909447275677021)
57.   (2,3,7,43,1807,3263443,10650070768501,8208594864385838099)
58:   (2,3,7,43,1807,3263443,10650070951271,8101435356694279945)
59:   (2,3,7,43,1807,3263443,10650072486971,7300635550436883325)
60:   (2,3,7,43,1807,3263443,10650073935703,6677926509977160761)
61:   (2,3,7,43,1807,3263443,10650075630037,6072193870844315891)
62:   (2,3,7,43,1807,3263443,10650075797891,6018114407503324837)
63:   (2,3,7,43,1807,3263443,10650075798763,6017835980166266861)
64:   (2,3,7,43,1807,3263443,10650080117897,4895908588016066791)
65:   (2,3,7,43,1807,3263443,10650080327161,4852080746332062743)
66:   (2,3,7,43,1807,3263443,10650082659077,4411964805021755491)
67:   (2,3,7,43,1807,3263443,10650083390441,4289922861606436807)
68:   (2,3,7,43,1807,3263443,10650086501387,3838301107202656141)
69:   (2,3,7,43,1807,3263443,10650089026489,3536138409375784151)
70:   (2,3,7,43,1807,3263443,10650107240345,2255424306869577271)
71:   (2,3,7,43,1807,3263443,10650107694601,2235233952835373159)
72:   (2,3,7,43,1807,3263443,10650112756565,2032483912456467571)
73:   (2,3,7,43,1807,3263443,10650119325763,1818428145004661861)
74:   (2,3,7,43,1807,3263443,10650125016751,1666390409500696049)
75:   (2,3,7,43,1807,3263443,10650126167623,1638683417332729961)
76:   (2,3,7,43,1807,3263443,10650141875291,1335593822040992797)
77:   (2,3,7,43,1807,3263443,10650150346961,1214447294214423823)
78:   (2,3,7,43,1807,3263443,10650151190591,1203575716078814017)
79:   (2,3,7,43,1807,3263443,10650170255905,1001057505462912671)
80:   (2,3,7,43,1807,3263443,10650171754619,987989272817600701)
81:   (2,3,7,43,1807,3263443,10650172786261,979190237648774003)
82:   (2,3,7,43,1807,3263443,10650174744829,962909362289669051)
83:   (2,3,7,43,1807,3263443,10650184347359,890330759859381601)
84:   (2,3,7,43,1807,3263443,10650185492161,882401481049911743)
85:   (2,3,7,43,1807,3263443,10650204703711,767668741486091873)
86:   (2,3,7,43,1807,3263443,10650215900875,713591422742310221)
87:   (2,3,7,43,1807,3263443,10650217329221,707236201920717475)
88:   (2,3,7,43,1807,3263443,10650273755813,523170675855668611)
89:   (2,3,7,43,1807,3263443,10650306159083,455146869530958061)
90:   (2,3,7,43,1807,3263443,10650308398501,451093381419476099)
91:   (2,3,7,43,1807,3263443,10650333494543,410158116370430161)
92:   (2,3,7,43,1807,3263443,10650335979601,406505302536854159)
93:   (2,3,7,43,1807,3263443,10650368825591,363694149046493017)
94:   (2,3,7,43,1807,3263443,10650399936763,330705568436396861)

 95:   (2,3,7,43,1807,3263443,10650403034891,327745203512106637)
 96:   (2,3,7,43,1807,3263443,10650473049925,272598857712855971)
 97:   (2,3,7,43,1807,3263443,10650527576309,241016954076584851)
 98:   (2,3,7,43,1807,3263443,10650623476301,200220021138143179)
 99:   (2,3,7,43,1807,3263443,10650630969871,197606374609080785)
100:   (2,3,7,43,1807,3263443,10650640675495,194320939124424521)
101:   (2,3,7,43,1807,3263443,10650645920921,192590392503494455)
102:   (2,3,7,43,1807,3263443,10650693933571,178074672017436965)
103:   (2,3,7,43,1807,3263443,10650753331583,162886646640370561)
104:   (2,3,7,43,1807,3263443,10650829717043,146786877620592661)
105:   (2,3,7,43,1807,3263443,10650851701151,142726804594022689)
106:   (2,3,7,43,1807,3263443,10651140975841,104642655216694367)
107:   (2,3,7,43,1807,3263443,10651268524627,93627490475138741)
108:   (2,3,7,43,1807,3263443,10651302992191,91037893951752257)
109:   (2,3,7,43,1807,3263443,10651372053809,86257668576922351)
110:   (2,3,7,43,1807,3263443,10651439669491,82040143319646677)
111:   (2,3,7,43,1807,3263443,10651568606639,75043412527964401)
112:   (2,3,7,43,1807,3263443,10651734418979,67626665001999301)
113:   (2,3,7,43,1807,3263443,10651756607257,66743967245067191)
114:   (2,3,7,43,1807,3263443,10651771880591,66149633732840017)
115:   (2,3,7,43,1807,3263443,10652118821905,55020739247502671)
116:   (2,3,7,43,1807,3263443,10652137446401,54528291588131839)
117:   (2,3,7,43,1807,3263443,10652344986925,49583168652060971)
118:   (2,3,7,43,1807,3263443,10652410078321,48211910860877615)
119:   (2,3,7,43,1807,3263443,10652614324043,44362296275187661)
120:   (2,3,7,43,1807,3263443,10652894840303,39978278527091761)
121:   (2,3,7,43,1807,3263443,10652911686593,39742422633566911)
122:   (2,3,7,43,1807,3263443,10652949446875,39223741570200221)
123:   (2,3,7,43,1807,3263443,10652975574251,38872707870445549)
124:   (2,3,7,43,1807,3263443,10653266722495,35347653446979521)
125:   (2,3,7,43,1807,3263443,10653538854691,32585849373634757)
126:   (2,3,7,43,1807,3263443,10653597727387,32044219659746141)
127:   (2,3,7,43,1807,3263443,10653920781991,29365895569679177)
128:   (2,3,7,43,1807,3263443,10654886505749,23495984501651251)
129:   (2,3,7,43,1807,3263443,10655608321979,20442307303554301)
130:   (2,3,7,43,1807,3263443,10656114819911,18734018140588393)
131:   (2,3,7,43,1807,3263443,10656217247519,18422703621847201)
132:   (2,3,7,43,1807,3263443,10656573903635,17415059146837621)
133:   (2,3,7,43,1807,3263443,10656632465821,17260053760945115)
134:   (2,3,7,43,1807,3263443,10657615229971,15017202551153525)
135:   (2,3,7,43,1807,3263443,10658444291671,13533853045960505)
136:   (2,3,7,43,1807,3263443,10658555233061,13357313494574083)
137:   (2,3,7,43,1807,3263443,10660274490163,11111533458755861)
138:   (2,3,7,43,1807,3263443,10660366306301,11012667895061179)
139:   (2,3,7,43,1807,3263443,10660366783285,11012158887132371)
140:   (2,3,7,43,1807,3263443,10660540618853,10829736711251011)
141:   (2,3,7,43,1807,3263443,10661395244023,10014246796303961)
142:   (2,3,7,43,1807,3263443,10661497131401,9925153775972839)
143:   (2,3,7,43,1807,3263443,10662843816991,8880979300598177)

144: (2,3,7,43,1807,3263443,10664119375043,8076379773522661)
145: (2,3,7,43,1807,3263443,10664203506947,8028411547797541)
146: (2,3,7,43,1807,3263443,10664246398291,8004175750978997)
147: (2,3,7,43,1807,3263443,10664330629741,7957004572274027)
148: (2,3,7,43,1807,3263443,10664519431151,7853268359600689)
149: (2,3,7,43,1807,3263443,10666105809251,7078050734956549)
150: (2,3,7,43,1807,3263443,10667602349407,6475234626089441)
151: (2,3,7,43,1807,3263443,10667760833711,6417363977509873)
152: (2,3,7,43,1807,3263443,10669352596429,5888852594015051)
153: (2,3,7,43,1807,3263443,10669526890387,5836231209041141)
154: (2,3,7,43,1807,3263443,10673989663055,4749925400059921)
155: (2,3,7,43,1807,3263443,10674204725521,4707716945890895)
156: (2,3,7,43,1807,3263443,10677813806671,4096981506271505)
157: (2,3,7,43,1807,3263443,10680582700979,3726323297169301)
158: (2,3,7,43,1807,3263443,10680858434371,3693060769930085)
159: (2,3,7,43,1807,3263443,10682641714951,3491531874928169)
160: (2,3,7,43,1807,3263443,10687918549705,3006395518149671)
161: (2,3,7,43,1807,3263443,10701144647591,2230826737311817)
162: (2,3,7,43,1807,3263443,10701606113201,2210951822987119)
163: (2,3,7,43,1807,3263443,10702008448127,2193911731866241)
164: (2,3,7,43,1807,3263443,10702475291041,2174467387810847)
165: (2,3,7,43,1807,3263443,10706748416891,2011369404821437)
166: (2,3,7,43,1807,3263443,10712034839959,1840717434292601)
167: (2,3,7,43,1807,3263443,10714490281387,1770976770356141)
168: (2,3,7,43,1807,3263443,10718833145899,1659821153171501)
169: (2,3,7,43,1807,3263443,10719742865297,1638291939587791)
170: (2,3,7,43,1807,3263443,10720369071991,1623796000265177)
171: (2,3,7,43,1807,3263443,10720789731511,1614202355120153)
172: (2,3,7,43,1807,3263443,10721557922767,1596974129821841)
173: (2,3,7,43,1807,3263443,10737002638549,1315185290379251)
174: (2,3,7,43,1807,3263443,10737783943811,1303566970778533)
175: (2,3,7,43,1807,3263443,10746535178921,1186290564363655)
176: (2,3,7,43,1807,3263443,10746539642689,1186236173401151)
177: (2,3,7,43,1807,3263443,10747406648711,1175766287368873)
178: (2,3,7,43,1807,3263443,10767101118073,979717680770711)
179: (2,3,7,43,1807,3263443,10768649289635,967066923127621)
180: (2,3,7,43,1807,3263443,10769720512051,958505125572629)
181: (2,3,7,43,1807,3263443,10781657590055,872528188414921)
182: (2,3,7,43,1807,3263443,10795228790627,791956632628741)
183: (2,3,7,43,1807,3263443,10801326374945,760462609626271)
184: (2,3,7,43,1807,3263443,10802685701671,753784704994505)
185: (2,3,7,43,1807,3263443,10814252372083,701434928863061)
186: (2,3,7,43,1807,3263443,10839364945301,609799149190579)
187: (2,3,7,43,1807,3263443,10877663168899,508983160526501)
188: (2,3,7,43,1807,3263443,10902629116039,459724534143401)
189: (2,3,7,43,1807,3263443,10907489100947,451246590807541)
190: (2,3,7,43,1807,3263443,10909814437411,447302391933893)
191: (2,3,7,43,1807,3263443,10935726631127,407695038861241)
192: (2,3,7,43,1807,3263443,10959946396571,376663532419165)

193:    (2,3,7,43,1807,3263443,10965186066515,370577805296821)
194:    (2,3,7,43,1807,3263443,10969354355689,365879104606151)
195:    (2,3,7,43,1807,3263443,10972223603711,362715399631873)
196:    (2,3,7,43,1807,3263443,10993937926271,340484276194945)
197:    (2,3,7,43,1807,3263443,10998486523261,336178433478203)
198:    (2,3,7,43,1807,3263443,11004377837029,330765854426051)
199:    (2,3,7,43,1807,3263443,11007561810611,327914871525013)
200:    (2,3,7,43,1807,3263443,11072760082445,278979540613771)
201:    (2,3,7,43,1807,3263443,11079887340733,274530219250811)
202:    (2,3,7,43,1807,3263443,11084785389521,271557103636495)
203:    (2,3,7,43,1807,3263443,11132470410221,245767280240875)
204:    (2,3,7,43,1807,3263443,11230065752587,206205215264141)
205:    (2,3,7,43,1807,3263443,11235277787141,204463581714787)
206:    (2,3,7,43,1807,3263443,11243018644951,201933430186169)
207:    (2,3,7,43,1807,3263443,11253044554543,198752950530161)
208:    (2,3,7,43,1807,3263443,11308060147051,183025683243629)
209:    (2,3,7,43,1807,3263443,11369418293447,168322851320041)
210:    (2,3,7,43,1807,3263443,11375916149911,166911372086393)
211:    (2,3,7,43,1807,3263443,11406404071501,160612567485899)
212:    (2,3,7,43,1807,3263443,11448324473627,152737402123741)
213:    (2,3,7,43,1807,3263443,11471034057191,148807031333257)
214:    (2,3,7,43,1807,3263443,11567632041437,134262515716891)
215:    (2,3,7,43,1807,3263443,11788088041271,110316677665945)
216:    (2,3,7,43,1807,3263443,11901612707899,101276233541501)
217:    (2,3,7,43,1807,3263443,11912917776971,100464952389325)
218:    (2,3,7,43,1807,3263443,11937217701511,98769363722153)
219:    (2,3,7,43,1807,3263443,12078405352411,90059053332893)
220:    (2,3,7,43,1807,3263443,12202382505665,83717019419071)
221:    (2,3,7,43,1807,3263443,12211597426295,83285838626521)
222:    (2,3,7,43,1807,3263443,12225702529351,82635606620009)
223:    (2,3,7,43,1807,3263443,12246543975221,81695866481875)
224:    (2,3,7,43,1807,3263443,12382881573515,76106024951821)
225:    (2,3,7,43,1807,3263443,12405802064689,75251525671151)
226:    (2,3,7,43,1807,3263443,12421661381921,74673216445855)
227:    (2,3,7,43,1807,3263443,12586861740613,69212341416611)
228:    (2,3,7,43,1807,3263443,12763572609001,64315953683399)
229:    (2,3,7,43,1807,3263443,12780068340967,63900341704841)
230:    (2,3,7,43,1807,3263443,12799208900441,63426089410807)
231:    (2,3,7,43,1807,3263443,12799308335473,63423647729711)
232:    (2,3,7,43,1807,3263443,13291823504627,53584854838741)
233:    (2,3,7,43,1807,3263443,13550100959711,49761088613473)
234:    (2,3,7,43,1807,3263443,13581596801207,49340887996441)
235:    (2,3,7,43,1807,3263443,13664994969491,48270635666677)
236:    (2,3,7,43,1807,3263443,14019739252817,44310118312591)
237:    (2,3,7,43,1807,3263443,14214830148269,42467989473451)
238:    (2,3,7,43,1807,3263443,14246863664011,42184615824653)
239:    (2,3,7,43,1807,3263443,14307679158979,41660288899301)
240:    (2,3,7,43,1807,3263443,14605847318009,39322889059351)
241:    (2,3,7,43,1807,3263443,14641394564911,39067525985393)

```
242:    (2,3,7,43,1807,3263443,15237932403961,35372548704023)
243:    (2,3,7,43,1807,3263443,16384623372515,30429008786821)
244:    (2,3,7,43,1807,3263443,16852103340505,28938167991671)
245:    (2,3,7,43,1807,3263443,16907835736271,28775292268945)
246:    (2,3,7,43,1807,3263443,17013643455335,28473922938121)
247:    (2,3,7,43,1807,3263443,17382069223163,27498468460861)
248:    (2,3,7,43,1807,3263443,18388223159933,25307756202811)
249:    (2,3,7,43,1807,3263443,18411684725101,25263449444459)
250:    (2,3,7,43,1807,3263443,18457759328251,25177213285949)
251:    (2,3,7,43,1807,3263443,19237016528393,23858889719911)
252:    (2,3,7,43,1807,3263443,19314180064351,23741250551009)
253:    (2,3,7,43,1807,3263443,19428782520221,23570350694875)
254:    (2,3,7,43,1807,3263443,20334080899841,22362513843967)
255:    (2,3,7,43,1807,3263443,21204775106587,21396313874141)
256:    (2,3,7,43,1807,3263443,21300113901611,21300113901613)
257:    (2,3,7,43,1807,3263447,2130014000915,2268479822046749809 0185209)
258:    (2,3,7,43,1807,3263453,968189962295,5427005175767543752 13351)
259:    (2,3,7,43,1807,3263453,968189962447,6097758625415225751631)
260:    (2,3,7,43,1807,3263453,968191675715,547088581335563051)
261:    (2,3,7,43,1807,3263453,968342456827,6148020023058931)
262:    (2,3,7,43,1807,3263467,426011650123,29087649458734661)
263:    (2,3,7,43,1807,3263467,431493708101,33492839204683)
264:    (2,3,7,43,1807,3263467,440049819613,13347917207171)
265:    (2,3,7,43,1807,3263467,440695252693,12780162328091)
266:    (2,3,7,43,1807,3263483,259760670451,55329996848759504 0585437)
267:    (2,3,7,43,1807,3263483,259760739379,978927350808388381)
268:    (2,3,7,43,1807,3263483,259767485441,9901316798999735)
269:    (2,3,7,43,1807,3263497,193640603263,738917082487556510171)
270:    (2,3,7,43,1807,3263573,81301383575,17297566015272971711)
271:    (2,3,7,43,1807,3263573,82889098211,4244463432803)
272:    (2,3,7,43,1807,3263591,71480133827,761302020256877140089593)
273:    (2,3,7,43,1807,3263671,46510048103,990738535140400014508189)
274:    (2,3,7,43,1807,3263683,44194462691,18581823597527605)
275:    (2,3,7,43,1807,3263773,33375205823,897552055711)
276:    (2,3,7,43,1807,3263843,26562000961,4898649404931690223)
277:    (2,3,7,43,1807,3263945,21176332403,11871780956994661586951)
278:    (2,3,7,43,1807,3264011,20287332797,242376931523)
279:    (2,3,7,43,1807,3264061,17208519803,192954128638848111 14963)
280:    (2,3,7,43,1807,3264167,14692992667,205671385394733992017)
281:    (2,3,7,43,1807,3264187,14298637519,152316021000302785506425)
282:    (2,3,7,43,1807,3264199,14072032361,34190638054365523)
283:    (2,3,7,43,1807,3264461,10843167823,291848881087)
284:    (2,3,7,43,1807,3264977,6941413295,11671343143607219)
285:    (2,3,7,43,1807,3266063,4066618531,47894463455777446945)
286:    (2,3,7,43,1807,3267031,2970678791,160775038921890887773)
287:    (2,3,7,43,1807,3267301,4840020859,6438862043)
288:    (2,3,7,43,1807,3268751,2009301055,404414650631470247053)
289:    (2,3,7,43,1807,3271393,1342724351,150893967753302743219)
290:    (2,3,7,43,1807,3272831,1137575263,24300186900453815 20285)
```

291: (2,3,7,43,1807,3276181,850534063,63446582375)
292: (2,3,7,43,1807,3276281,833444231,1031685689123)
293: (2,3,7,43,1807,3276773,802157275,23893916024601906811)
294: (2,3,7,43,1807,3288647,425800789,31785447208059535)
295: (2,3,7,43,1807,3289787,407516761,2587292602794911023)
296: (2,3,7,43,1807,3332107,158365099,1082703262718405)
297: (2,3,7,43,1807,3354583,120117325,10431144828791)
298: (2,3,7,43,1807,3404603,95947769,438097871)
299: (2,3,7,43,1807,3409363,84824905,754136231)
300: (2,3,7,43,1807,3419881,71341483,114256603062479)
301: (2,3,7,43,1807,3483107,55865009,701928095)
302: (2,3,7,43,1807,3558181,39397291,549162395020523)
303: (2,3,7,43,1807,3599423,34961825,561529103817011)
304: (2,3,7,43,1807,3667523,29619721,19819217116471)
305: (2,3,7,43,1807,5778053,7498711,431098798655023)
306: (2,3,7,43,1811,654137,112603355159,2343968949833659663)
307: (2,3,7,43,1811,654167,12660248233,1341879276045695)
308: (2,3,7,43,1811,654167,12660392623,607509924211097)
309: (2,3,7,43,1811,654433,1427906867,4841683965821195)
310: (2,3,7,43,1811,656011,228620033,532822850387)
311: (2,3,7,43,1811,658087,108880649,2974997080463)
312: (2,3,7,43,1811,658313,104250607,8764202183)
313: (2,3,7,43,1811,676567,19738273,36468880087)
314: (2,3,7,43,1811,730717,6248239,5660284603)
315: (2,3,7,43,1811,830113,3087647,4669841615)
316: (2,3,7,43,1811,1033163,1786303,976842557)
317: (2,3,7,43,1817,303479,17543063,156178151522731)
318: (2,3,7,43,1819,267097,4693135,4776988367)
319: (2,3,7,43,1823,193667,637617223447,40655572363562390933836 1)
320: (2,3,7,43,1823,193667,637617223451,8131114472763487564642 9)
321: (2,3,7,43,1823,193667,637617223483,10987992531312922415741)
322: (2,3,7,43,1823,193667,637617223631,2197598506772678261905)
323: (2,3,7,43,1823,193667,637617223895,905469318310192484281)
324: (2,3,7,43,1823,193667,637617225691,181093864172132275613)
325: (2,3,7,43,1823,193667,637617240059,24472144358497635901)
326: (2,3,7,43,1823,193667,637617260377,11008522574065905911)
327: (2,3,7,43,1823,193667,637617306511,4894429381793305937)
328: (2,3,7,43,1823,193667,637617408101,2201705024906959939)
329: (2,3,7,43,1823,193667,637618589893,297528257521241891)
330: (2,3,7,43,1823,193667,637620676465,1177398460128 29231)
331: (2,3,7,43,1823,193667,637624055681,59506161598027135)
332: (2,3,7,43,1823,193667,637633805465,24518503845394231)
333: (2,3,7,43,1823,193667,637634488541,23548479296344603)
334: (2,3,7,43,1823,193667,637655604065,10593372550315231)
335: (2,3,7,43,1823,193667,637700133541,4904210862857603)
336: (2,3,7,43,1823,193667,637744985149,3182778384672251)
337: (2,3,7,43,1823,193667,637809126541,2119184603841803)
338: (2,3,7,43,1823,193667,638230758149,663282650417251)
339: (2,3,7,43,1823,193667,638256031961,637065770713207)

340: (2,3,7,43,1823,193667,639037306349,286927750550251)
341: (2,3,7,43,1823,193667,639167628977,262863025676911)
342: (2,3,7,43,1823,193667,640684896961,133166623862207)
343: (2,3,7,43,1823,193667,644717637961,57895643888807)
344: (2,3,7,43,1823,193667,645369251101,53082698914139)
345: (2,3,7,43,1823,193667,654850121377,24229454490911)
346: (2,3,7,43,1823,193667,694982228093,7724790424891)
347: (2,3,7,43,1823,193667,723781713101,5355984676939)
348: (2,3,7,43,1823,193667,765140668135,3825703340681)
349: (2,3,7,43,1823,193667,924442246681,2055051863735)
350: (2,3,7,43,1823,193667,1275234446891,1275234446893)
351: (2,3,7,43,1823,193727,624696145,1249029542707580051)
352: (2,3,7,43,1823,193747,468685325,2558779870050491)
353: (2,3,7,43,1823,195227,24235643,71694375412021)
354: (2,3,7,43,1831,138683,2861051,1456230512169437)
355: (2,3,7,43,1831,238123,297557,17623382424931)
356: (2,3,7,43,1843,89959,11976990133,34817546746755863675)
357: (2,3,7,43,1843,89959,11976990545,344728197469850263)
358: (2,3,7,43,1853,71203,10360092227,36845145540307875671)
359: (2,3,7,43,1853,71203,10382523559,4795251910195)
360: (2,3,7,43,1867,55277,1961928211,4520530841303023)
361: (2,3,7,43,1873,50639,16834739,45310242714479)
362: (2,3,7,43,1873,51911,2163883,12322207)
363: (2,3,7,43,1913,32311,46532123,273389846882848007)
364: (2,3,7,43,1919,30671,975201683,1494685587100417)
365: (2,3,7,43,1997,18883,959198423,514365153707602931)
366: (2,3,7,43,1997,18883,959225723,33700641672431)
367: (2,3,7,43,2003,18371,39817607,529220481881695069)
368: (2,3,7,43,2033,32287,32411,4601387132471)
369: (2,3,7,43,2075,13939,24410963,45271367153)
370: (2,3,7,43,2087,13739,565721,67188075943)
371: (2,3,7,43,2147,11371,1520374063,13406916288849768157)
372: (2,3,7,43,2161,21803,22175,19209741287)
373: (2,3,7,43,2225,9599,10677439,2083854611)
374: (2,3,7,43,2243,12491,35995,25573817)
375: (2,3,7,43,2441,6943,129692807,248849507)
376: (2,3,7,43,2593,5951,58180667,6322831502951)
377: (2,3,7,43,3203,4141,66724871,12730559229347)
378: (2,3,7,43,3263,4045,40608343,727261435685959)
379: (2,3,7,43,3263,4051,2558951,61088439723561977)
380: (2,3,7,43,3307,3979,642937961,627272420323)
381: (2,3,7,43,3307,3979,740369317,4847851055)
382: (2,3,7,43,3407,5123,15385,25524552402311)
383: (2,3,7,43,3559,3667,33816127,797040720326433785)
384: (2,3,7,43,3719,5347,10225,10069719847)
385: (2,3,7,43,3791,4177,34063,47251)
386: (2,3,7,47,395,779731,607979652631,36963925801270344556 9529)
387: (2,3,7,47,395,779731,607979652643,2843378907846147703 0853)
388: (2,3,7,47,395,779731,607979652661,1192384703325622048 8659)

389: (2,3,7,47,395,779731,607979652701,5206186733172704524699)
390: (2,3,7,47,395,779731,607979653033,917219003119382793863)
391: (2,3,7,47,395,779731,607979653553,400475903112958488943)
392: (2,3,7,47,395,779731,607979654831,167941508110067551729)
393: (2,3,7,47,395,779731,607979658569,62239309921348143271)
394: (2,3,7,47,395,779731,607979681243,12918578108140260253)
395: (2,3,7,47,395,779731,607979699141,7947351944482747939)
396: (2,3,7,47,395,779731,607979729837,4787639785931074987)
397: (2,3,7,47,395,779731,607979836739,2007720263249603941)
398: (2,3,7,47,395,779731,607980074299,876610598308786301)
399: (2,3,7,47,395,779731,607980257273,611335326172198423)
400: (2,3,7,47,395,779731,607981094471,256366780124913769)
401: (2,3,7,47,395,779731,607982046047,154440581461956577)
402: (2,3,7,47,395,779731,607982954911,111935133845893409)
403: (2,3,7,47,395,779731,607985134327,67432145697278297)
404: (2,3,7,47,395,779731,607992724369,28278349603173071)
405: (2,3,7,47,395,779731,607998396563,19721082760057333)
406: (2,3,7,47,395,779731,608022582283,8610956123209613)
407: (2,3,7,47,395,779731,608082023341,3611399136628139)
408: (2,3,7,47,395,779731,608149585237,2175818873769587)
409: (2,3,7,47,395,779731,608255881459,1338771195093461)
410: (2,3,7,47,395,779731,609310471873,278361145573823)
411: (2,3,7,47,395,779731,611570627407,103543611609617)
412: (2,3,7,47,395,779731,616542746329,43774534989431)
413: (2,3,7,47,395,779731,627591899489,19455348884191)
414: (2,3,7,47,395,779731,654747318217,8511715136807)
415: (2,3,7,47,395,779731,719299870717,3928483909307)
416: (2,3,7,47,395,779731,862938861797,2057777285827)
417: (2,3,7,47,395,779731,1215959305259,1215959305261)
418: (2,3,7,47,395,779747,35764242853,113844911673252232819)
419: (2,3,7,47,395,779759,21057289541,4813266433751)
420: (2,3,7,47,395,779759,38765626621,45659592271)
421: (2,3,7,47,395,779767,16432641181,12695263971863917607)
422: (2,3,7,47,395,779767,16436722861,66173085861047)
423: (2,3,7,47,395,779771,19900801759,58194426121)
424: (2,3,7,47,395,779783,14885469661,50028641623)
425: (2,3,7,47,395,779831,6020372531,3660733426607933569529)
426: (2,3,7,47,395,779837,5682995267,189977953601291)
427: (2,3,7,47,395,779959,3643426013,9796210087)
428: (2,3,7,47,395,780007,2197019071,3521690089637)
429: (2,3,7,47,395,780071,1783709071,33505194320147509)
430: (2,3,7,47,395,780499,791389997,1140143708160043)
431: (2,3,7,47,395,780809,564244859,266091145461825119)
432: (2,3,7,47,395,784751,121867639,21195582322021)
433: (2,3,7,47,395,805453,24415343,51283314802521631)
434: (2,3,7,47,395,810307,20696909,12711787483)
435: (2,3,7,47,395,841727,10617301,3627426007)
436: (2,3,7,47,395,852997,12808339,31168163)
437: (2,3,7,47,395,1036747,5630039,7126513)

438:    (2,3,7,47,395,1465487,1668001,1644259747)
439:    (2,3,7,47,395,1506907,1615811,4344500161018957)
440:    (2,3,7,47,397,71275,160970303,196784088475439)
441:    (2,3,7,47,403,19403,15435513367,238255072887400163321)
442:    (2,3,7,47,403,19403,15435513371,47651014589828443357)
443:    (2,3,7,47,403,19403,15435513467,2358961132979717821)
444:    (2,3,7,47,403,19403,15435513871,471792238944354257)
445:    (2,3,7,47,403,19403,15466078739,7810369763701)
446:    (2,3,7,47,403,19403,15588340231,1574422363433)
447:    (2,3,7,47,403,19403,18522616039,92613080201)
448:    (2,3,7,47,403,19403,30871026731,30871026733)
449:    (2,3,7,47,403,20017,632531,2486398615943)
450:    (2,3,7,47,407,13171,813986039,297015603735318229)
451:    (2,3,7,47,407,13171,813986779,892679069216489)
452:    (2,3,7,47,407,13171,813986909,759631511994559)
453:    (2,3,7,47,407,13171,813988649,253643515029019)
454:    (2,3,7,47,407,13171,814276249,2283878577419)
455:    (2,3,7,47,407,13171,814855189,763135297079)
456:    (2,3,7,47,407,13171,815007419,649516546249)
457:    (2,3,7,47,407,13171,1153824559,2763656909)
458:    (2,3,7,47,415,8111,6644612311,44150872756848148409)
459:    (2,3,7,47,415,8111,6644612561,175899898079686159)
460:    (2,3,7,47,415,8111,6644612579,164129645121205381)
461:    (2,3,7,47,415,8111,6644679829,653909586750131)
462:    (2,3,7,47,415,8111,6644710721,448644223238719)
463:    (2,3,7,47,415,8111,6669313471,1794045323969)
464:    (2,3,7,47,415,8111,6671084869,1674442302371)
465:    (2,3,7,47,415,8111,13289224619,13289224621)
466:    (2,3,7,47,451,4219,12721,809846028175)
467:    (2,3,7,47,475,2339,7588853,1097355724147)
468:    (2,3,7,47,475,2339,10937161,24788231)
469:    (2,3,7,47,583,1223,1407479767,1980999293106894521)
470:    (2,3,7,47,583,1223,1407479771,396199859747362717)
471:    (2,3,7,47,583,1223,1407479795,68310321810907861)
472:    (2,3,7,47,583,1223,1407479911,13662065488165385)
473:    (2,3,7,47,583,1223,1407480007,8219915480921321)
474:    (2,3,7,47,583,1223,1407480971,1643984222168077)
475:    (2,3,7,47,583,1223,1407486755,283446720357061)
476:    (2,3,7,47,583,1223,1407514711,56690470055225)
477:    (2,3,7,47,583,1223,1407520043,49185787937581)
478:    (2,3,7,47,583,1223,1407681151,9838283571329)
479:    (2,3,7,47,583,1223,1408647799,1697420599001)
480:    (2,3,7,47,583,1223,1413319931,340610103613)
481:    (2,3,7,47,583,1223,1417186523,205492045981)
482:    (2,3,7,47,583,1223,1456013551,42224393009)
483:    (2,3,7,47,583,1223,1688975719,8444878601)
484:    (2,3,7,47,583,1223,2814959531,2814959533)
485:    (2,3,7,55,179,24323,10057317271,101149630679497570169)
486:    (2,3,7,55,179,24323,10057317337,1509695990198216567)

487:    (2,3,7,55,179,24323,10057317373,982035259275183803)
488:    (2,3,7,55,179,24323,10057324171,14657252582957069)
489:    (2,3,7,55,179,24323,10057326451,11017288204516949)
490:    (2,3,7,55,179,24323,10057932397,164447044588907)
491:    (2,3,7,55,179,24323,10058262913,106973922824063)
492:    (2,3,7,55,179,24323,10058412719,92346287182321)
493:    (2,3,7,55,179,24323,10058774641,69415603790639)
494:    (2,3,7,55,179,24323,10120675351,1606532921849)
495:    (2,3,7,55,179,24323,10130712353,1388210001823)
496:    (2,3,7,55,179,24323,10154961127,1045960995977)
497:    (2,3,7,55,179,24323,10170148517,906525568387)
498:    (2,3,7,55,179,24323,10207426483,683897574293)
499:    (2,3,7,55,179,24323,17617010819,23437440421)
500:    (2,3,7,55,179,24323,20114634539,20114634541)
501:    (2,3,7,55,197,1963,23731,5851217083)
502:    (2,3,7,65,121,6233,108365629,3657689751023669)
503:    (2,3,7,67,113,32159,285907,55656282215)
504:    (2,3,7,67,115,5779,64691,12184647841)
505:    (2,3,7,67,187,283,334651,49836124516793)
506:    (2,3,7,71,103,61463,111704965,86118313000187)
507:    (2,3,7,71,103,61583,24596843,3122473822388077)
508:    (2,3,7,71,103,61679,15167713,18655073941247)
509:    (2,3,7,71,103,62453,3747683,32674646527)
510:    (2,3,7,71,103,65635,958667,68866718993)
511:    (2,3,7,97,101,313,2473,63700207645)
512:    (2,3,11,17,101,149,3109,52495396601)
513:    (2,3,11,23,31,47059,2214502423,4904020979258368505)
514:    (2,3,11,23,31,47059,2214502429,700574427506483291)
515:    (2,3,11,23,31,47059,2214502441,258106369427337479)
516:    (2,3,11,23,31,47059,2214502471,100082062970499689)
517:    (2,3,11,23,31,47059,2214502555,36872340387764573)
518:    (2,3,11,23,31,47059,2214503353,5267479096397015)
519:    (2,3,11,23,31,47059,2259696349,110725121051)
520:    (2,3,11,23,31,47059,2331055181,44290048459)
521:    (2,3,11,23,31,47059,2530859911,17716019369)
522:    (2,3,11,23,31,47059,3030371735,8225294713)
523:    (2,3,11,23,31,47059,3073187035,7925587613)
524:    (2,3,11,23,31,47059,4429004843,4429004845)
525:    (2,3,11,23,31,47063,442938131,980970939025927673)
526:    (2,3,11,23,31,47065,316397839,6803428301729243)
527:    (2,3,11,23,31,47095,59897203,132743972247361529)
528:    (2,3,11,23,31,47107,47322265,1028152391)
529:    (2,3,11,23,31,47131,30382063,67384091875543673)
530:    (2,3,11,23,31,47137,28282147,3892535183)
531:    (2,3,11,23,31,47183,17789723,11693897561)
532:    (2,3,11,23,31,47243,12017087,26715920281613177)
533:    (2,3,11,23,31,47423,6114059,13644326865136505)
534:    (2,3,11,23,31,47735,3318041,58687879710227)
535:    (2,3,11,23,31,48175,2676569,8395963)

536: (2,3,11,23,31,49759,866923,2029951372029305)
537: (2,3,11,23,31,50027,792917,266665980383003)
538: (2,3,11,23,31,50299,731935,331193033)
539: (2,3,11,23,31,53899,371089,420055991)
540: (2,3,11,23,31,55195,445661,1124947)
541: (2,3,11,23,31,60563,211031,601432790177273)
542: (2,3,11,23,31,69745,144667,24989779178951)
543: (2,3,11,23,31,90721,97775,414514424707)
544: (2,3,11,25,29,1097,2753,144508961849)
545: (2,3,11,31,35,67,369067,1770735487289)
546: (2,3,13,17,37,257,504565,1225587899)
547: (2,3,13,17,37,257,507175,90784289)
548: (2,3,13,17,37,257,521749,15130715)
549: (2,3,13,17,53,83,253805,333362609)
550: (2,3,13,25,29,67,2981,11294561849)

References

[1] Michel Abdalla, Mihir Bellare, and Phillip Rogaway. The oracle Diffie-Hellman assumptions and an analysis of DHIES. In David Naccache, editor, *CT-RSA*, volume 2020 of *Lecture Notes in Computer Science*, pages 143–158. Springer, 2001.

[2] Leonard M. Adleman. Molecular computation of solutions to combinatorial problems. *Science*, 266:1021–1024, November 1994.

[3] Jae Hyun Ahn, Matthew Green, and Susan Hohenberger. Synchronized aggregate signatures: new definitions, constructions, and applications. In Ehab Al-Shaer, Angelos D. Keromytis, and Vitaly Shmatikov, editors, *ACM Conference on Computer and Communications Security*, pages 473–484. ACM, 2010.

[4] Dario Alpern. Factorization using the elliptic curve method. http://www.alpertron.com.ar/ECM.HTM.

[5] Iris Anshel, Michael Anshel, Benji Fisher, and Dorian Goldfeld. New key agreement protocols in braid group cryptography. In David Naccache, editor, *CT-RSA*, volume 2020 of *Lecture Notes in Computer Science*, pages 13–27. Springer, 2001.

[6] Iris Anshel, Michael Anshel, and Dorian Goldfeld. An algebraic method for public-key cryptography. *Math. Res. Lett.*, 6(3-4):287–291, 1999.

[7] Giuseppe Ateniese, Karyn Benson, and Susan Hohenberger. Key-private proxy re-encryption. In Marc Fischlin, editor, *CT-RSA*, volume 5473 of *Lecture Notes in Computer Science*, pages 279–294. Springer, 2009.

[8] Giuseppe Ateniese, Jan Camenisch, Marc Joye, and Gene Tsudik. A practical and provably secure coalition-resistant group signature scheme. In Mihir Bellare, editor, *CRYPTO*, volume 1880 of *Lecture Notes in Computer Science*, pages 255–270. Springer, 2000.

[9] Giuseppe Ateniese, Kevin Fu, Matthew Green, and Susan Hohenberger. Improved proxy re-encryption schemes with applications to secure distributed storage. In *NDSS*. The Internet Society, 2005.

[10] Giuseppe Ateniese, Kevin Fu, Matthew Green, and Susan Hohenberger. Improved proxy re-encryption schemes with applications to secure distributed storage. *ACM Trans. Inf. Syst. Secur.*, 9(1):1–30, 2006.

[11] Giuseppe Ateniese and Susan Hohenberger. Proxy re-signatures: new definitions, algorithms, and applications. In Vijay Atluri, Catherine Meadows, and Ari Juels, editors, *ACM Conference on Computer and Communications Security*, pages 310–319. ACM, 2005.

[12] Michael Atiyah. Mathematics in the 20th century. *Bull. London Math. Soc.*, 34:1–15, 2002.

[13] Nuttapong Attrapadung and Hideki Imai. Graph-decomposition-based frameworks for subset-cover broadcast encryption and efficient instantiations. In Bimal K. Roy, editor, *ASIACRYPT*, volume 3788 of *Lecture Notes in Computer Science*, pages 100–120. Springer, 2005.

[14] Man Ho Au, Patrick P. Tsang, Willy Susilo, and Yi Mu. Dynamic universal accumulators for ddh groups and their application to attribute-based anonymous credential systems. In Marc Fischlin, editor, *CT-RSA*, volume 5473 of *Lecture Notes in Computer Science*, pages 295–308. Springer, 2009.

[15] Randolph Baden, Adam Bender, Neil Spring, Bobby Bhattacharjee, and Daniel Starin. Persona: an online social network with user-defined privacy. In Pablo Rodriguez, Ernst W. Biersack, Konstantina Papagiannaki, and Luigi Rizzo, editors, *SIGCOMM*, pages 135–146. ACM, 2009.

[16] Joonsang Baek, Willy Susilo, and Jianying Zhou. New constructions of fuzzy identity-based encryption. In Feng Bao and Steven Miller, editors, *ASIACCS*, pages 368–370. ACM, 2007.

[17] G. Baumslag, B. Fine, and X. Xu. A proposed public key cryptosystem using the modular group. *Contemporary Mathematics*, 421:35, 2006.

[18] Gilbert Baumslag, Nelly Fazio, Antonio Nicolosi, Vladimir Shpilrain, and William E. Skeith III. Generalized learning problems and applications to noncommutative cryptography. In Xavier Boyen and Xiaofeng Chen, editors, *ProvSec*, volume 6980 of *Lecture Notes in Computer Science*, pages 324–339. Springer, 2011.

[19] Gilbert Baumslag, Benjamin Fine, and Xiaowei Xu. Cryptosystems using linear groups. *Appl. Algebr. Eng. Comm.*, 17(3-4):205–217, 2006.

[20] Amos Beimel. *Secure Schemes for Secret Sharing and Key Distribution*. PhD thesis, Israel Institute of Technology, Technion, Haifa, Israel, 1996.

[21] Mihir Bellare, Juan A. Garay, and Tal Rabin. Fast batch verification for modular exponentiation and digital signatures. In Kaisa Nyberg, editor, *EUROCRYPT*, volume 1403 of *Lecture Notes in Computer Science*, pages 236–250. Springer, 1998.

[22] Mihir Bellare, Chanathip Namprempre, David Pointcheval, and Michael Semanko. The one-more-rsa-inversion problems and the security of chaum's blind signature scheme. *J. Cryptol.*, 16(3):185–215, 2003.

[23] Mihir Bellare and Gregory Neven. Multi-signatures in the plain public-key model and a general forking lemma. In Ari Juels, Rebecca N. Wright, and Sabrina De Capitani di Vimercati, editors, *ACM Conference on Computer and Communications Security*, pages 390–399. ACM, 2006.

[24] Mitchell A. Berger. The genus of closed 3-braids. *J. Knot. Theor. Ramif.*, 1(3):303–326, 1992.

[25] Mitchell A. Berger. Minimum crossing numbers for 3-braids. *J. Phys. A: Math. Gen.*, 27:6205–6213, 1994.

[26] David Bernard and Alper Epstein. *Word processing in groups.* Jones and Bartlett Publishers, 1992.

[27] John Bethencourt, Amit Sahai, and Brent Waters. Ciphertext-policy attribute-based encryption. In *IEEE Symposium on Security and Privacy*, pages 321–334. IEEE Computer Society, 2007.

[28] Jean-Camille Birget, Spyros S. Magliveras, and Michal Sramka. On public-key cryptosystems based on combinatorial group theory. *IACR Cryptology ePrint Archive*, 2005:70, 2005.

[29] Matt Blaze, Gerrit Bleumer, and Martin Strauss. Divertible protocols and atomic proxy cryptography. In Kaisa Nyberg, editor, *EUROCRYPT*, volume 1403 of *Lecture Notes in Computer Science*, pages 127–144. Springer, 1998.

[30] Avrim Blum, Adam Kalai, and Hal Wasserman. Noise-tolerant learning, the parity problem, and the statistical query model. *J. ACM*, 50(4):506–519, 2003.

[31] Rakeshbabu Bobba, Omid Fatemieh, Fariba Khan, Arindam Khan, Carl A. Gunter, Himanshu Khurana, and Manoj Prabhakaran. Attribute-based messaging: Access control and confidentiality. *ACM Trans. Inf. Syst. Secur.*, 13(4):31, 2010.

[32] Rakeshbabu Bobba, Himanshu Khurana, Musab AlTurki, and Farhana Ashraf. Pbes: a policy based encryption system with application to data sharing in the power grid. In Wanqing Li, Willy Susilo, Udaya Kiran Tupakula, Reihaneh Safavi-Naini, and Vijay Varadharajan, editors, *ASIACCS*, pages 262–275. ACM, 2009.

[33] Rudolf M. Bolle, Jonathan Connell, Sharathchandra Pankanti, Nalini K. Ratha, and Andrew W. Senior. Biometrics 101. *Report RC22481, IBM Research*, pages 545–546, 2002.

[34] Dan Boneh and Xavier Boyen. Efficient selective-id secure identity-based encryption without random oracles. In Christian Cachin and Jan Camenisch, editors, *EUROCRYPT*, volume 3027 of *Lecture Notes in Computer Science*, pages 223–238. Springer, 2004.

[35] Dan Boneh, Xavier Boyen, and Eu-Jin Goh. Hierarchical identity based encryption with constant size ciphertext. In Ronald Cramer, editor, *EUROCRYPT*, volume 3494 of *Lecture Notes in Computer Science*, pages 440–456. Springer, 2005.

[36] Dan Boneh, Ran Canetti, Shai Halevi, and Jonathan Katz. Chosen-ciphertext security from identity-based encryption. *SIAM J. Comput.*, 36(5):1301–1328, 2007.

[37] Dan Boneh and Matthew K. Franklin. Identity-based encryption from the weil pairing. In Joe Kilian, editor, *CRYPTO*, volume 2139 of *Lecture Notes in Computer Science*, pages 213–229. Springer, 2001.

[38] Dan Boneh, Craig Gentry, Ben Lynn, and Hovav Shacham. Aggregate and verifiably encrypted signatures from bilinear maps. In Eli Biham, editor, *EUROCRYPT*, volume 2656 of *Lecture Notes in Computer Science*, pages 416–432. Springer, 2003.

[39] Dan Boneh, Craig Gentry, and Brent Waters. Collusion resistant broadcast encryption with short ciphertexts and private keys. In Victor Shoup, editor, *CRYPTO*, volume 3621 of *Lecture Notes in Computer Science*, pages 258–275. Springer, 2005.

[40] Dan Boneh and Michael Hamburg. Generalized identity based and broadcast encryption schemes. In Josef Pieprzyk, editor, *ASIACRYPT*, volume 5350 of *Lecture Notes in Computer Science*, pages 455–470. Springer, 2008.

[41] Dan Boneh, Ben Lynn, and Hovav Shacham. Short signatures from the weil pairing. *J. Cryptol.*, 17(4):297–319, 2004.

[42] Dan Boneh, Amit Sahai, and Brent Waters. Fully collusion resistant traitor tracing with short ciphertexts and private keys. In Serge Vaudenay, editor, *EUROCRYPT*, volume 4004 of *Lecture Notes in Computer Science*, pages 573–592. Springer, 2006.

[43] Dan Boneh and Hovav Shacham. Fast variants of RSA. *In RSA Laboratories' Cryptobytes*, 5(1):1–8, 2002.

[44] David Borwein, J. M. Borwein, P. B. Borwein, and R. Girgensohn. Giuga's conjecture on primality. *Am. Math. Mon.*, 103:40–50, 1996.

[45] Colin Boyd and Chris Pavlovski. Attacking and repairing batch verification schemes. In Tatsuaki Okamoto, editor, *ASIACRYPT*, volume 1976 of *Lecture Notes in Computer Science*, pages 58–71. Springer, 2000.

[46] Xavier Boyen, Yevgeniy Dodis, Jonathan Katz, Rafail Ostrovsky, and Adam Smith. Secure remote authentication using biometric data. In Ronald Cramer, editor, *EUROCRYPT*, volume 3494 of *Lecture Notes in Computer Science*, pages 147–163. Springer, 2005.

[47] Xavier Boyen and Brent Waters. Anonymous hierarchical identity-based encryption (without random oracles). In Cynthia Dwork, editor, *CRYPTO*, volume 4117 of *Lecture Notes in Computer Science*, pages 290–307. Springer, 2006.

[48] Xavier Boyen and Brent Waters. Compact group signatures without random oracles. In Serge Vaudenay, editor, *EUROCRYPT*, volume 4004 of *Lecture Notes in Computer Science*, pages 427–444. Springer, 2006.

[49] Lawrence Brenton and Daniel Drucker. Perfect graphs and complex surface singularities with perfect local fundamental group. *Tohoku Math. J.*, 41(4):507–525, 1989.

[50] Lawrence Brenton and Daniel Drucker. On the number of solutions of $\sum_{j=1}^{s}\left(\frac{1}{x_j}\right) + \frac{1}{(x_1 \ldots x_s)} = 1$. *J. Number Theory*, 44:25–29, 1993.

[51] Lawrence Brenton and Richard Hill. On the diophantine equation $1 = \sum(1/n_i) + 1/(\prod n_j)$ and a class of homologically trivial complex surface singularities. *Pac. J. Math.*, 133(1):41–67, 1988.

[52] Lawrence Brenton and Lynda Jaje. Perfectly weighted graphs. *Graph. Combinator.*, 17(3):389–407, 2001.

[53] Lawrence Brenton and Ana Vasiliu. Znám's problem. *Mathematics Magazine*, 75(1):3–11, 2002.

[54] Emmanuel Bresson, Dario Catalano, and David Pointcheval. A simple public-key cryptosystem with a double trapdoor decryption mechanism and its applications. In Chi-Sung Laih, editor, *ASIACRYPT*, volume 2894 of *Lecture Notes in Computer Science*, pages 37–54. Springer, 2003.

[55] Achim D. Brucker, Helmut Petritsch, and Stefan G. Weber. Attribute-based encryption with break-glass. In Pierangela Samarati, Michael Tunstall, Joachim Posegga, Konstantinos Markantonakis, and Damien Sauveron, editors, *WISTP*, volume 6033 of *Lecture Notes in Computer Science*, pages 237–244. Springer, 2010.

[56] J. Buhler, H. Lenstra, and Carl Pomerance. Factoring integers with the number field sieve. In Arjen Lenstra and Hendrik Lenstra, editors, *The development of the number field sieve*, volume 1554 of *Lecture Notes in Mathematics*, pages 50–94. Springer Berlin / Heidelberg, 1993. 10.1007/BFb0091539.

[57] Nechemia Burshtein. On distinct unit fractions whose sum equals 1. *Discrete Math.*, 5:201–206, 1973.

[58] William Butske, Lynda M. Jaje, and Daniel R. Mayernik. On the equation $\sum_{p|N} \frac{1}{p} + \frac{1}{N} = 1$, pseudoperfect numbers, and perfectly weighted graphs. *Math. Comput.*, 69(229):407–420, 2000.

[59] Christian Cachin and Jan Camenisch, editors. *Advances in Cryptology - EU-ROCRYPT 2004, International Conference on the Theory and Applications of Cryptographic Techniques, Interlaken, Switzerland, May 2-6, 2004, Proceedings*, volume 3027 of *Lecture Notes in Computer Science*. Springer, 2004.

[60] Jan Camenisch, Susan Hohenberger, and Michael Østergaard Pedersen. Batch verification of short signatures. In Moni Naor, editor, *EUROCRYPT*, volume 4515 of *Lecture Notes in Computer Science*, pages 246–263. Springer, 2007.

[61] Jan Camenisch and Markus Stadler. Efficient group signature schemes for large groups (extended abstract). In Burton S. Kaliski Jr., editor, *CRYPTO*, volume 1294 of *Lecture Notes in Computer Science*, pages 410–424. Springer, 1997.

[62] Ran Canetti, Shai Halevi, and Jonathan Katz. A forward-secure public-key encryption scheme. In Eli Biham, editor, *EUROCRYPT*, volume 2656 of *Lecture Notes in Computer Science*, pages 255–271. Springer, 2003.

[63] Ran Canetti, Shai Halevi, and Jonathan Katz. Chosen-ciphertext security from identity-based encryption. In Cachin and Camenisch [59], pages 207–222.

[64] Ran Canetti, Shai Halevi, and Jonathan Katz. A forward-secure public-key encryption scheme. *J. Cryptol.*, 20(3):265–294, 2007.

[65] Ran Canetti and Susan Hohenberger. Chosen-ciphertext secure proxy re-encryption. In Peng Ning, Sabrina De Capitani di Vimercati, and Paul F. Syverson, editors, *ACM Conference on Computer and Communications Security*, pages 185–194. ACM, 2007.

[66] Ning Cao, Cong Wang, Ming Li, Kui Ren, and Wenjing Lou. Privacy-preserving multi-keyword ranked search over encrypted cloud data. In *INFOCOM*, pages 829–837. IEEE, 2011.

[67] Tianjie Cao, Dongdai Lin, and Rui Xue. Security analysis of some batch verifying signatures from pairings. *I. J. Network Security*, 3(2):138–143, 2006.

[68] Zhenfu Cao. Mordell's problem on unit fractions. *J. Mathematics (PRC)*, 7(3):245–250, 1987.

[69] Zhenfu Cao. On the number of solution of the diophantine equation $\sum_{j=1}^{s} \frac{1}{x_j} + \frac{1}{x_1 \cdots x_s} = 1$. In *International Conference on Number Theory and Analysis in Memory of Loo-Keng Hua*, Beijing, 1988.

[70] Zhenfu Cao. *Introduction to Diophantine Equations*. Harbin Institute of Technology Press, Harbin, 1989. (in Chinese).

[71] Zhenfu Cao. Two new world records were born in our lab, 2008. `http://tdt.sjtu.edu.cn/English\%20Version/notes/20081210/index.htm`.

[72] Zhenfu Cao. Application oriented cryptographic technology. *Bulletin of Chinese Association for Cryptologic Research*, 1:6–11, 2011.

[73] Zhenfu Cao, Xiaolei Dong, and Licheng Wang. New public key cryptosystems using polynomials over non-commutative rings. *IACR Cryptology ePrint Archive*, 2007:9, 2007.

[74] Zhenfu Cao and Chenming Jin. On the number of solutions of Znám's problem. *J. Harbin Instisute of Technology*, 30(1):46–49, 1998. (in Chinese).

[75] Zhenfu Cao, Rui Liu, and Liangrui Zhang. On the equation $\sum_{j=1}^{s} \frac{1}{x_j} + \frac{1}{x_1 \ldots x_s} = 1$ and Znám's problem. *J. Number Theory*, 27:206–211, 1987.

[76] Zhenfu Cao, Rui Liu, and Liangrui Zhang. On the equation $\sum_{1 \leq j \leq s} \frac{1}{x_j} + \frac{1}{x_1 \ldots x_s} = 1$ and Znám's problem. *Chinese Journal of Nature*, 12(7):554–555, 1989. (in Chinese).

[77] David Cash, Eike Kiltz, and Victor Shoup. The twin Diffie-Hellman problem and applications. *J. Cryptol.*, 22(4):470–504, 2009.

[78] Melissa Chase. Multi-authority attribute based encryption. In Salil P. Vadhan, editor, *TCC*, volume 4392 of *LNCS*, pages 515–534. Springer, 2007.

[79] Melissa Chase and Sherman S. M. Chow. Improving privacy and security in multi-authority attribute-based encryption. In Ehab Al-Shaer, Somesh Jha, and Angelos D. Keromytis, editors, *ACM Conference on Computer and Communications Security*, pages 121–130. ACM, 2009.

[80] Le Chen and Zhenfu Cao. Methods for resist length-based attacks. *J. Shanghai Jiao Tong University (in Chinese)*, 44(7):968–974, 2010.

[81] Yi-Ruei Chen, J. D. Tygar, and Wen-Guey Tzeng. Secure group key management using uni-directional proxy re-encryption schemes. In *INFOCOM*, pages 1952–1960. IEEE, 2011.

[82] Jung Hee Cheon and Byungheup Jun. A polynomial time algorithm for the braid Diffie-Hellman conjugacy problem. In Dan Boneh, editor, *CRYPTO*, volume 2729 of *Lecture Notes in Computer Science*, pages 212–225. Springer, 2003.

[83] Jung Hee Cheon and Dong Hoon Lee. Use of sparse and/or complex exponents in batch verification of exponentiations. *IEEE T. Comput.*, 55(12):1536–1542, 2006.

[84] Ling Cheung and Calvin C. Newport. Provably secure ciphertext policy abe. In Peng Ning, Sabrina De Capitani di Vimercati, and Paul F. Syverson, editors, *ACM Conference on Computer and Communications Security*, pages 456–465. ACM, 2007.

[85] Sherman S. M. Chow and Raphael C.-W. Phan. Proxy re-signatures in the standard model. In Tzong-Chen Wu, Chin-Laung Lei, Vincent Rijmen, and Der-Tsai Lee, editors, *ISC*, volume 5222 of *Lecture Notes in Computer Science*, pages 260–276. Springer, 2008.

[86] Sherman S. M. Chow, Jian Weng, Yanjiang Yang, and Robert H. Deng. Efficient unidirectional proxy re-encryption. In Daniel J. Bernstein and Tanja Lange, editors, *AFRICACRYPT*, volume 6055 of *Lecture Notes in Computer Science*, pages 316–332. Springer, 2010.

[87] Clifford Cocks. An identity based encryption scheme based on quadratic residues. In Bahram Honary, editor, *IMA Int. Conf.*, volume 2260 of *Lecture Notes in Computer Science*, pages 360–363. Springer, 2001.

[88] Don Coppersmith. Modifications to the number field sieve. *J. Cryptol.*, 6(3):169–180, 1993.

[89] Ronald Cramer and Victor Shoup. A practical public key cryptosystem provably secure against adaptive chosen ciphertext attack. In Hugo Krawczyk, editor, *CRYPTO*, volume 1462 of *Lecture Notes in Computer Science*, pages 13–25. Springer, 1998.

[90] Ronald Cramer and Victor Shoup. Universal hash proofs and a paradigm for adaptive chosen ciphertext secure public-key encryption. In Lars R. Knudsen, editor, *EUROCRYPT*, volume 2332 of *Lecture Notes in Computer Science*, pages 45–64. Springer, 2002.

[91] Ronald Cramer and Victor Shoup. Design and analysis of practical public-key encryption schemes secure against adaptive chosen ciphertext attack. *SIAM J. Comput.*, 33(1):167–226, 2003.

[92] George I. Davida, Yair Frankel, and Brian J. Matt. On enabling secure applications through off-line biometric identification. In *IEEE Symposium on Security and Privacy*, pages 148–157. IEEE Computer Society, 1998.

[93] Patrick Dehornoy. Braid-based cryptography. *Contemporary Mathematics*, 360:5–33, 2004.

[94] Patrick Dehornoy. Using shifted conjugacy in braid-based cryptography. *CoRR*, abs/cs/0609091, 2006.

[95] Cécile Delerablée. Identity-based broadcast encryption with constant size ciphertexts and private keys. In Kaoru Kurosawa, editor, *ASIACRYPT*, volume 4833 of *Lecture Notes in Computer Science*, pages 200–215. Springer, 2007.

[96] Robert H. Deng, Jian Weng, Shengli Liu, and Kefei Chen. Chosen-ciphertext secure proxy re-encryption without pairings. In Matthew K. Franklin, Lucas Chi Kwong Hui, and Duncan S. Wong, editors, *CANS*, volume 5339 of *Lecture Notes in Computer Science*, pages 1–17. Springer, 2008.

[97] Somnath Dey and Debasis Samanta. Improved feature processing for iris biometric authentication system. *I. J. Electrical Comput. System. Engineering*, 4(2), 2010.

[98] Steven DiBenedetto, Pado Gasti, Tene Tsudik, and Ersin Uzun. Andana: Anonymous named data networking application. *Arxiv preprint arXiv:1112.2205*, 2011.

[99] Whitfield Diffie and Martin E. Hellman. New directions in cryptography. *IEEE T Inform. Theory.*, 22(6):644–654, 1976.

[100] Jeong-Min Do, You-Jin Song, and Namje Park. Attribute based proxy re-encryption for data confidentiality in cloud computing environments. In *Computers, Networks, Systems and Industrial Engineering (CNSI), 2011 First ACIS/JNU International Conference on*, pages 248–251. IEEE, 2011.

[101] Yevgeniy Dodis and Nelly Fazio. Public key broadcast encryption for stateless receivers. In Joan Feigenbaum, editor, *Digital Rights Management Workshop*, volume 2696 of *Lecture Notes in Computer Science*, pages 61–80. Springer, 2002.

[102] Yevgeniy Dodis, Leonid Reyzin, and Adam Smith. Fuzzy extractors: How to generate strong keys from biometrics and other noisy data. In Christian Cachin and Jan Camenisch, editors, *EUROCRYPT*, volume 3027 of *Lecture Notes in Computer Science*, pages 523–540. Springer, 2004.

[103] Michael Domaratzki, Keith Ellul, Jeffrey Shallit, and Ming wei Wang. Non-uniqueness and radius of cyclic unary nfas. *Int. J. Found. Comput. Sci.*, 16(5):883–896, 2005.

[104] Changyu Dong, Giovanni Russello, and Naranker Dulay. Shared and searchable encrypted data for untrusted servers. *Journal of Computer Security*, 19(3):367–397, 2011.

[105] Xiaolei Dong, Lifei Wei, Haojin Zhu, Zhenfu Cao, and Licheng Wang. EP^2DF: An efficient privacy preserving data forwarding scheme for service-oriented vehicular ad hoc networks. *IEEE T Veh. Technol.*, 60(2):580–591, 2011.

[106] B. Eick and D. Kahrobaei. Polycyclic groups: A new platform for cryptology? *Arxiv preprint math/0411077*, 2004.

[107] Taher ElGamal. A public key cryptosystem and a signature scheme based on discrete logarithms. *IEEE T Inform. Theory.*, 31(4):469–472, 1985.

[108] Elsayed A. Elrifai and Hugh R. p. Algorithms for positive braids. *Q. J. MATH.*, 45(180):479–498, 1994.

[109] Paul Erdős and Sherman Stein. Sums of distinct unit fractions. *P. Am. Math. Soc.*, 14:126–131, 1963.

[110] Anna Lisa Ferrara, Matthew Green, Susan Hohenberger, and Michael Østergaard Pedersen. Practical short signature batch verification. In Marc Fischlin, editor, *CT-RSA*, volume 5473 of *Lecture Notes in Computer Science*, pages 309–324. Springer, 2009.

[111] Amos Fiat. Batch RSA. *J. Cryptol.*, 10(2):75–88, 1997.

[112] Benjamin Fine, Maggie Habeeb, Delaram Kahrobaei, and Gerhard Rosenberger. Aspects of nonabelian group based cryptography: A survey and open problems. *CoRR*, abs/1103.4093, 2011.

[113] Nuno Franco and Juan González-Meneses. Conjugacy problem for braid groups and garside groups. *J. Algebra*, 266(1):112–132, 2003.

[114] Keith B. Frikken, Mikhail J. Atallah, and Jiangtao Li. Attribute-based access control with hidden policies and hidden credentials. *IEEE T. Comput.*, 55(10):1259–1270, 2006.

[115] Eiichiro Fujisaki and Tatsuaki Okamoto. How to enhance the security of public-key encryption at minimum cost. In Hideki Imai and Yuliang Zheng, editors, *Public Key Cryptography*, volume 1560 of *Lecture Notes in Computer Science*, pages 53–68. Springer, 1999.

[116] Eiichiro Fujisaki and Tatsuaki Okamoto. Secure integration of asymmetric and symmetric encryption schemes. In Michael J. Wiener, editor, *CRYPTO*, volume 1666 of *Lecture Notes in Computer Science*, pages 537–554. Springer, 1999.

[117] David Garber. Braid group cryptography. *CoRR*, abs/0711.3941, 2007.

[118] V. Gebhardt. A new approach to the conjugacy problem in garside groups. *J. Algebra*, 292(1):282–302, 2005.

[119] Craig Gentry. Fully homomorphic encryption using ideal lattices. In Michael Mitzenmacher, editor, *STOC*, pages 169–178. ACM, 2009.

[120] Craig Gentry and Shai Halevi. Hierarchical identity based encryption with polynomially many levels. In Omer Reingold, editor, *TCC*, volume 5444 of *Lecture Notes in Computer Science*, pages 437–456. Springer, 2009.

[121] Craig Gentry and Zulfikar Ramzan. Identity-based aggregate signatures. In Moti Yung, Yevgeniy Dodis, Aggelos Kiayias, and Tal Malkin, editors, *Public Key Cryptography*, volume 3958 of *Lecture Notes in Computer Science*, pages 257–273. Springer, 2006.

[122] Craig Gentry and Alice Silverberg. Hierarchical id-based cryptography. In Yuliang Zheng, editor, *ASIACRYPT*, volume 2501 of *Lecture Notes in Computer Science*, pages 548–566. Springer, 2002.

[123] Craig Gentry and Brent Waters. Adaptive security in broadcast encryption systems (with short ciphertexts). In Antoine Joux, editor, *EUROCRYPT*, volume 5479 of *Lecture Notes in Computer Science*, pages 171–188. Springer, 2009.

[124] G. Giuga. Su una presumibile proprietà caratteristica dei numeri primi. *Ist. Lombardo Sci. Lett. Rend A*, 83:511–528, 1950.

[125] Juan Gonzalez-Meneses. Improving an algorithm to solve multiple simultaneous conjugacy problems in braid groups. *Contemporary Mathematics*, 372:35–42, 2005.

[126] Michael T. Goodrich, Jonathan Z. Sun, and Roberto Tamassia. Efficient tree-based revocation in groups of low-state devices. In Matthew K. Franklin, editor, *CRYPTO*, volume 3152 of *Lecture Notes in Computer Science*, pages 511–527. Springer, 2004.

[127] Vipul Goyal, Abhishek Jain, Omkant Pandey, and Amit Sahai. Bounded ciphertext policy attribute based encryption. In Luca Aceto, Ivan Damgård, Leslie Ann Goldberg, Magnús M. Halldórsson, Anna Ingólfsdóttir, and Igor Walukiewicz, editors, *ICALP (2)*, volume 5126 of *Lecture Notes in Computer Science*, pages 579–591. Springer, 2008.

[128] Vipul Goyal, Omkant Pandey, Amit Sahai, and Brent Waters. Attribute-based encryption for fine-grained access control of encrypted data. In Ari Juels, Rebecca N. Wright, and Sabrina De Capitani di Vimercati, editors, *ACM Conference on Computer and Communications Security*, pages 89–98. ACM, 2006.

[129] Vipul Goyal, Omkant Pandey, Amit Sahai, and Brent Waters. Attribute-based encryption for fine-grained access control of encrypted data. In Ari Juels, Rebecca N. Wright, and Sabrina De Capitani di Vimercati, editors, *ACM Conference on Computer and Communications Security*, pages 89–98. ACM, 2006.

[130] Matthew Green and Giuseppe Ateniese. Identity-based proxy re-encryption. In Jonathan Katz and Moti Yung, editors, *ACNS*, volume 4521 of *Lecture Notes in Computer Science*, pages 288–306. Springer, 2007.

[131] Dima Grigoriev and Ilia V. Ponomarenko. On non-abelian homomorphic public-key cryptosystems. *CoRR*, cs.CR/0207079, 2002.

[132] Dima Grigoriev and Ilia V. Ponomarenko. Homomorphic public-key cryptosystems over groups and rings. *CoRR*, cs.CR/0309010, 2003.

[133] Dimitri Grigoriev and Ilia V. Ponomarenko. Constructions in public-key cryptography over matrix groups. *CoRR*, abs/math/0506180, 2005.

[134] Dani Halevy and Adi Shamir. The LSD broadcast encryption scheme. In Moti Yung, editor, *CRYPTO*, volume 2442 of *Lecture Notes in Computer Science*, pages 47–60. Springer, 2002.

[135] Chris Hall, Ian Goldberg, and Bruce Schneier. Reaction attacks against several public-key cryptosystems. In Vijay Varadharajan and Yi Mu, editors, *ICICS*, volume 1726 of *Lecture Notes in Computer Science*, pages 2–12. Springer, 1999.

[136] Goichiro Hanaoka, Yutaka Kawai, Noboru Kunihiro, Takahiro Matsuda, Jian Weng, Rui Zhang, and Yunlei Zhao. Generic construction of chosen ciphertext secure proxy re-encryption. In Orr Dunkelman, editor, *CT-RSA*, volume 7178 of *Lecture Notes in Computer Science*, pages 349–364. Springer, 2012.

[137] Lein Harn. Batch verifying multiple DSA-type digital signatures. *Electron. Lett.*, 34(9):870–871, 1998.

[138] Lein Harn. Batch verifying multiple RSA digital signatures. *Electron. Lett.*, 34(12):1219–1220, 1998.

[139] Javier Herranz, Fabien Laguillaumie, Benoît Libert, and Carla Ràfols. Short attribute-based signatures for threshold predicates. In Orr Dunkelman, editor, *CT-RSA*, volume 7178 of *Lecture Notes in Computer Science*, pages 51–67. Springer, 2012.

[140] Thomas S. Heydt-Benjamin, Hee-Jin Chae, Benessa Defend, and Kevin Fu. Privacy for public transportation. In George Danezis and Philippe Golle, editors, *Privacy Enhancing Technologies*, volume 4258 of *Lecture Notes in Computer Science*, pages 1–19. Springer, 2006.

[141] Christopher James Hill. Risk of masquerade arising from the storage of biometrics. Master's thesis, Australian National University, 2001.

[142] M. Jason Hinek, Shaoquan Jiang, Reihaneh Safavi-Naini, and Siamak Fayyaz Shahandashti. Attribute-based encryption with key cloning protection. *IACR Cryptology ePrint Archive*, 2008:478, 2008.

[143] Jeffrey Hoffstein, Jill Pipher, and Joseph Silverman. NTRU: A ring based public-key cryptosystem. In *Third International Algorithmic Number Theory Symposium (ANTS-III)*, volume 1423 of *Lecture Notes in Computer Science*, pages 268–288. Springer-Verlag, Berlin Germany, 1998.

[144] Dennis Hofheinz and Rainer Steinwandt. A practical attack on some braid group based cryptographic primitives. In Yvo Desmedt, editor, *Public Key Cryptography*, volume 2567 of *Lecture Notes in Computer Science*, pages 187–198. Springer, 2003.

[145] Susan Hohenberger, Guy N. Rothblum, Abhi Shelat, and Vinod Vaikuntanathan. Securely obfuscating re-encryption. *J. Cryptol.*, 24(4):694–719, 2011.

[146] Susan Hohenberger and Brent Waters. Realizing hash-and-sign signatures under standard assumptions. In Antoine Joux, editor, *EUROCRYPT*, volume 5479 of *Lecture Notes in Computer Science*, pages 333–350. Springer, 2009.

[147] Xiaoyan Hong, Dijiang Huang, Mario Gerla, and Zhen Cao. Sat: situation-aware trust architecture for vehicular networks. In *Proceedings of the 3rd international workshop on Mobility in the evolving internet architecture*, MobiArch '08, pages 31–36, New York, NY, USA, 2008. ACM.

[148] Fumitaka Hoshino, Masayuki Abe, and Tetsutaro Kobayashi. Lenient/strict batch verification in several groups. In George I. Davida and Yair Frankel, editors, *ISC*, volume 2200 of *Lecture Notes in Computer Science*, pages 81–94. Springer, 2001.

[149] Chun-Ying Huang, Yun-Peng Chiu, Kuan-Ta Chen, and Chin-Laung Lei. Secure multicast in dynamic environments. *Comput. Netw.*, 51(10):2805–2817, 2007.

[150] Dijiang Huang and Mayank Verma. Aspe: attribute-based secure policy enforcement in vehicular ad hoc networks. *Ad Hoc Netw.*, 7(8):1526–1535, 2009.

[151] Junbcom Hur. Improving security and efficiency in attribute-based data sharing. *IEEE T Knowl. Data En.*, (99):1–1, 2011.

[152] Min-Shiang Hwang, Cheng-Chi Lee, and Yuan-Liang Tang. Two simple batch verifying multiple digital signatures. In Sihan Qing, Tatsuaki Okamoto, and Jianying Zhou, editors, *ICICS*, volume 2229 of *Lecture Notes in Computer Science*, pages 233–237. Springer, 2001.

[153] Min-Shiang Hwang, Iuon-Chang Lin, and Kuo-Feng Hwang. Cryptanalysis of the batch verifying multiple RSA digital signatures. *Informatica, Lith. Acad. Sci.*, 11(1):15–19, 2000.

[154] Luan Ibraimi, Muhammad Asim, and Milan Petkovic. Secure management of personal health records by applying attribute-based encryption. In *Wearable Micro and Nano Technologies for Personalized Health (pHealth), 2009 6th International Workshop on*, pages 71–74. IEEE, 2009.

[155] The informed dialogue about consumer acceptability of DRM solutions in Europe (INDICARE). Consumer survey on digital music and drm. www.indicare.org/survey.

[156] Anca-Andreea Ivan and Yevgeniy Dodis. Proxy cryptography revisited. In *NDSS*. The Internet Society, 2003.

[157] Sonia Jahid, Prateek Mittal, and Nikita Borisov. Easier: encryption-based access control in social networks with efficient revocation. In Bruce S. N. Cheung, Lucas Chi Kwong Hui, Ravi S. Sandhu, and Duncan S. Wong, editors, *ASIACCS*, pages 411–415. ACM, 2011.

[158] Jaroslav Janak and Ladislav Skula. On the integers x_i for which $x_i | x_1 \ldots x_{i-1} x_{x+1} \ldots x_n + 1$ holds. *Mathematica Slovaca*, 28(3):305–310, 1978.

[159] Nam-Su Jho, Jung Yeon Hwang, Jung Hee Cheon, Myung-Hwan Kim, Dong Hoon Lee, and Eun Sun Yoo. One-way chain based broadcast encryption schemes. In Ronald Cramer, editor, *EUROCRYPT*, volume 3494 of *Lecture Notes in Computer Science*, pages 559–574. Springer, 2005.

[160] Xiaoqi Jia, Jun Shao, Jiwu Jing, and Peng Liu. Cca-secure type-based proxy re-encryption with invisible proxy. In *CIT*, pages 1299–1305. IEEE Computer Society, 2010.

[161] Yixin Jiang, Haojin Zhu, Minghui Shi, Xuemin (Sherman) Shen, and Chuang Lin. An efficient dynamic-identity based signature scheme for secure network coding. *Comput. Netw.*, 54(1):28–40, 2010.

[162] Antoine Joux and Kim Nguyen. Separating decision Diffie-Hellman from Diffie-Hellman in cryptographic groups. Cryptology ePrint Archive, Report 2001/003, 2001. http://eprint.iacr.org/.

[163] Marc Joye and Tancrède Lepoint. Traitor tracing schemes for protected software implementations. In Yan Chen, Stefan Katzenbeisser, and Ahmad-Reza Sadeghi, editors, *Digital Rights Management Workshop*, pages 15–22. ACM, 2011.

[164] Ari Juels and Madhu Sudan. A fuzzy vault scheme. *Design. Code. Cryptogr.*, 38(2):237–257, 2006.

[165] Ari Juels and Martin Wattenberg. A fuzzy commitment scheme. In Juzar Motiwalla and Gene Tsudik, editors, *ACM Conference on Computer and Communications Security*, pages 28–36. ACM, 1999.

[166] Seny Kamara and Kristin Lauter. Cryptographic cloud storage. In Radu Sion, Reza Curtmola, Sven Dietrich, Aggelos Kiayias, Josep M. Miret, Kazue Sako, and Francesc Sebé, editors, *Financial Cryptography Workshops*, volume 6054 of *Lecture Notes in Computer Science*, pages 136–149. Springer, 2010.

[167] Eun Song Kang, Ki Hyoung Ko, and Sang Jin Lee. Band-generator presentation for the 4-braids. *Topology Appl.*, 78:39–60, 1997.

[168] Apu Kapadia, Patrick P. Tsang, and Sean W. Smith. Attribute-based publishing with hidden credentials and hidden policies. In *NDSS*. The Internet Society, 2007.

[169] Zhao Ke and Qi Sun. On the representation of 1 by unit fractions. *J. Sichuan University (Natural Science Edition)*, 31:13–29, 1964. (in Chinese).

[170] Florian Kerschbaum and Alessandro Sorniotti. RFID-based supply chain partner authentication and key agreement. In David A. Basin, Srdjan Capkun, and Wenke Lee, editors, *WISEC*, pages 41–50. ACM, 2009.

[171] Himanshu Khurana and Hyung-Seok Hahm. Certified mailing lists. In Ferng-Ching Lin, Der-Tsai Lee, Bao-Shuh Paul Lin, Shiuhpyng Shieh, and Sushil Jajodia, editors, *ASIACCS*, pages 46–58. ACM, 2006.

[172] Alexei Kitaev. Quantum measurements and the abelian stabilizer problem. *Electronic Colloquium on Computational Complexity (ECCC)*, 3(3), 1996.

[173] Ki Hyoung Ko, Doo Ho Choi, Mi Sung Cho, and Jang-Won Lee. New signature scheme using conjugacy problem. *IACR Cryptology ePrint Archive*, 2002:168, 2002.

[174] Ki Hyoung Ko, Jang-Won Lee, and Tony Thomas. Towards generating secure keys for braid cryptography. *Design, Codes, and Cryptography*, 45(3):317–333, 2007.

[175] Ki Hyoung Ko, Sangjin Lee, Jung Hee Cheon, Jae Woo Han, Ju-Sung Kang, and Choonsik Park. New public-key cryptosystem using braid groups. In Mihir Bellare, editor, *CRYPTO*, volume 1880 of *Lecture Notes in Computer Science*, pages 166–183. Springer, 2000.

[176] Rob H. Koenen, Jack Lacy, Michael MacKay, and Steve Mitchell. The long march to interoperable digital rights management. *P. IEEE*, 92(6):883–897, 2004.

[177] Kaoru Kurosawa and Toshihiko Matsuo. How to remove mac from dhies. In Huaxiong Wang, Josef Pieprzyk, and Vijay Varadharajan, editors, *ACISP*, volume 3108 of *Lecture Notes in Computer Science*, pages 236–247. Springer, 2004.

[178] Laurie Law and Brian J. Matt. Finding invalid signatures in pairing-based batches. In Steven D. Galbraith, editor, *IMA Int. Conf.*, volume 4887 of *Lecture Notes in Computer Science*, pages 34–53. Springer, 2007.

[179] Eonkyung Lee. Braid groups in cryptography. *IEICE Transactions on Fundamentals*, E87-A(5):986–992, 2004.

[180] Seungwon Lee, Seongje Cho, Jongmoo Choi, and Yookun Cho. Efficient identification of bad signatures in RSA-type batch signature. *IEICE Transactions*, 89-A(1):74–80, 2006.

[181] Allison B. Lewko, Tatsuaki Okamoto, Amit Sahai, Katsuyuki Takashima, and Brent Waters. Fully secure functional encryption: Attribute-based encryption and (hierarchical) inner product encryption. In Henri Gilbert, editor, *EUROCRYPT*, volume 6110 of *LNCS*, pages 62–91. Springer, 2010.

[182] Allison B. Lewko and Brent Waters. New techniques for dual system encryption and fully secure hibe with short ciphertexts. *IACR Cryptology ePrint Archive*, 2009:482, 2009.

[183] Allison B. Lewko and Brent Waters. New techniques for dual system encryption and fully secure hibe with short ciphertexts. In Daniele Micciancio, editor, *TCC*, volume 5978 of *LNCS*, pages 455–479. Springer, 2010.

[184] Allison B. Lewko and Brent Waters. Decentralizing attribute-based encryption. In Kenneth G. Paterson, editor, *EUROCRYPT*, volume 6632 of *LNCS*, pages 568–588. Springer, 2011.

[185] Jin Li, Man Ho Au, Willy Susilo, Dongqing Xie, and Kui Ren. Attribute-based signature and its applications. In Dengguo Feng, David A. Basin, and Peng Liu, editors, *ASIACCS*, pages 60–69. ACM, 2010.

[186] Jin Li, Qiong Huang, Xiaofeng Chen, Sherman S. M. Chow, Duncan S. Wong, and Dongqing Xie. Multi-authority ciphertext-policy attribute-based encryption with accountability. In Bruce S. N. Cheung, Lucas Chi Kwong Hui, Ravi S. Sandhu, and Duncan S. Wong, editors, *ASIACCS*, pages 386–390. ACM, 2011.

[187] Jin Li, Kui Ren, and Kwangjo Kim. A2be: Accountable attribute-based encryption for abuse free access control. *IACR Cryptology ePrint Archive*, 2009:118, 2009.

[188] Jin Li, Kui Ren, Bo Zhu, and Zhiguo Wan. Privacy-aware attribute-based encryption with user accountability. In Pierangela Samarati, Moti Yung, Fabio Martinelli, and Claudio Agostino Ardagna, editors, *ISC*, volume 5735 of *Lecture Notes in Computer Science*, pages 347–362. Springer, 2009.

[189] Ming Li, Wenjing Lou, and Kui Ren. Data security and privacy in wireless body area networks. *IEEE Wireless Communications*, 17(1):51–58, 2010.

[190] Ming Li, Shucheng Yu, Kui Ren, and Wenjing Lou. Securing personal health records in cloud computing: Patient-centric and fine-grained data access control in multi-owner settings. In Sushil Jajodia and Jianying Zhou, editors, *SecureComm*, volume 50 of *Lecture Notes of the Institute for Computer Sciences, Social Informatics and Telecommunications Engineering*, pages 89–106. Springer, 2010.

[191] Xiaohui Liang, Zhenfu Cao, Huang Lin, and Jun Shao. Attribute based proxy re-encryption with delegating capabilities. In Wanqing Li, Willy Susilo, Udaya Kiran Tupakula, Reihaneh Safavi-Naini, and Vijay Varadharajan, editors, *ASIACCS*, pages 276–286. ACM, 2009.

[192] Xiaohui Liang, Zhenfu Cao, Huang Lin, and Dongsheng Xing. Provably secure and efficient bounded ciphertext policy attribute based encryption. In Wanqing Li, Willy Susilo, Udaya Kiran Tupakula, Reihaneh Safavi-Naini, and Vijay Varadharajan, editors, *ASIACCS*, pages 343–352. ACM, 2009.

[193] Benoît Libert and Damien Vergnaud. Multi-use unidirectional proxy re-signatures. In Peng Ning, Paul F. Syverson, and Somesh Jha, editors, *ACM Conference on Computer and Communications Security*, pages 511–520. ACM, 2008.

[194] Benoît Libert and Damien Vergnaud. Unidirectional chosen-ciphertext secure proxy re-encryption. In Ronald Cramer, editor, *Public Key Cryptography*, volume 4939 of *Lecture Notes in Computer Science*, pages 360–379. Springer, 2008.

[195] Benoît Libert and Damien Vergnaud. Unidirectional chosen-ciphertext secure proxy re-encryption. *IEEE T Inform. Theory.*, 57(3):1786–1802, 2011.

[196] Chae Hoon Lim and Pil Joong Lee. On the security of interactive dsa batch verification. *Electron. Lett.*, 30(19):1592–1593, 1994.

[197] Hsiao-Ying Lin and Wen-Guey Tzeng. A secure erasure code-based cloud storage system with secure data forwarding. *IEEE T PARALL. DISTR.*, 23(6):995–1003, 2012.

[198] Huang Lin, Zhenfu Cao, Xiaohui Liang, and Jun Shao. Secure threshold multi authority attribute based encryption without a central authority. In Dipanwita Roy Chowdhury, Vincent Rijmen, and Abhijit Das, editors, *INDOCRYPT*, volume 5365 of *Lecture Notes in Computer Science*, pages 426–436. Springer, 2008.

[199] Huang Lin, Zhenfu Cao, Xiaohui Liang, Muxin Zhou, Haojin Zhu, and Dongsheng Xing. How to construct interval encryption from binary tree encryption. In Jianying Zhou and Moti Yung, editors, *ACNS*, volume 6123 of *Lecture Notes in Computer Science*, pages 19–34, 2010.

[200] Yi-Ru Liu and Wen-Guey Tzeng. Public key broadcast encryption with low number of keys and constant decryption time. In Ronald Cramer, editor, *Public Key Cryptography*, volume 4939 of *Lecture Notes in Computer Science*, pages 380–396. Springer, 2008.

[201] Zhen Liu, Zhenfu Cao, Qiong Huang, Duncan S. Wong, and Tsz Hon Yuen. Fully secure multi-authority ciphertext-policy attribute-based encryption without random oracles. In Vijay Atluri and Claudia Díaz, editors, *ESORICS*, volume 6879 of *Lecture Notes in Computer Science*, pages 278–297. Springer, 2011.

[202] Jonathan Longrigg and Alexander Ushakov. Cryptanalysis of shifted conjugacy authentication protocol. *CoRR*, abs/0708.1768, 2007.

[203] Steve Lu, Rafail Ostrovsky, Amit Sahai, Hovav Shacham, and Brent Waters. Sequential aggregate signatures and multisignatures without random oracles. In Serge Vaudenay, editor, *EUROCRYPT*, volume 4004 of *Lecture Notes in Computer Science*, pages 465–485. Springer, 2006.

[204] Junzhou Luo, Xiaogang Wang, and Ming Yang. A resilient p2p anonymous routing approach employing collaboration scheme. *J. UCS*, 15(9):1797–1811, 2009.

[205] Anna Lysyanskaya, Silvio Micali, Leonid Reyzin, and Hovav Shacham. Sequential aggregate signatures from trapdoor permutations. In Christian Cachin and Jan Camenisch, editors, *EUROCRYPT*, volume 3027 of *Lecture Notes in Computer Science*, pages 74–90. Springer, 2004.

[206] Samuel Maffre. A weak key test for braid based cryptography. *Designs, Codes, and Cryptography*, 39(3):347–373, June 2006.

[207] Spyros S. Magliveras, Douglas R. Stinson, and Tran van Trung. New approaches to designing public key cryptosystems using one-way functions and trapdoors in finite groups. *J. Cryptol.*, 15(4):285–297, 2002.

[208] Ayan Mahalanobis. The Diffie-Hellman key exchange protocol and non-abelian nilpotent groups. *CoRR*, abs/math/0602282, 2006.

[209] Wenbo Mao. *Modern Cryptography: Theory and Practice*. Prentice-Hall PTR, 2004.

[210] Toshihide Matsuda, Ryo Nishimaki, and Keisuke Tanaka. Cca proxy re-encryption without bilinear maps in the standard model. In Phong Q. Nguyen and David Pointcheval, editors, *Public Key Cryptography*, volume 6056 of *Lecture Notes in Computer Science*, pages 261–278. Springer, 2010.

[211] Tsutomu Matsumoto, Hiroyuki Matsumoto, Koji Yamada, and Satoshi Hoshino. Impact of artificial "gummy" fingers on fingerprint systems. *Datenschutz und Datensicherheit*, 26(8), 2002.

[212] Ueli M. Maurer. Abstract models of computation in cryptography. In Nigel P. Smart, editor, *IMA Int. Conf.*, volume 3796 of *Lecture Notes in Computer Science*, pages 1–12. Springer, 2005.

[213] Dmitriy N. Moldovyan and Nikolay A. Moldovyan. A new hard problem over non-commutative finite groups for cryptographic protocols. In Igor V. Kotenko and Victor A. Skormin, editors, *MMM-ACNS*, volume 6258 of *Lecture Notes in Computer Science*, pages 183–194. Springer, 2010.

[214] Fabian Monrose, Michael K. Reiter, and Susanne Wetzel. Password hardening based on keystroke dynamics. *International Journal of Information Security*, 1(2):69–83, 2002.

[215] H.R. Morton. The multivariable alexander polynomial for a closed braid. *Contemporary Mathematics*, 233:167–172, 1999.

[216] Ritesh Mukherjee and J. William Atwood. Scalable solutions for secure group communications. *Comput. Netw.*, 51(12):3525–3548, 2007.

[217] Sascha Müller, Stefan Katzenbeisser, and Claudia Eckert. Distributed attribute-based encryption. In Pil Joong Lee and Jung Hee Cheon, editors, *ICISC*, volume 5461 of *LNCS*, pages 20–36. Springer, 2008.

[218] Alexei Myasnikov, Vladimir Shpilrain, and Alexander Ushakov. *Group-based Cryptography*. Advanced Courses in Mathematics – CRM Barcelona. Birkhäuser, 2008.

[219] David Naccache, David M'Raïhi, Serge Vaudenay, and Dan Raphaeli. Can d.s.a. be improved? complexity trade-offs with the digital signature standard. In Alfredo De Santis, editor, *EUROCRYPT*, volume 950 of *Lecture Notes in Computer Science*, pages 77–85. Springer, 1994.

[220] Deholo Nali, Carlisle M. Adams, and Ali Miri. Using threshold attribute-based encryption for practical biometric-based access control. *International Journal of Network Security*, 1(3):173–182, 2005.

[221] Dalit Naor, Moni Naor, and Jeffery Lotspiech. Revocation and tracing schemes for stateless receivers. In Joe Kilian, editor, *CRYPTO*, volume 2139 of *Lecture Notes in Computer Science*, pages 41–62. Springer, 2001.

[222] National Bureau of Standards. Data encryption standard, fips-pub.46. U.S. Department of Commerce, Washington D.C., January 1977.

[223] Rafail Ostrovsky, Amit Sahai, and Brent Waters. Attribute-based encryption with non-monotonic access structures. In Peng Ning, Sabrina De Capitani di Vimercati, and Paul F. Syverson, editors, *ACM Conference on Computer and Communications Security*, pages 195–203. ACM, 2007.

[224] Seong-Hun Paeng, Kil-Chan Ha, Jae Heon Kim, Seongtaek Chee, and Choonsik Park. New public key cryptosystem using finite non abelian groups. In Joe Kilian, editor, *CRYPTO*, volume 2139 of *Lecture Notes in Computer Science*, pages 470–485. Springer, 2001.

[225] Pascal Paillier. Public-key cryptosystems based on composite degree residuosity classes. In Jacques Stern, editor, *EUROCRYPT*, volume 1592 of *Lecture Notes in Computer Science*, pages 223–238. Springer, 1999.

[226] Mike Paterson and Alexander A. Razborov. The set of minimal braids is co-np-complete. *J. Algorithm*, 12(3):393–408, 1991.

[227] Matthew Pirretti, Patrick Traynor, Patrick McDaniel, and Brent Waters. Secure attribute-based systems. *Journal of Computer Security*, 18(5):799–837, 2010.

[228] Salil Prabhakar, Sharath Pankanti, and Anil K. Jain. Biometric recognition: Security and privacy concerns. *IEEE Security & Privacy*, 1(2):33–42, 2003.

[229] John Proos and Christof Zalka. Shor's discrete logarithm quantum algorithm for elliptic curves. *Quantum Information & Computation*, 3(4):317–344, 2003.

[230] Jean-Jacques Quisquater and Chantal Couvreur. Fast decipherment algorithm for rsa public-key cryptosystem. *Electron. Lett.*, 18(21):905–907, 1982.

[231] P.Vasudeva Reddy, G.S.G.N.Anjaneyulu, D. V. Rama Koti Reddy, and M. Padmavathamma. New digital signature scheme using polynomials over non-commutative groups. *International Journal of Computer Science and Network Security (IJCSNS)*, 8(1):245–250, 2008.

[232] Oded Regev. On lattices, learning with errors, random linear codes, and cryptography. *J. ACM*, 56(6), 2009.

[233] Ronald L. Rivest, Adi Shamir, and Leonard M. Adleman. A method for obtaining digital signatures and public-key cryptosystems. *Commun. ACM*, 21(2):120–126, 1978.

[234] Martin Rötteler. Quantum algorithms: A survey of some recent results. *Inform., Forsch. Entwickl.*, 21(1-2):3–20, 2006.

[235] Dima Ruinskiy, Adi Shamir, and Boaz Tsaban. Length-based cryptanalysis: The case of thompson's group. *J. Mathematical Cryptology*, 1(4):359–372, 2007.

[236] Sushmita Ruj, Amiya Nayak, and Ivan Stojmenovic. Dacc: Distributed access control in clouds. *International Joint Conference of IEEE TrustCom/IEEE ICESS/FCST*, 0:91–98, 2011.

[237] Sushmita Ruj, Amiya Nayak, and Ivan Stojmenovic. Distributed fine-grained access control in wireless sensor networks. In *IPDPS*, pages 352–362. IEEE, 2011.

[238] Sushmita Ruj, Amiya Nayak, and Ivan Stojmenovic. Improved access control mechanism in vehicular ad hoc networks. In Hannes Frey, Xu Li, and Stefan Rührup, editors, *ADHOC-NOW*, volume 6811 of *Lecture Notes in Computer Science*, pages 191–205. Springer, 2011.

[239] Sushmita Ruj, Amiya Nayak, and Ivan Stojmenovic. A security architecture for data aggregation and access control in smart grids. *CoRR*, abs/1111.2619, 2011.

[240] Amit Sahai and Brent Waters. Revocation systems with very small private keys.

[241] Amit Sahai and Brent Waters. Fuzzy identity-based encryption. In Ronald Cramer, editor, *EUROCRYPT*, volume 3494 of *Lecture Notes in Computer Science*, pages 457–473. Springer, 2005.

[242] Hovav Shacham and Dan Boneh. Improving ssl handshake performance via batching. In David Naccache, editor, *CT-RSA*, volume 2020 of *Lecture Notes in Computer Science*, pages 28–43. Springer, 2001.

[243] Claude Elwood Shannon. A mathematical theory of communication. *Bell system technical journal*, 27:379–423 and 623–656, 1948.

[244] Claude Elwood Shannon. Communication Theory of Secrecy Systems. *Bell system technical journal*, 28(4):656–715, 1949.

[245] Jun Shao. Bibliography on proxy re-cryptography. http://tdt.sjtu.edu.cn/~jshao/prcbib.htm.

[246] Jun Shao and Zhenfu Cao. Cca-secure proxy re-encryption without pairings. In Stanislaw Jarecki and Gene Tsudik, editors, *Public Key Cryptography*, volume 5443 of *Lecture Notes in Computer Science*, pages 357–376. Springer, 2009.

[247] Jun Shao, Zhenfu Cao, Xiaohui Liang, and Huang Lin. Proxy re-encryption with keyword search. *Inf. Sci.*, 180(13):2576–2587, 2010.

[248] Jun Shao, Zhenfu Cao, and Peng Liu. Cca-secure pre scheme without random oracles. *IACR Cryptology ePrint Archive*, 2010:112, 2010.

[249] Jun Shao, Zhenfu Cao, and Peng Liu. SCCR: a generic approach to simultaneously achieve CCA security and collusion-resistance in proxy re-encryption. *SECUR. COMMUN. NETW.*, 4(2):122–135, 2011.

[250] Jun Shao, Zhenfu Cao, Licheng Wang, and Xiaohui Liang. Proxy re-signature schemes without random oracles. In K. Srinathan, C. Pandu Rangan, and Moti Yung, editors, *INDOCRYPT*, volume 4859 of *Lecture Notes in Computer Science*, pages 197–209. Springer, 2007.

[251] Jun Shao, Min Feng, Bin Zhu, Zhenfu Cao, and Peng Liu. The security model of unidirectional proxy re-signature with private re-signature key. In Ron Steinfeld and Philip Hawkes, editors, *ACISP*, volume 6168 of *Lecture Notes in Computer Science*, pages 216–232. Springer, 2010.

[252] Jun Shao, Peng Liu, Zhenfu Cao, and Guiyi Wei. Multi-use unidirectional proxy re-encryption. In *ICC*, pages 1–5. IEEE, 2011.

[253] Jun Shao, Peng Liu, Guiyi Wei, and Yun Ling. Anonymous proxy re-encryption. *Secur. Commun. Netw.*, 5(5):439–449, 2012.

[254] Jun Shao, Peng Liu, and Yuan Zhou. Achieving key privacy without losing cca security in proxy re-encryption. *J. Syst. Software*, 85(3):655–665, 2012.

[255] Jun Shao, Guiyi Wei, Yun Ling, and Mande Xie. Identity-based conditional proxy re-encryption. In *ICC*, pages 1–5. IEEE, 2011.

[256] Emily Shen, Elaine Shi, and Brent Waters. Predicate privacy in encryption systems. In Omer Reingold, editor, *TCC*, volume 5444 of *Lecture Notes in Computer Science*, pages 457–473. Springer, 2009.

[257] Peter W. Shor. Algorithms for quantum computation: Discrete logarithms and factoring. In *FOCS*, pages 124–134. IEEE Computer Society, 1994.

[258] Victor Shoup and Rosario Gennaro. Securing threshold cryptosystems against chosen ciphertext attack. In Kaisa Nyberg, editor, *EUROCRYPT*, volume 1403 of *Lecture Notes in Computer Science*, pages 1–16. Springer, 1998.

[259] Vladimir Shpilrain and Alexander Ushakov. Thompson's group and public key cryptography. In John Ioannidis, Angelos D. Keromytis, and Moti Yung, editors, *ACNS*, volume 3531 of *Lecture Notes in Computer Science*, pages 151–163, 2005.

[260] Hervé Sibert, Patrick Dehornoy, and Marc Girault. Entity authentication schemes using braid word reduction. *Discrete Appl. Math.*, 154(2):420–436, 2006.

[261] Neil J. A. Sloane. The on-line encyclopedia of integer sequences. In Manuel Kauers, Manfred Kerber, Robert Miner, and Wolfgang Windsteiger, editors, *Calculemus/MKM*, volume 4573 of *Lecture Notes in Computer Science*, page 130. Springer, 2007.

[262] Martin Stanek. Attacking lccc batch verification of rsa signatures. Cryptology ePrint Archive, Report 2006/111, 2006. http://eprint.iacr.org/.

[263] Qi Sun. On the number of the representations of 1 by unit fractions. *J. Sichuan University (Natural Science Edition)*, 2:15–18, 1978. (in Chinese).

[264] Qi Sun. On a problem of Znám. *J. Sichuan University (Natural Science Edition)*, 4:9–11, 1983.

[265] Qi Sun. Number theory entered the applied science. *J. Mathematical Research and Exposition*, 4:149–154, 1986.

[266] Qi Sun and Zhenfu Cao. On the equation $\frac{1}{x_1} + \cdots + \frac{1}{x_s} - \frac{1}{x_1 \ldots x_s} = 1$ and its application. *Chinese Sci. Bull.*, 2:155–157, 1985.

[267] Qi Sun and Zhenfu Cao. On the equation $\sum_{j=1}^{s} \frac{1}{x_j} + \frac{1}{x_1 \ldots x_s} = n$ and the number of solutions of Znám's problem. *Advances in Mathematics (China)*, 3:329–330, 1986.

[268] Qi Sun and Zhenfu Cao. On the equation $\sum_{j=1}^{s} \frac{1}{x_j} + \frac{1}{x_1 \ldots x_s} = 1$. *J. Mathematical Research and Exposition*, 1:125–128, 1987. (in Chinese).

[269] Yipin Sun, Rongxing Lu, Xiaodong Lin, Xuemin Shen, and Jinshu Su. An efficient pseudonymous authentication scheme with strong privacy preservation for vehicular communications. *IEEE T Veh. Technol.*, 59(7):3589–3603, 2010.

[270] James Joseph Sylvester. On a point in the theory of vulgar fractions. *Am. J. Math.*, 3(4):332–335, 1880.

[271] Gelareh Taban, Alvaro A. Cárdenas, and Virgil D. Gligor. Towards a secure and interoperable drm architecture. In Moti Yung, Kaoru Kurosawa, and Reihaneh Safavi-Naini, editors, *Digital Rights Management Workshop*, pages 69–78. ACM, 2006.

[272] Qiang Tang. Type-based proxy re-encryption and its construction. In Dipanwita Roy Chowdhury, Vincent Rijmen, and Abhijit Das, editors, *INDOCRYPT*, volume 5365 of *Lecture Notes in Computer Science*, pages 130–144. Springer, 2008.

[273] Patrick Traynor, Kevin R. B. Butler, William Enck, and Patrick McDaniel. Realizing massive-scale conditional access systems through attribute-based cryptosystems. In *NDSS*. The Internet Society, 2008.

[274] Pim Tuyls and Jasper Goseling. Capacity and examples of template-protecting biometric authentication systems. *Biometric Authentication*, pages 158–170, 2004.

[275] Ton van der Putte and Jeroen Keuning. Biometrical fingerprint recognition: Don't get your fingers burned. In Josep Domingo-Ferrer, David Chan, and Anthony Watson, editors, *CARDIS*, volume 180 of *IFIP Conference Proceedings*, pages 289–306. Kluwer, 2000.

[276] Maria Isabel Gonzalez Vasco, Consuelo Martínez, and Rainer Steinwandt. Towards a uniform description of several group based cryptographic primitives. *Design. Code. Cryptogr.*, 33(3):215–226, 2004.

[277] Nitin Vats. NNRU, a noncommutative analogue of NTRU. *CoRR*, abs/0902.1891, 2009.

[278] Evgeny Verbitskiy, Pim Tuyls, Dee Denteneer, and Jean-Paul Linnartz. Reliable biometric authentication with privacy protection. In *Proc. 24th Benelux Symposium on Information theory*, page 19, 2003.

[279] Neal R. Wagner and Marianne R. Magyarik. A public key cryptosystem based on the word problem. In G. R. Blakley and David Chaum, editors, *CRYPTO*, volume 196 of *Lecture Notes in Computer Science*, pages 19–36. Springer, 1984.

[280] Cong Wang, Ning Cao, Jin Li, Kui Ren, and Wenjing Lou. Secure ranked keyword search over encrypted cloud data. In *ICDCS*, pages 253–262. IEEE Computer Society, 2010.

[281] Cong Wang, Qian Wang, Kui Ren, and Wenjing Lou. Privacy-preserving public auditing for data storage security in cloud computing. In *INFOCOM*, pages 525–533. IEEE, 2010.

[282] Guojun Wang, Qin Liu, and Jie Wu. Hierarchical attribute-based encryption for fine-grained access control in cloud storage services. In Ehab Al-Shaer, Angelos D. Keromytis, and Vitaly Shmatikov, editors, *ACM Conference on Computer and Communications Security*, pages 735–737. ACM, 2010.

[283] Hongbing Wang, Zhenfu Cao, and Licheng Wang. Multi-use and unidirectional identity-based proxy re-encryption schemes. *Inf. Sci.*, 180(20):4042–4059, 2010.

[284] Licheng Wang, Zhenfu Cao, Peng Zeng, and Xiangxue Li. One-more matching conjugate problem and security of braid-based signatures. In Steven Miller (Eds.) Feng Bao, editor, *Proceedings of the 2007 ACM Symposium on Information, Computer and Communications Security (ASIACCS 2007)*, pages 295–301. ACM Press, 2007.

[285] Licheng Wang, Zhenfu Cao, Shihui Zheng, Xiaofang Huang, and Yixian Yang. Transitive signatures from braid groups. In K. Srinathan, C. Pandu Rangan, and Moti Yung, editors, *INDOCRYPT*, volume 4859 of *Lecture Notes in Computer Science*, pages 183–196. Springer, 2007.

[286] Licheng Wang, Lihua Wang, Zhenfu Cao, Yixian Yang, and Xinxin Niu. Conjugate adjoining problem in braid groups and new design of braid-based signatures. *Sci. China Inform. Sci.*, 53(3):524–536, 2010.

[287] Lihua Wang, Licheng Wang, Zhenfu Cao, Eiji Okamoto, and Jun Shao. New constructions of public-key encryption schemes from conjugacy search problems. In Xuejia Lai, Moti Yung, and Dongdai Lin, editors, *Inscrypt*, volume 6584 of *Lecture Notes in Computer Science*, pages 1–17. Springer, 2010.

[288] Pan Wang, Peng Ning, and Douglas S. Reeves. Storage-efficient stateless group key revocation. In Kan Zhang and Yuliang Zheng, editors, *ISC*, volume 3225 of *Lecture Notes in Computer Science*, pages 25–38. Springer, 2004.

[289] Brent Waters. Efficient identity-based encryption without random oracles. In Ronald Cramer, editor, *EUROCRYPT*, volume 3494 of *Lecture Notes in Computer Science*, pages 114–127. Springer, 2005.

[290] Brent Waters. Dual system encryption: Realizing fully secure ibe and hibe under simple assumptions. In Shai Halevi, editor, *CRYPTO*, volume 5677 of *Lecture Notes in Computer Science*, pages 619–636. Springer, 2009.

[291] Brent Waters. Ciphertext-policy attribute-based encryption: An expressive, efficient, and provably secure realization. In Dario Catalano, Nelly Fazio, Rosario Gennaro, and Antonio Nicolosi, editors, *Public Key Cryptography*, volume 6571 of *LNCS*, pages 53–70. Springer, 2011.

[292] Lifei Wei, Haojin Zhu, Zhenfu Cao, Weiwei Jia, and Athanasios V. Vasilakos. Seccloud: Bridging secure storage and computation in cloud. In *ICDCS Workshops*, pages 52–61. IEEE Computer Society, 2010.

[293] Jian Weng, Min-Rong Chen, Yanjiang Yang, Robert H. Deng, Kefei Chen, and Feng Bao. Cca-secure unidirectional proxy re-encryption in the adaptive corruption model without random oracles. *Sci. China Inform. Sci.*, 53(3):593–606, 2010.

[294] Jian Weng, Robert H. Deng, Xuhua Ding, Cheng-Kang Chu, and Junzuo Lai. Conditional proxy re-encryption secure against chosen-ciphertext attack. In Wanqing Li, Willy Susilo, Udaya Kiran Tupakula, Reihaneh Safavi-Naini, and Vijay Varadharajan, editors, *ASIACCS*, pages 322–332. ACM, 2009.

[295] Jian Weng, Yunlei Zhao, and Goichiro Hanaoka. On the security of a bidirectional proxy re-encryption scheme from pkc 2010. In Dario Catalano, Nelly Fazio, Rosario Gennaro, and Antonio Nicolosi, editors, *Public Key Cryptography*, volume 6571 of *Lecture Notes in Computer Science*, pages 284–295. Springer, 2011.

[296] Michael J. Wiener. Cryptanalysis of short rsa secret exponents. *IEEE T Inform. Theory.*, 36(3):553–558, 1990.

[297] Bert Wiest. An algorithm for the word problem in braid groups, November 11 2002.

[298] Yacov Yacobi and Michael J. Beller. Batch Diffie-Hellman key agreement systems. *J. Cryptol.*, 10(2):89–96, 1997.

[299] Akihiro Yamamura. Public-key cryptosystems using the modular group. In Hideki Imai and Yuliang Zheng, editors, *Public Key Cryptography*, volume 1431 of *Lecture Notes in Computer Science*, pages 203–216. Springer, 1998.

[300] Piyi Yang, Zhenfu Cao, and Xiaolei Dong. Fuzzy identity based signature. Cryptology ePrint Archive, Report 2008/002, 2008. `http://eprint.iacr.org/`.

[301] Piyi Yang, Zhenfu Cao, and Xiaolei Dong. Fuzzy identity based signature with applications to biometric authentication. *Computers & Electrical Engineering*, 37(4):532–540, 2011.

[302] Sung-Ming Yen and Chi-Sung Laih. Improved digital signature suitable for batch verification. *IEEE T. Comput.*, 44(7):957–959, 1995.

[303] Eun Sun Yoo, Nam-Su Iho, Jung Hee Cheon, and Myung-Hwan Kim. Efficient broadcast encryption using multiple interpolation methods. In Choonsik Park and Seongtaek Chee, editors, *ICISC*, volume 3506 of *Lecture Notes in Computer Science*, pages 87–103. Springer, 2004.

[304] HyoJin Yoon, Jung Hee Cheon, and Yongdae Kim. Batch verifications with ID-based signatures. In Choonsik Park and Seongtaek Chee, editors, *ICISC*, volume 3506 of *Lecture Notes in Computer Science*, pages 233–248. Springer, 2004.

[305] Shucheng Yu, Kui Ren, and Wenjing Lou. Attribute-based on-demand multicast group setup with membership anonymity. *Comput. Netw.*,.

[306] Shucheng Yu, Cong Wang, Kui Ren, and Wenjing Lou. Achieving secure, scalable, and fine-grained data access control in cloud computing. In *INFOCOM*, pages 534–542. IEEE, 2010.

[307] Chenxi Zhang, Rongxing Lu, Xiaodong Lin, Pin-Han Ho, and Xuemin Shen. An efficient identity-based batch verification scheme for vehicular sensor networks. In *INFOCOM*, pages 246–250. IEEE, 2008.

[308] Fangguo Zhang and Kwangjo Kim. Efficient ID-based blind signature and proxy signature from bilinear pairings. In Reihaneh Safavi-Naini and Jennifer Seberry, editors, *ACISP*, volume 2727 of *Lecture Notes in Computer Science*, pages 312–323. Springer, 2003.

[309] Fangguo Zhang, Reihaneh Safavi-Naini, and Willy Susilo. Efficient verifiably encrypted signature and partially blind signature from bilinear pairings. In Thomas Johansson and Subhamoy Maitra, editors, *INDOCRYPT*, volume 2904 of *Lecture Notes in Computer Science*, pages 191–204. Springer, 2003.

[310] Lixia Zhang, Deborah Estrin, Jeffrey Burke, Van Jacobson, James D. Thornton, Diana K. Smetters, Beichuan Zhang, Gene Tsudik, Dan Massey, Christos Papadopoulos, et al. Named data networking (ndn) project. Technical report, PARC, Tech. report ndn-0001, 2010.

[311] Haojin Zhu, Xiaodong Lin, Rongxing Lu, Xuemin Shen, and Pin-Han Ho. Bba: An efficient batch bundle authentication scheme for delay tolerant networks. In *GLOBECOM*, pages 1882–1886. IEEE, 2008.

[312] Haojin Zhu, Xiaodong Lin, Rongxing Lu, Xuemin Shen, Dongsheng Xing, and Zhenfu Cao. An opportunistic batch bundle authentication scheme for energy constrained dtns. In *INFOCOM*, pages 605–613. IEEE, 2010.

Index